TRIGONOMETRY
IN HIGH SCHOOL

TRIGONOMETRY

·············· *TECHNIQUES OF PROBLEM SOLVING* ··············

IN HIGH SCHOOL

MIHAI ROSU

Library of Congress Control Number:		2015903274
ISBN:	Hardcover	978-1-5035-2676-1
	Softcover	978-1-5035-2677-8
	eBook	978-1-5035-2678-5

Print information available on the last page.

Corrections by
Emanuel Teodorescu, PhD., Mathematics, Univ. of Kansas
Vlad Rosu, Student
Alexandru Rosu, Student

Rev. date: 03/30/2015

To order additional copies of this book, contact:
Xlibris
1-888-795-4274
www.Xlibris.com
Orders@Xlibris.com
701491

CONTENT

I dedicate this book to my truly inspiring sons

Alexandru and **Vlad,**

whose help has been essential in finalizing this book.
I hope they come to seeing it as an important professional experience.

FOREWORD

High school mathematics professor Mihai Rosu's second volume "Trigonometry in High School: Techniques of Problem Solving" is a continuation of the series of high school mathematics problem books which started with "Methods of Solving Problems in Middle and High School Mathematics" a fruit of his decades-long professional activity in teaching mathematics. A mathematics professor formed—as a mathematics major—in the Romanian education system, at standards set earlier in the 20th century by a number of eminent Romanian mathematicians with doctorates from French and German universities, Mihai Rosu offers to parents and students another solid piece of work, in the strong tradition of the Romanian school of mathematics, well aligned to the mathematics curricula in North America.

The present volume is mainly addressed to high school students—but also to college students taking introductory mathematic courses—with an offer of over 1700 problems and exercises, many with full solutions, all with final answers. They cover the Trigonometry part of the (former) grades 9 and 10 Romanian mathematics education, at a depth mostly unattained in modern day schools. The chosen material increases in difficulty gradually—from "introductory", all the way to "very advanced", facilitating a student's path to the mastery of the subject. A rich bibliography further indicates the broad reach of this work.

As with the previous volume, this volume is another concrete response to the call of many Canadians (university professors at "WISE Math", petitioners at "change.org" from multiple Canadian provinces (Alberta, Manitoba, British Columbia, Ontario) and many, many others, for raising the standards of Canadian mathematical (but not only) primary and secondary education, both of students and of their teachers.

9

Those students who will be putting in the effort of carefully going through the material (including carefully reading the given solutions at the back of the book) will benefit enormously, towards gaining of a good, clear image of what modern high school mathematics might be. Most importantly, this work (together with others of the same type and similar caliber, covering other branches of high school mathematics) will prepare students for the significantly harder time of university mathematics, on their path to becoming highly qualified experts in their fields of interest.

Em. Theodorescu, Ph.D.

PREFACE

Trigonometry in High School: Techniques of Problem Solving includes a range of trigonometry exercises and problems developed in conformity with the Ontario Secondary Curriculum. The material is divided into chapters and graded from basic to more challenging levels of complexity, in order to facilitate understanding and to encourage the development of problem solving skills. The book has two parts: the first part consists of 11 chapters of exercises and problems, while the second part offers answers, hints, and solutions.

This new book is a natural and necessary sequel to my earlier *Methods of Solving Problems in Elementary, Middle, and High School Mathematics*, published with Xlibris in 2010. I wrote *Trigonometry in High School* with the intention of providing teachers and especially high school students with a useful collection of problems and exercises. Many of these problems have been used as test and exam questions, solved in class, and discussed with my students and collaborators for more than three decades. I have gathered this personal experience into a single-author volume, containing the kind of information that many of my colleagues keep on their personal book shelves, in folders, or in notebooks revisited with generations of students.

Answers are provided to every problem in this book, with a degree of detail depending on the complexity of the problem. Wherever I considered that supplementary clarifications or elaborations were necessary, I often gave two, three, or even four detailed solutions. I clearly indicated the steps of problem solving, with the purpose of enabling prospective students to develop a strong conceptual basis and good mathematical thinking skills. By creating the premises of an enjoyable and intellectually stimulating dialogue between the reader and the text, I attempted to reduce mathematical anxiety, demystifying the idea that mathematics is an impossibly difficult discipline.

This does not mean that I encourage an oversimplifying attitude towards mathematics. To offer an efficient and enjoyable learning experience, a mathematical problem should not have a self-evident solution. As the famous mathematician David Hilbert put it more than a hundred years ago, "a mathematical problem should be difficult in order to entice us, yet not completely inaccessible, lest it mock at our efforts." Observant of Hilbert's idea, my book contains a few difficult exercises and problems. However, due to the organization of chapters along an increasing scale of difficulty, the reader can determine his/her own "comfort zone" and build up from there, at a pace that suits him/her and his/her future endeavours. This book will also be very useful to those university students who take mathematics in their first and second academic years.

Unlike high school textbooks, providing solutions to trigonometric equations only within the interval $[0, 2\pi)$, often through rather confusing approaches, my book gives very clear solutions within the general interval $[2k\pi, 2\pi + 2k\pi)_{k \in \mathbb{Z}}$. It also contains a few trigonometric equations, which I consider exemplary, with solutions very clearly represented on the unit circle. Last, but not least, inverse trigonometric functions, which are succinctly treated in the Ontario Secondary Curriculum, benefit from a generous space of exposition in Unit VI of my book. Its rich material includes trigonometric expressions, identities, and equations containing inverse trigonometric functions. I mention that most of the problems in this chapter are given almost-complete solutions in the second part of the book.

MIHAI ROSU

rosumihai@yahoo.ca

UNIT I

Trigonometric Circle

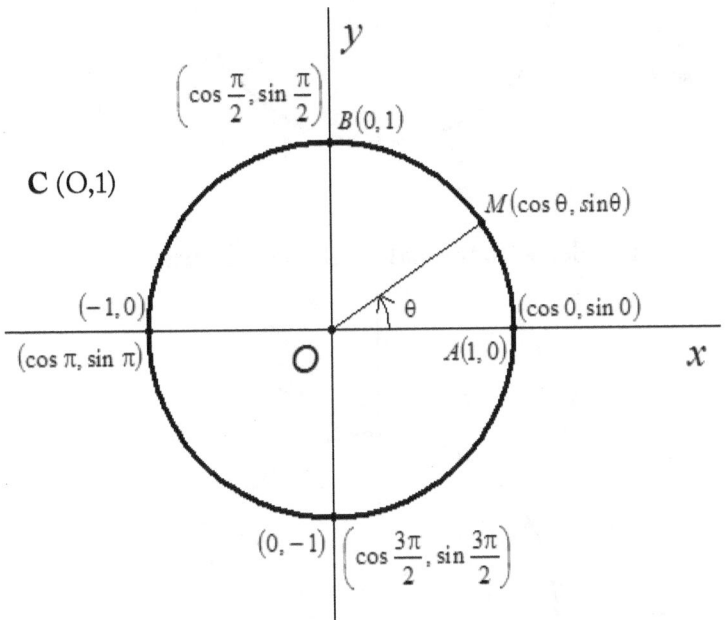

Geometric representations of trigonometric functions of the angle θ on the unit circle

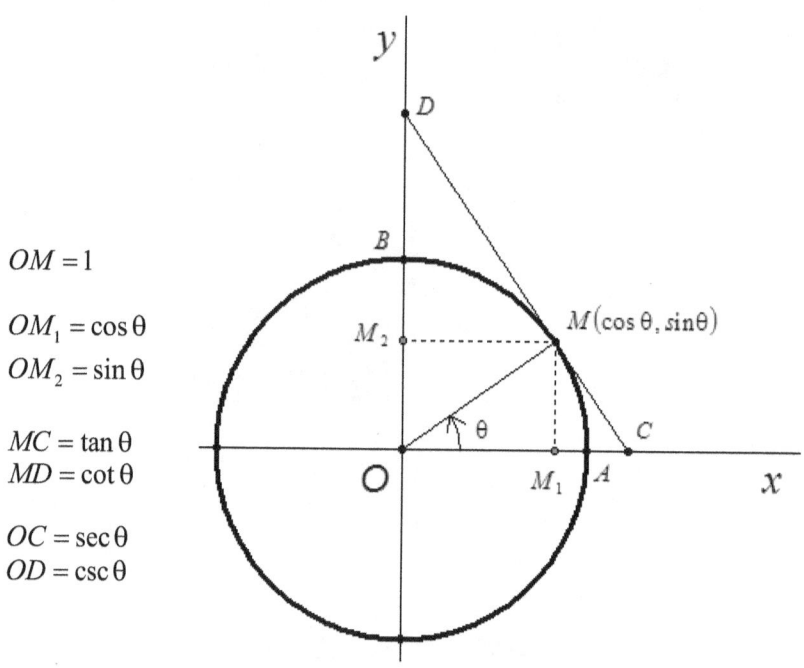

$OM = 1$

$OM_1 = \cos\theta$

$OM_2 = \sin\theta$

$MC = \tan\theta$

$MD = \cot\theta$

$OC = \sec\theta$

$OD = \csc\theta$

Unit Circle Labeled with Special Angles and Values

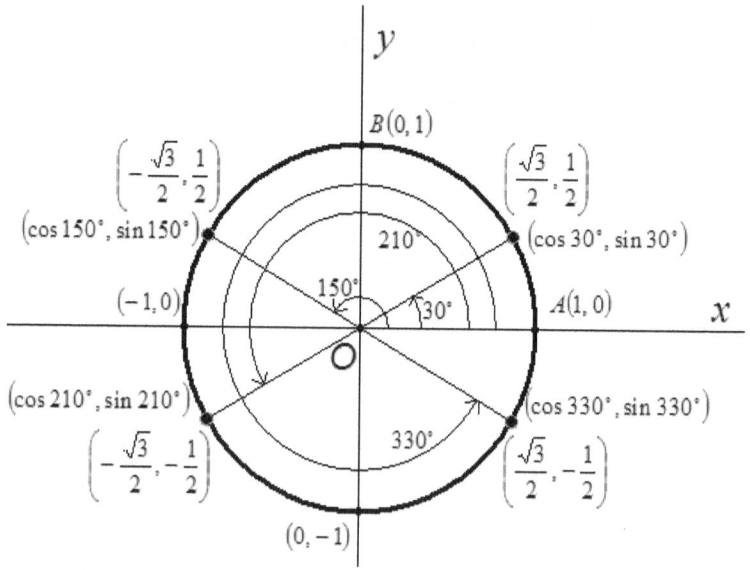

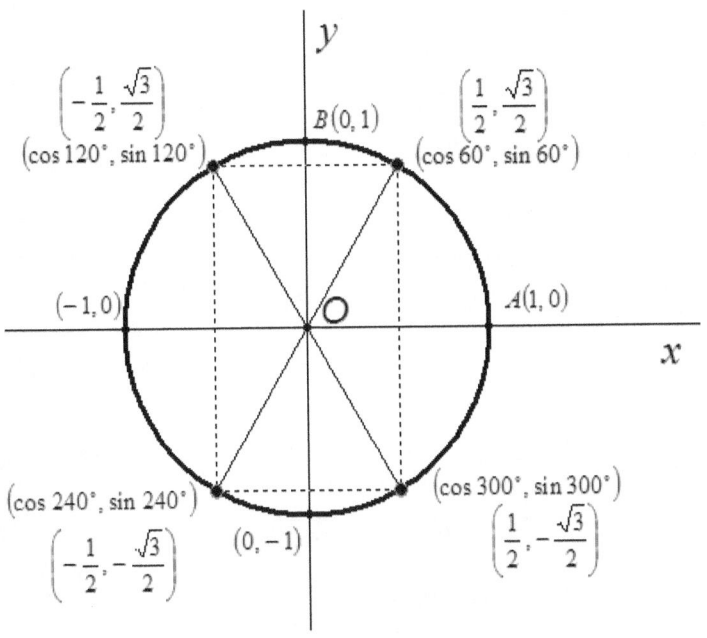

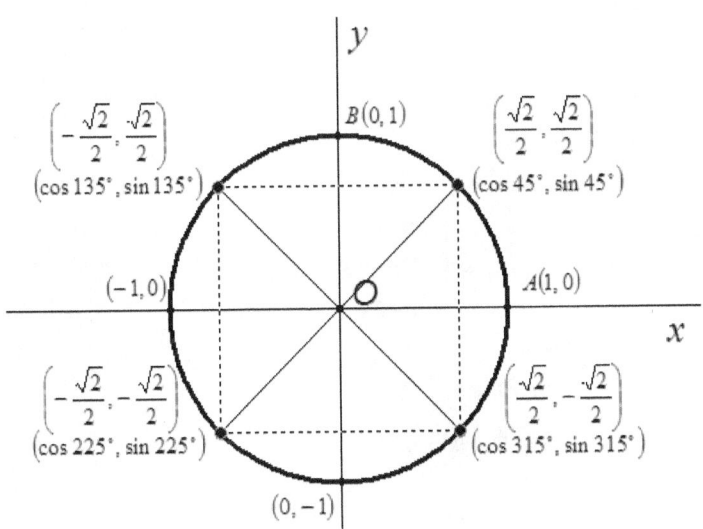

Pythagorean formula $\sin^2 t + \cos^2 t = 1, \quad \forall t \in \mathbf{R}$,

Fundamental Proprieties:

$\sin(t + 2k\pi) = \sin t, \; \cos(t + 2k\pi) = \cos t, \; \forall k \in \mathbf{Z}$,

$-1 \le \sin t \le 1, \qquad -1 \le \cos t \le 1, \quad \forall t \in \mathbf{R}$.

$\sin(-t) = -\sin t, \quad \cos(-t) = \cos t, \qquad \forall t \in \mathbf{R}$.

$\sin t = \sin(\pi - t) \qquad \cos t = -\cos(\pi - t) \qquad t \in \left(\dfrac{\pi}{2}, \pi \right)$

$\sin t = -\sin(t - \pi) \qquad \cos t = -\cos(t - \pi), \qquad t \in \left(\pi, \dfrac{3\pi}{2} \right)$

$\sin t = -\sin(2\pi - t) \qquad \cos t = \cos(2\pi - t) \qquad t \in \left(\dfrac{3\pi}{2}, 2\pi \right)$

Complementary formulas

$\sin\left(\dfrac{\pi}{2} - x\right) = \cos x \quad \cos\left(\dfrac{\pi}{2} - x\right) = \sin x \quad \tan\left(\dfrac{\pi}{2} - x\right) = \cot x$

Sum-Difference Formulas (Ptolemy's identities)

$\sin(a + b) = \sin a \cos b + \sin b \cos a$

$\sin(a - b) = \sin a \cos b - \sin b \cos a$

$\cos(a + b) = \cos a \cos b - \sin a \sin b$

$\cos(a - b) = \cos a \cos b + \sin a \sin b$

Double angle formulas

$$\sin 2x = 2\sin x \cos x \quad \cos 2x = \begin{cases} \cos^2 x - \sin^2 x \\ 2\cos^2 x - 1 \\ 1 - 2\sin^2 x \end{cases}$$

Product-to-Sum Formulas

$\sin x \cos y = \dfrac{1}{2}[\sin(x + y) + \sin(x - y)]$

$\cos x \cos y = \dfrac{1}{2}[\cos(x + y) + \cos(x - y)]$

$\sin x \sin y = \dfrac{1}{2}[\cos(x - y) - \cos(x + y)]$

Sum-to-Product Formulas

$$\sin a + \sin b = 2\sin\frac{a+b}{2}\cos\frac{a-b}{2}$$

$$\sin a - \sin b = 2\sin\frac{a-b}{2}\cos\frac{a+b}{2}$$

$$\cos a + \cos b = 2\cos\frac{a+b}{2}\cos\frac{a-b}{2}$$

$$\cos a - \cos b = -2\sin\frac{a+b}{2}\sin\frac{a-b}{2}$$

Sum, difference, and double angle formulas for tangent

$$\tan(x+y)=\frac{\tan x + \tan y}{1-\tan x\cdot\tan y}\quad \tan(x-y)=\frac{\tan x - \tan y}{1+\tan x\cdot\tan y}$$

$$\tan 2x = \frac{2\tan x}{1-\tan^2 x}\quad \tan x\pm\tan y = \frac{\sin(x\pm y)}{\cos x\cos y}$$

Half-angle formulas

$$1-\cos t = 2\sin^2\frac{t}{2}\quad 1+\cos t = 2\cos^2\frac{t}{2}$$

$$\sin t = \frac{2\tan\frac{t}{2}}{1+\tan^2\frac{t}{2}}\quad \cos t = \frac{1-\tan^2\frac{t}{2}}{1+\tan^2\frac{t}{2}}\quad \tan t = \frac{2\tan\frac{t}{2}}{1+\tan^2\frac{t}{2}}$$

Triple angle formulas

$$\sin 3x = 3\sin x - 4\sin^3 x\quad \cos 3x = 4\cos^3 x - 3\cos x$$

$$\tan 3x = \frac{3\tan x - \tan^3 x}{1-3\tan^2 x}\quad;$$

I. 1. Determine the following (assume a related angle calculation):

a) $\sec 300°$; b) $\cos 135°$; c) $\sin 330°$; d) $\csc(-120°)$;

e) $\sec 750°$; f) $\tan(-300°)$; g) $\sin(2190°)$; h) $\cos(-690°)$.

I. 2. Calculate:

a) $(\sin 20° + \sin 70°)^2 - 2\cos 20°\cos 70°$;

b) $(\cos 10° + \cos 80°)^2 - 2\sin 10°\sin 80°$.

I. 3. Find $\alpha \in \left(0, \dfrac{\pi}{2}\right)$ such that:

a) $\tan \alpha = \dfrac{1 - \tan 10°}{1 + \tan 10°}$; b) $\tan \alpha = \dfrac{\left(1 + \tan 10°\right)\left(1 + \tan 1°\right) - 2}{\left(1 - \tan 10°\right)\left(1 - \tan 1°\right) - 2}$.

I. 4. $M(-6, 7)$ lies on the terminal arm of an angle in standard position. What is the value of the principal angle θ to the nearest degree?

I. 5. $N(6, -7)$ lies on the terminal arm of an angle in standard position. What is the value of the principal angle θ to the nearest degree?

I. 6. Show that for any $x \in \mathbf{R}$

a) $\sin\left[(2k+1)\dfrac{\pi}{2} + x\right] = (-1)^k \cos x$;

b) $\cos\left[(2k+1)\dfrac{\pi}{2} + x\right] = (-1)^{k+1} \sin x$, $k \in \mathbf{Z}$.

I. 7. If $\alpha \in \left(\dfrac{\pi}{2}, \pi\right)$ and $\cos \alpha = -\dfrac{5}{13}$, calculate $\sin \alpha$, $\tan \alpha$, $\cot \alpha$, $\sec \alpha$, and $\csc \alpha$.

I. 8. If $\alpha \in \left(\dfrac{\pi}{2}, \pi\right)$ and $\sin \alpha = \dfrac{5}{13}$, calculate $\cos \alpha$, $\tan \alpha$, $\cot \alpha$, $\sec \alpha$, and $\csc \alpha$.

I. 9. If $\alpha \in \left(\dfrac{3\pi}{2}, 2\pi\right)$ and $\tan \alpha = -\dfrac{1}{2}$, calculate $\sin \alpha$, $\cos \alpha$, $\cot \alpha$, $\sec \alpha$, and $\csc \alpha$.

I. 10. If $\dfrac{3\pi}{2} < a < 2\pi$, and $\cos a = \dfrac{\sqrt{2 + \sqrt{2}}}{2}$ find $\tan a$.

I. 11. Find $\alpha \in \left(0, \dfrac{\pi}{2}\right)$ if $2\sin^2 \alpha - 7\sin \alpha + 3 = 0$.

I. 12. If $\cos x < 0$ and $\sin y > 0$, $\cos y = \dfrac{4}{5}$, and $\sin x = -\dfrac{12}{13}$, then calculate $\cos(x + y)$.

I. 13. If $\cos y > 0$, $\sin x < 0$, $\sin y = -\dfrac{3}{5}$, and $\cos x = \dfrac{5}{13}$, then calculate $\sin(x + y)$.

I. 14. If $\sin a = -\dfrac{3}{5}$ and $\pi < a < \dfrac{3\pi}{2}$, calculate $\sin\dfrac{a}{2}$ and $\tan\dfrac{a}{2}$.

I. 15. If $\sin \alpha = \dfrac{5}{13}$, $\alpha \in \left(0, \dfrac{\pi}{2}\right)$, calculate $\dfrac{\cot \alpha + \tan \alpha}{\cot \alpha - \tan \alpha}$.

I. 16. If $\sin \alpha = \dfrac{5}{13}$, $\alpha \in \left(\dfrac{\pi}{2}, \pi\right)$ calculate
$\cos \alpha$, $\tan \alpha$, and $\cot \alpha$.

I. 17. Find $\sin \alpha$ if $\alpha \in \left(\pi, \dfrac{3\pi}{2}\right)$ and $4\sin^2 \alpha - 4\sin \alpha - 3 = 0$.

I. 18. Calculate:

a) $\cos x$, if $\sin x = -\dfrac{1}{5}$ and $x \in \left(\pi, \dfrac{3\pi}{2}\right)$;

b) $\tan x$, if $\sin^2 x - 2\cos^2 x = \sin x \cos x$ and $x \in \left(0, \dfrac{\pi}{2}\right)$;

c) $\tan 2x$, if $\cos(x - 90°) = \dfrac{1}{4}$ and $x \in (90°, 180°)$;

d) $\tan\dfrac{x}{2}$, if $\sin x - \cos x = \dfrac{1 + 2\sqrt{2}}{3}$;

e) $x + y$, if $1 + \cot\left(\dfrac{\pi}{4} - x\right) = \dfrac{2}{1 - \cot y}$ and $x, y \in \left(0, \dfrac{\pi}{2}\right)$.

I. 19. If a and b are acute angles, such that $\cot a = \dfrac{3}{5}$ and $\cot b = \dfrac{1}{4}$, show that $a + b = \dfrac{3\pi}{4}$.

I. 20. If $\sin x + \sin y = \dfrac{\sqrt{2}}{3}$ and $\cos x + \cos y = \dfrac{\sqrt{2}}{2\sqrt{3}}$, determine $\cos(x - y)$.

I. 21. If $\cos\alpha = \dfrac{2}{5}$, $\cos\beta = \dfrac{1}{2}$, $\cos\gamma = \dfrac{1}{5}$, calculate

$\tan^2\dfrac{\alpha}{2} + \tan^2\dfrac{\beta}{2} + \tan^2\dfrac{\gamma}{2}$.

I. 22. Find $\sin\dfrac{x+y}{2}$ and $\cos\dfrac{x+y}{2}$ if $\sin x + \sin y = -\dfrac{21}{65}$ and

$\cos x + \cos y = -\dfrac{27}{65}$ where $\dfrac{\pi}{2} < x < \pi$ and $-\dfrac{\pi}{2} < y < 0$.

I. 23. Find $\cos\dfrac{x-y}{2}$ if $\sin x + \sin y = -\dfrac{27}{65}$ and $\tan\dfrac{x+y}{2} = \dfrac{7}{9}$ where

$\dfrac{\pi}{2} < x < \pi$ and $-\dfrac{\pi}{2} < y < 0$.

I. 24. If $\sin\dfrac{\alpha}{2} + \cos\dfrac{\alpha}{2} = 1.4$ find $\sin\alpha$.

I. 25. Calculate $\dfrac{2\sin\alpha - 3\cos\alpha}{4\sin\alpha - 5\cos\alpha}$ if $\tan\alpha = \sqrt{3}$.

I. 26. Calculate the value of the expressions for indicated value:

a) $A = \dfrac{\cos x + 3\sin x}{3\cos x - 5\sin x}$ if $\tan x = 3$; b) $B = \dfrac{2\sin x - 5\cos x}{-3\sin x + 4\cos x}$ if $\tan x = \sqrt{6}$;

c) $C = \dfrac{4\cos x - 3\sin x}{2\cos x + 5\sin x}$ if $\cot x = 2$; d) $D = \dfrac{\sin\alpha + 2\cos\alpha}{\cos\alpha - 3\sin\alpha}$ if $\cot\alpha = \dfrac{2}{5}$;

e) $E = \dfrac{3 + \sin x + \cos x}{2 + \sin x - \cos x}$ if $\tan\dfrac{x}{2} = \dfrac{1}{2}$; f) $F = \sqrt{3}\sin 2x - \cos 2x$, if

$\tan x = \dfrac{\sqrt{3}}{2}$;

g) $G = \sqrt{2}\sin x + \sqrt{3}\cos x$ if $\tan\dfrac{x}{2} = \dfrac{\sqrt{6}}{3}$; h) $H = \dfrac{5\sin x - \cos x}{5\sin x + \cos x}$ if $\cot\dfrac{x}{2} = 2$.

I. 27. If $a \in \left(\pi, \dfrac{3\pi}{2}\right)$ and $\tan a + \cot a = 3$ calculate:

a) $A = \sin a + \cos a$;

b) $B = \tan^2 a + \cot^2 a$;

c) $C = \dfrac{1}{\sin^3 a} + \dfrac{1}{\cos^3 a}$.

I. 28. If $\tan x + \cot x = 5$, calculate:

a) $\tan^2 x + \cot^2 x$; b) $\tan^3 x + \cot^3 x$; c) $\tan^4 x + \cot^4 x$.

I. 29. If $\sin x + \cos x = p$, calculate each expression in terms of p:

a) $\sin^3 x + \cos^3 x$; b) $\sin^6 x + \cos^6 x$; c) $\dfrac{(\sin x - \cos x)^2 (\sin x + \cos x)}{\sin x \cos x}$.

I. 30. If $\cos 2x = p$, calculate $\cos^8 x - \sin^8 x$.

I. 31. If $\sin \alpha + \cos \alpha = p$ and $\tan \alpha + \cot \alpha = q$, prove that $\left(p^2 - 1\right)q = 2$.

I. 32. If $3\sin \alpha + 4\cos \alpha + 5 = 0$, prove that $4 \tan \alpha + 3 \cot \alpha = 7$.

I. 33. If $a \in \left(0, \dfrac{\pi}{2}\right)$ and $\sin a \cos a = \dfrac{1}{4}$ calculate
$(1 + \cot a)\sin^3 a + (1 + \tan a)\cos^3 a$.

I. 34. If $\sin t = \dfrac{a^2 - b^2}{a^2 + b^2}$, $t \in \left[-\dfrac{\pi}{2}, \dfrac{\pi}{2}\right]$ and $0 < a < b$,
find $\cos t$ and $\tan t$.

I. 35. Prove that if $\dfrac{\sin(x - \alpha)}{\sin(x - \beta)} = \dfrac{a}{b}$ and $\dfrac{\cos(x - \beta)}{\cos(x - \alpha)} = \dfrac{A}{B}$, then
$\sec(\alpha - \beta) = \dfrac{aA + bB}{aB + bA}$.

I. 36. If $x \sin \alpha + y \cos \alpha = p$ and $x \cos \alpha - y \sin \alpha = q$, find $x^2 + y^2$.

I. 37. Prove that if $\cos 2\alpha = \cos 2\beta \cos 2\chi$ then
$1 + \cot(\alpha + \beta) \cot(\alpha - \beta) = \csc^2 \chi$

I. 38. Show that, if $\dfrac{\cos \alpha}{\cos \beta} + \dfrac{\sin \alpha}{\sin \beta} = -1$, then $\dfrac{\cos^3 \beta}{\cos \alpha} + \dfrac{\sin^3 \beta}{\sin \alpha} = 1$.

I. 39. Show that, if $a \cos \alpha + b \sin \alpha = c$ and $a \cos \beta + b \sin \beta = c$, then
$\cos^2 \dfrac{\alpha - \beta}{2} = \dfrac{c^2}{a^2 + b^2}$.

ELIMINATIONS

I. 40. Eliminate θ between the relations $\sin\theta - \cos\theta = a$ and $\sin^3\theta - \cos^3\theta = b$.

I. 41. Eliminate θ between the relations $\sin\theta + \cos\theta = a$ and $\sin^3\theta + \cos^3\theta = b$.

I. 42. Eliminate θ between the relations $\sin\theta + \cos\theta = a$ and $\sin^5\theta + \cos^5\theta = b$.

I. 43. Eliminate θ between the relations $\tan^2\theta + \cot^2\theta = a$ and $\tan^4\theta + \cot^4\theta = b$.

I. 44. Eliminate θ between the relations $\dfrac{a}{\sin\theta} = \dfrac{b}{\sin 2\theta} = \dfrac{c}{\sin 3\theta}$.

I. 45. Eliminate θ between the relations $a\sin^2\theta + b\sin\theta = c$ and $b\sin^2\theta + a\sin\theta = c$, $b \neq a$.

I. 46. Eliminate θ between the relations $x = \sin\theta$ and $y = \dfrac{\sin\theta}{2 + \cos\theta}$.

I. 47. Eliminate θ between the relations $x = 2a\cos\theta(1 - \cos\theta)$ and $y = 2a\sin\theta(1 - \cos\theta)$.

I. 48. Eliminate θ between the relations $\tan\theta + \cot\theta = a$ and $\tan 2\theta + \cot 2\theta = b$.

I. 49. Eliminate α and β between the relations $\tan\alpha + \tan\beta = a$, $\cot\alpha + \cot\beta = b$, and $\alpha + \beta = c$.

I. 50. Eliminate α and β between the relations

$$x\cot^2\alpha + y\cot^2\beta = 1 \qquad (1)$$
$$x\cos^2\alpha + y\cos^2\beta = 1 \qquad (2)$$
$$x\sin\alpha = y\sin\beta \qquad (3)$$

I. 51. Eliminate α and β between the relations

$x\cos 2\alpha + y\sin 2\alpha = 2a$ (1)

$x\cos 2\beta + y\sin 2\beta = 2a$ (2)

$2\sin\alpha\sin\beta = 1$ (3)

I. 52. Eliminate α and β between the relations

$\sin\alpha + \sin\beta = a$ (1)

$\cos\alpha + \cos\beta = b$ (2)

$\tan\alpha + \tan\beta = c$ (3)

I. 53. Eliminate α and β between the relations

$x = a\cos\alpha\cos\beta$, $y = b\cos\alpha\cos\beta$, $z = c\sin\alpha$.

I. 54. Eliminate a and b between the relations $\tan x = \sqrt{\dfrac{a-b}{a+b}}$ and $\cos y = \sqrt{\dfrac{a+b}{2a}}$.

I. 55. Consider $a_{n+1} = \sin^n x + \cos^n x$, $n \in \mathbf{N}^\bullet$, $x \in \left(0, \dfrac{\pi}{2}\right)$. Show that $a_{n+1} = a_n\sqrt{1 + \dfrac{2}{t} - \dfrac{1}{t}a_{n-1}}$, where $t = \tan x + \cot x$.

I. 56. Consider $a_n = \sin^n x + \cos^n x$, $n \in \mathbf{N}^\bullet$. Evaluate a_3 and a_4 in terms of t, where $\sin x + \cos x = t$.

Prove that $2a_n = 2t\,a_{n-1} - \left(t^2 - 1\right)a_{n-2}$, $n \geq 3$.

I. 57. Consider $a_n = \tan^n x + \cot^n x$, $n \in \mathbf{N}^\bullet$. Evaluate a_1 and a_2 in terms of t, where $\sin x + \cos x = t$.

Prove that $\left(t^2 - 1\right)a_n = 2a_{n-1} - \left(t^2 - 1\right)a_{n-2}$.

I. 58. Consider $a_n = \tan^n x + \cot^n x$, $n \in \mathbf{N}^\bullet$. Show that $|t| \geq 2$ and prove that $a_n = t\,a_{n-1} - a_{n-2}$, where $\tan x + \cot x = t$.

I. 59. Prove that $a_n = \sin(2n+1)x$ and $b_n = \cos 2nx$ can be written in terms of $\sin x = t$. Also, prove the equalities $a_n = \left(1 - 2t^2\right)a_{n-1} + tb_n + tb_{n-1}$ and $b_n = 2\left(1 - 2t^2\right)b_{n-1} - b_{n-2}$.

UNIT II

Functions, graphs, transformations

Trigonometric functions

$\sin : \mathbf{R} \to [-1, 1]$

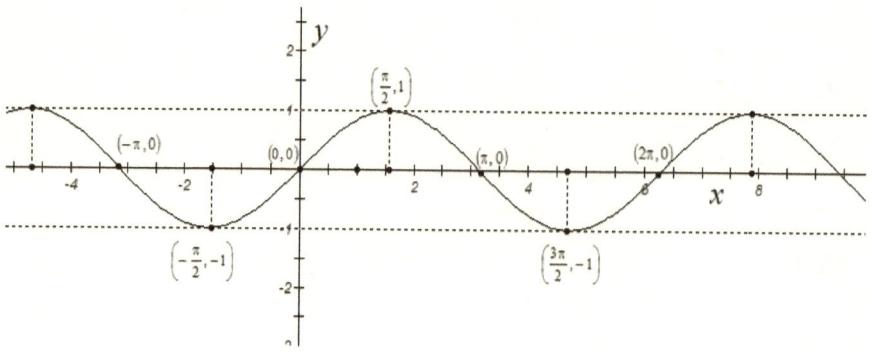

$\cos : \mathbf{R} \to [-1, 1], \quad y = \cos x$

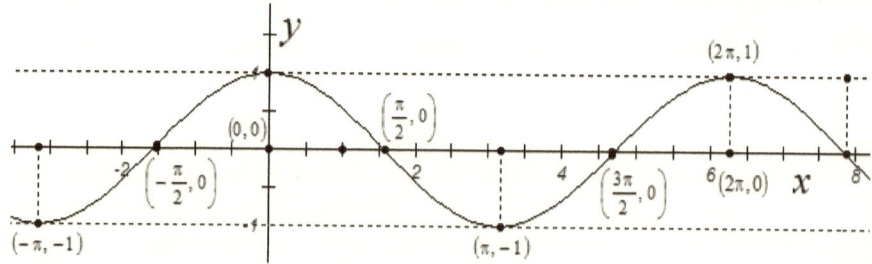

$$\tan : \mathbf{R} - \left\{ (2k+1)\frac{\pi}{2} \,/\, k \in Z \right\} \to \mathbf{R},$$

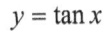

$$y = \tan x$$

$$\cot : \mathbf{R} - \left\{ k\pi \,/\, k \in Z \right\} \to \mathbf{R}, \quad y = \cot x$$

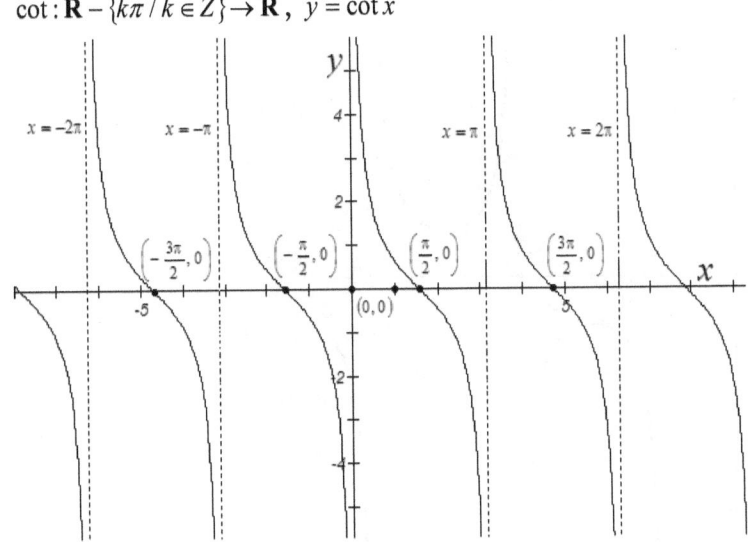

$\csc : \mathbf{R} - \{k\pi / k \in Z\} \to (-\infty, 1] \cup [-1, \infty), \ y = \csc x$

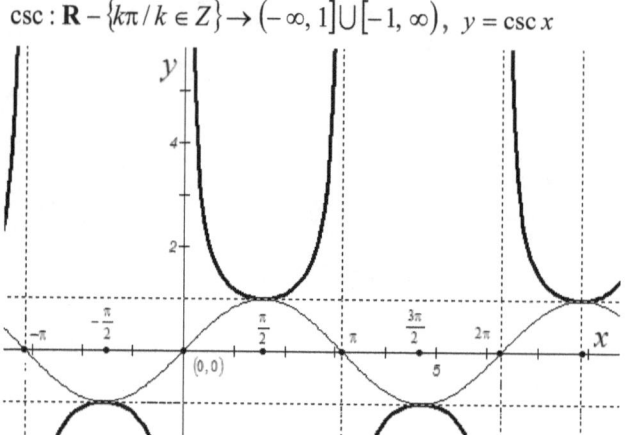

$\sec : \mathbf{R} - \left\{ (2k+1)\dfrac{\pi}{2} / k \in Z \right\} \to (-\infty, 1] \cup [-1, \infty), \ y = \sec x$

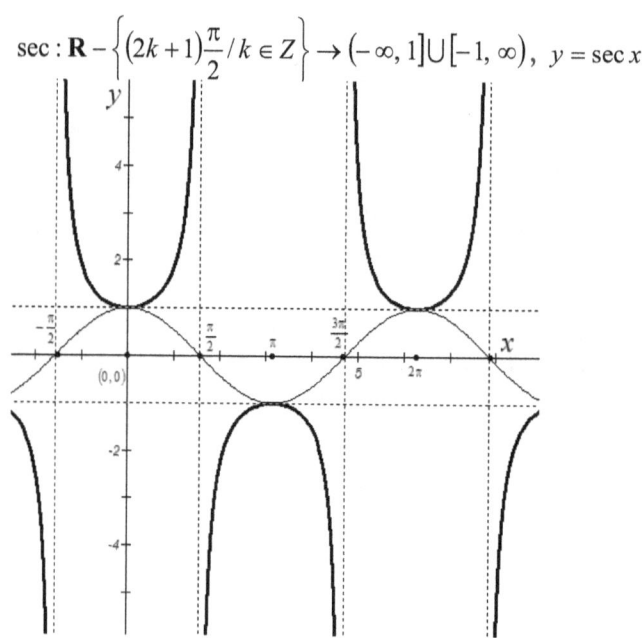

II. 1. Graph each function.

a) $y = 2\sin x + 2$;   b) $y = -2\sin\left(x - \dfrac{3\pi}{4}\right) - 1$;   c) $y = \sin\left(3x + \dfrac{\pi}{2}\right) + \dfrac{2}{3}$;

d) $y = \cos(3x + \pi) + \dfrac{1}{2}$;   e) $y = -3\sin\left(\dfrac{3}{2}x - \dfrac{\pi}{4}\right) - 6$;   f) $y = \tan(x + \pi) - 7$;

g) $y = \tan\dfrac{x}{2} - 5$;   h) $y = 3\tan\left(\dfrac{2}{3}x + \dfrac{\pi}{2}\right) + 8$;   i) $y = 5 + \csc\left(x - \dfrac{\pi}{2}\right)$;

j) $y = 2\sec\left(\dfrac{x}{2} + \pi\right) - \dfrac{1}{2}$;   k) $y = \cos\dfrac{3x}{2} - \dfrac{1}{2}$;   l) $y = -2\sin\left(x - \dfrac{3\pi}{4}\right) - 1$;

m) $y = \sin\left(3x + \dfrac{\pi}{2}\right) + \dfrac{2}{3}$;   n) $y = -3\cos\left(\dfrac{1}{2}x + \pi\right) - 1$.

II. 2. Let $f : D \to \mathbf{R}$, $f(x) = \sqrt{\sin^4 x + 4\cos^2 x} + \sqrt{\cos^4 x + 4\sin^2 x} + 1$ be a function. Find D and show that $f(x) = 4$. Draw the graph of f.

II. 3. Show that the function $f : \mathbf{R} \to \mathbf{R}$, $f(x) = 2\{x\} - \cos 2\pi x$, where $\{x\}$ is the fractional part of the real number x, is a periodic function.

II. 4. If the roots of the equation $x^2 - 2x\cos a + \sin^2 a = 0$, $a \in \mathbf{R}$ are real and distinct, then the equation $x^2 - 2x\sin a + \cos^2 a = 0$ has no real roots.

II. 5. Show that the function $f : \mathbf{R} \to \mathbf{R}$,

$f(x) = (1 - \sin a)x^2 - 2x\cos a + 1 + \sin a$ is positive on its domain of definition.

II. 6. Prove that the function $f : \mathbf{R} \setminus \left\{\dfrac{n\pi}{2}, n \in \mathbf{Z}\right\} \to \mathbf{R}$

$f(x) = \dfrac{m\cos^3 x - \cos 3x}{\cos x} + \dfrac{m\sin^3 x + \sin 3x}{\sin x}$ is a constant function for any $m \in \mathbf{R}$.

II. 7. Find the function f, such that:

a) $f(x) + 3f(-x) = \sin x + \cos x$;

b) $f(x)\sin x + f(-x)\cos x = \sin x + \cos x$.

II. 8. Find the range of the functions:

a) $f(x) = \cos 2x - 4\sin x$, $x \in \mathbf{R}$;

b) $f(x) = \sin^2 2x - \sin x \sin 3x$, $x \in \mathbf{R}$;

c) $f(x) = \tan x + 3 \cot x$, $x \in \left(0, \dfrac{\pi}{2}\right)$;

d) $f(x) = (\sin x + \csc x)^2 + (\cos x + \sec x)^2$, $x \in \left(0, \dfrac{\pi}{2}\right)$;

e) $f(x) = \sec\left(\dfrac{\pi}{6} - x\right) + \sec\left(\dfrac{\pi}{6} + x\right)$, $x \in \left[0, \dfrac{\pi}{3}\right]$.

II. 9. Find the domain (D) and range (R) of the function:

a) $f(x) = \dfrac{\sin x + \tan x}{\cos x + \cot x}$; **b)** $f(x) = \dfrac{5}{2\cos\left(2x - \dfrac{\pi}{3}\right) + 3}$.

II. 10. Find the range of the functions:

$f(x) = \dfrac{x(1 - x^2)}{(1 + x^2)^2}$ and $g(x) = \dfrac{x(1 - x^2)(x^4 - 6x^2 + 1)}{(1 + x^2)^4}$ for any $x \in \mathbf{R}$.

II. 11. Find the period and range of the function

$f(x) = \cos 3x + \sqrt{3} \sin 3x + 1$.

II. 12. State whether following functions are odd, even or neither:

a) $f(x) = 2x^4 - 3\cos x$;

b) $f(x) = 3\tan^2 x - \cos x + 2x^2$;

c) $f(x) = \cos x - \sin^2 x + 2\cot^2 x$;

d) $f(x) = x - \sin^3 x + 2\cot^5 x + \tan x$;

e) $f(-x) = \sin x - 2\tan x + 4\cos x$.

II. 13. Show that the principal period of the function

$f(x) = \sin 6x + \cos 15x$, $x \in \mathbf{R}$ is $\dfrac{2\pi}{3}$.

II. 14. Find the period of the functions:

a) $f(x) = \sin 20x + \cos 30x$, $x \in \mathbf{R}$;

b) $f(x) = |3\cos 2x|$, $x \in \mathbf{R}$;

c) $f(x) = \sin 2x + \tan x$, $x \in \mathbf{R}$;

d) $f(x) = 4\tan\dfrac{x}{2} - \tan\dfrac{x}{3}$, $x \in \mathbf{R}$;

e) $f(x) = 3\tan 2x - 2\cot 3x$, $x \in \mathbf{R}$;

f) $f(x) = 2\sin\dfrac{\pi x}{3} + \cos\dfrac{\pi x}{4}$, $x \in \mathbf{R}$;

g) $f(x) = \sin^2 2x + \cot 2x$, $x \in \mathbf{R}$;

h) $f(x) = \sin 6x \cdot \cos 15x$, $x \in \mathbf{R}$;

i) $f(x) = 2(\sin x + \cos x)^2$, $x \in \mathbf{R}$;

j) $f(x) = \dfrac{\sin 2x}{2\cos\dfrac{x}{2} + \tan\dfrac{x}{2}}$.

II. 15. Show that the function $f(x) = 2\sin 4x + \cos\left(x\sqrt{2}\right)$, $x \in \mathbf{R}$ is not a periodic function.

II. 16. Consider the function $f : \left(0, \dfrac{\pi}{2}\right) \to \mathbf{R}$ $f(x) = \tan x + 3\cot x$. Find the minimum value of $f(x)$.

II. 17. Find $\alpha \in \left(0, \dfrac{\pi}{2}\right)$ such as the expression

$\dfrac{\cot \alpha - \tan \alpha}{1 + \cos 4\alpha}$ has a minimum value.

II. 18. Consider the function $f : \mathbf{R} \to \mathbf{R}$

$f(x) = 2\sin^2 x + 3\sin x \cos x + 4\cos^2 x$.

Find the maximum and the minimum value of $f(x)$.

II. 19. The graph of a periodic function is shown below. What is the period of the function?

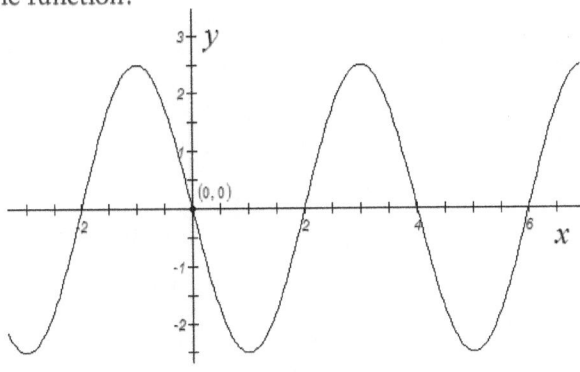

II. 20. The graph of a periodic function is shown below. What is the equation of the axis of the function?

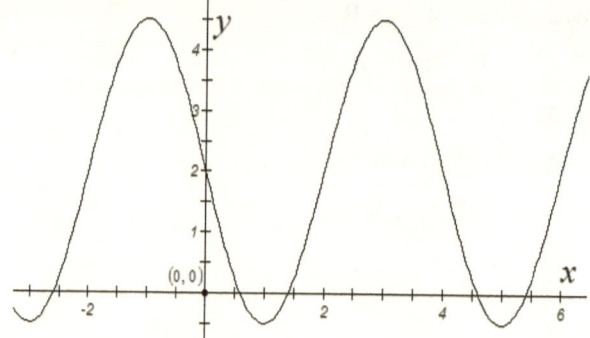

II. 21. Identify the function with a period of 4 and amplitude of 2.5.

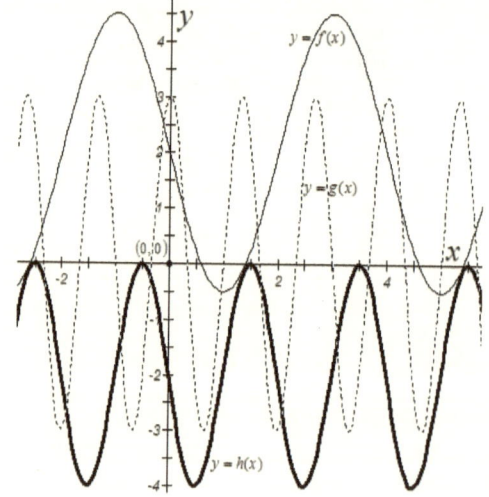

II. 22. Determine the equation of the axis of the function

$$y = -3.1\cos 4\left(x - \frac{\pi}{3}\right) + 2.$$

II. 23. Determine the coordinates of the point $(0,5)$ after a rotation of $1500°$ about $(0,0)$.

II. 24. The height of a saw tooth in inches after time t in seconds is given by the function $y = 0.4\sin(16t + 2.4) - 0.2$. Determine the maximum height the saw tooth reaches.

II. 25. The height of a saw tooth in inches after time t in seconds is given by the function $y = 0.8\cos(10t + 1.4) - 0.5$. Determine the minimum height the saw tooth reaches.

II. 26. State the transformations for the functions:

a) $y = 0.64\cos(6x + 32.4) - 2$;
b) $y = 4\sin(2x + 4) + 3$;
c) $y = -3\sin(4x + \pi) + 5$;
d) $y = -2\sin(-4x + \pi) - 3$.

II. 27. Which of the equations of a sine function would have a range of $\{y \in \mathbf{R} \mid 3 \le y \le 8\}$ and a period of $60°$?

a) $y = 2.5\sin(6x + 32.4) + 5.5$;
b) $y = 3\sin(6x + 60°) + 5$;
c) $y = -3.5\sin(6x + \pi) + 4.5$;
d) $y = -2.5\sin(6x - \pi) + 4.5$.

II. 28. Determine a correct equation of the following graph:

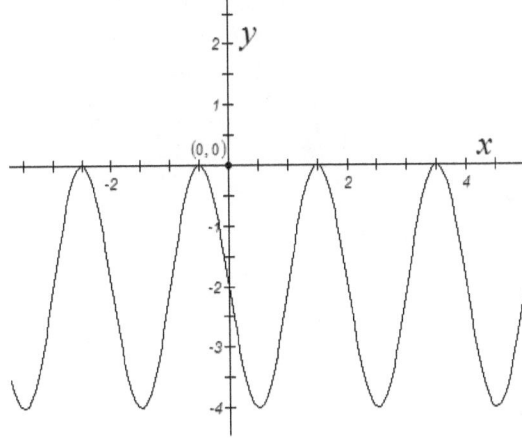

31

II. 29. Determine an equation of the following graph as a sine and a cosine function:

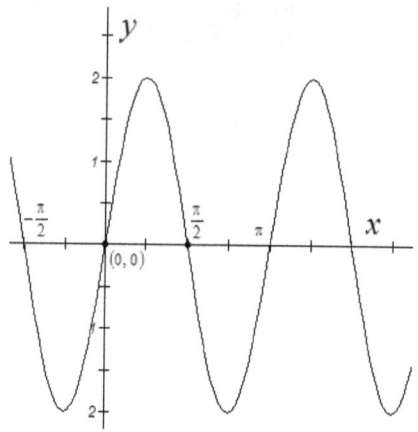

II. 30. Find a sine and cosine equations for each of the following:

a)

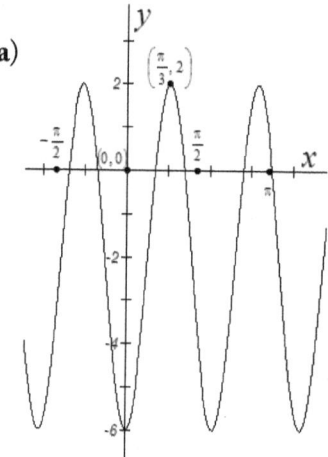

b)

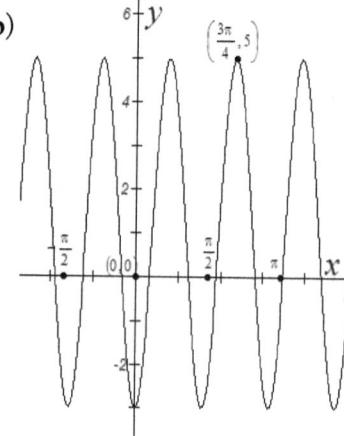

c)

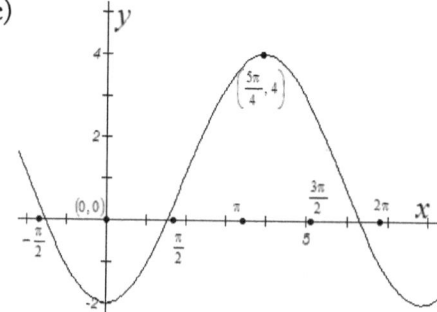

d)

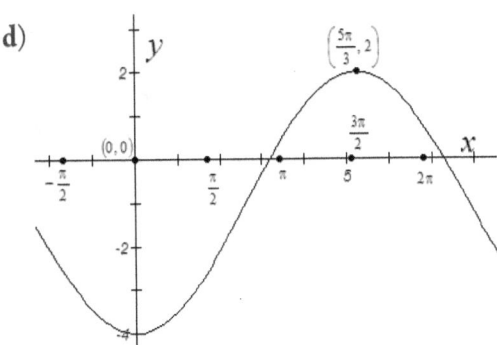

II. 31. Consider the function $y = 2\cos\left(\dfrac{1}{2}x - \dfrac{3\pi}{4}\right) + 1$. Find the phase shift and the period.

II. 32. Consider the function $y = 2\cos\left(2x + \dfrac{\pi}{4}\right) + 1.25$,

a) What is the amplitude?
b) What is the period?
c) Describe the phase shift;
d) Describe the vertical translation;
e) If the base function is $y = \cos x$, what are the mapping transformations?
f) Sketch a graph of the function over two cycles;
g) Write an equivalent sine function.

II. 33. A sinusoidal function has an amplitude of 1.75 units, a period of $1440°$, and a maximum at $(0, -2)$. Determine an equation for the function.

II. 34. A part of the graph of a sinusoidal function is given below.

The point $A\left(\dfrac{\pi}{4}, 3\right)$ is a maximum point

and $B\left(\dfrac{3\pi}{4}, -4\right)$ is a minimum point.

Find a sine and a cosine
equation of the graph
function passing
through these points.

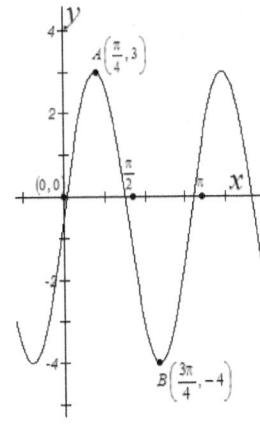

33

II. 35. The graph of

$y = a \cos px$
is given below.
Find the value of a, p.

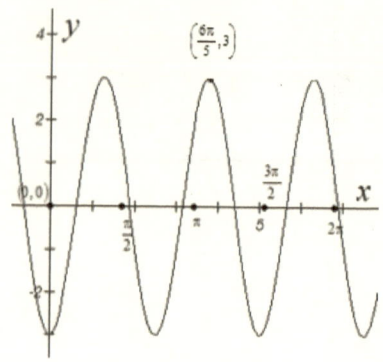

II. 36. The graph of

$y = a \cos(x - b)$
is given below.
Find the value of a,
and one of the value of b.

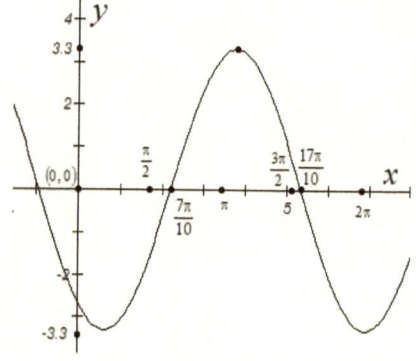

II. 37. The graph of $y = a \cos px + h$
is given below.
The point

$A(4, 5)$ is a maximum point and

$B(0, -3)$ is a minimum point.
Find the value of a, p, and h.

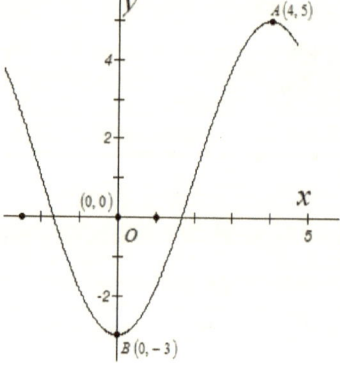

II. 38. Consider the function $f : \mathbf{R} \to [-1,1]$, $f(x) = \sin x$. The following transformations are applied:

a) an amplitude of 4;

b) a horizontal compression by a factor of $\dfrac{1}{2}$;

c) a horizontal translation of $\dfrac{\pi}{8}$ to the right;

d) a vertical translation of 1 unit down.

Use the transformation above to sketch the graph of the transformed function.

II. 39. Consider the function $f : \mathbf{R} \to [-1,1]$, $f(x) = \sin x$. Determine the equation of the transformed function in each case:

a) A period of π, an amplitude 5, and the equation of the axis $y = -1$;

b) A horizontal compression by a factor of $\dfrac{1}{3}$, an amplitude 3, a horizontal translation of 2π to the left, a reflection in the x-axis, and a vertical translation of 1 unit down;

c) A period of $\dfrac{\pi}{6}$, an amplitude 6, a horizontal translation of $\dfrac{2\pi}{5}$ to the right, a reflection in x-axis, and the equation of the axis $y = -3$;

d) A period of 2, an amplitude 6, a horizontal translation of $\dfrac{\pi}{2}$ to the left, a reflection in x-axis, a reflection in y-axis, and a vertical translation of 3 units down;

e) A vertical stretch by a factor of 4, a reflection in the x-axis, a horizontal compression by a factor of $\dfrac{1}{3}$, a horizontal translation of $\dfrac{\pi}{6}$ units to the left, and a vertical translation of 5 units down.

II. 40. The points $\left(\dfrac{\pi}{2}, 0\right)$ and $(\pi, -1)$ are on the curve $y = \cos\theta$. State the new coordinates of these points under the transformations given by $y = 3\cos\dfrac{2}{5}\left(\theta - \dfrac{\pi}{4}\right) - 2$.

II. 41. A sinusoidal function has a period of $\dfrac{\pi}{5}$, an amplitude of 6 units, a horizontal translation of $\dfrac{\pi}{4}$ to the right, a reflection in y-axis, and a vertical translation of 3 units down. Three points on the transformed curve are $\left(\dfrac{7\pi}{30},0\right)$, $\left(\dfrac{\pi}{5},3\right)$, and $\left(\dfrac{3\pi}{20},-3\right)$. Determine the points of the initial function.

II. 42. A cosine function has a period of π, a maximum value of 5, a minimum value of –3, and a phase shift of $\dfrac{\pi}{5}$ to the left. Write an equation for the function and draw the graph over two periods.

II. 43. A sine function has two consecutive points of maximum at $\left(-\dfrac{1}{6}\dfrac{\pi}{},1\right)$ and $\left(-\dfrac{\pi}{2},1\right)$. One of the minimums points is at $\left(-\dfrac{7\pi}{6},-5\right)$. Find an equation of the graph function passing through these points and draw the graph.

II. 44. A sinusoidal function has two consecutive points of minimum at $\left(\dfrac{\pi}{3},-4\right)$ and $(\pi,-4)$. One of the maximum points is at $\left(\dfrac{2\pi}{3},2\right)$.
Find an equation of the graph function passing through these points.

II. 45. A sinusoidal function has two consecutive points of minimum at $\left(\dfrac{19\pi}{12},-3\right)$ and $\left(\dfrac{17\pi}{6},-3\right)$. One of the maximum points is at $\left(\dfrac{23\pi}{24},2\right)$

Find an equation of the graph function passing through these points.

II. 46. A sine function has two consecutive points of minimum at $\left(\dfrac{\pi}{3},-1\right)$ and $\left(\dfrac{19\pi}{12},-1\right)$. The range of the function is the set $\{y \in R \,|-1 \le y \le 5\}$. Write an equation of the function.

II. 47. The graph of the function $f(x)=\sin x$ is vertically compressed by a factor of $\dfrac{1}{3}$, reflected in the x-axis and y-axis, horizontally compressed by a factor of $\dfrac{1}{4}$, horizontally translated $\dfrac{\pi}{4}$ units to the left, and vertically translated 6 units up. Write the equation of the transformed function.

II. 48. The graph of the function $y = \cos x$ is reflected in y-axis, vertically compressed by a factor of $\dfrac{1}{7}$, horizontally compressed by a factor of $\dfrac{1}{3}$, horizontally translated $\dfrac{\pi}{6}$ units to the left, and vertically translated 10 units down. Determine the value of the new function when $x = 0$.

II. 49. Create a flow chart that summarises the steps used for the transformed function $y = 3\cos\left(2x + \dfrac{\pi}{3}\right) + 1$.

II. 50. Determine the average rate of change of the function $y = f(x) = 3\cos\left(2x + \dfrac{\pi}{3}\right) + 1$ for each interval

a) $0 \le x \le \dfrac{\pi}{6}$; b) $0 \le x \le \dfrac{\pi}{2}$; c) $\dfrac{\pi}{3} \le x \le \dfrac{\pi}{2}$;

d) $\dfrac{\pi}{2} \le x \le \dfrac{2\pi}{3}$; e) $\dfrac{\pi}{6} \le x \le \dfrac{2\pi}{3}$; f) $\dfrac{\pi}{12} \le x \le \dfrac{7\pi}{12}$.

II. 51. For the graph of the function below, state two points where the function has an instantaneous rate of change in $f(x)$ that is
a) zero;
b) a negative value;
c) a positive value.

$$y = -3\cos\dfrac{4}{5}x + 1$$

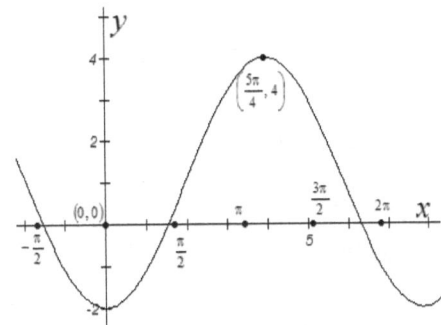

II. 52. Consider the function $f : \mathbf{R} \rightarrow [-5, 1]$, $f(x) = 3\cos\dfrac{3}{2}\left(x + \dfrac{\pi}{2}\right) - 2$ and its graph below:

State a point(s) or an interval (s) in which the function has an *instantaneous rate of change* in $f(x)$ that is

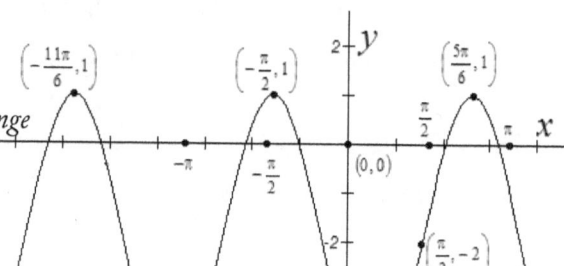

a) zero;
b) a negative value;
c) a positive value;
d) the greatest value;
e) the least value.

II. 53. a) Estimate the instantaneous rate of change of the function $f(x) = \sin x$ for the following values of x: 0, $\dfrac{\pi}{2}$, π, $\dfrac{3\pi}{2}$. 2π.

Use the points that represent the instantaneous rate of change to find the equation of a cosine function (this function is called *the derivative* of $f(x) = \sin x$);

b) Determine the derivative of the function $f(x) = 3\cos 2\left(x + \dfrac{\pi}{3}\right) - 1$

using the following values of x: $-\dfrac{\pi}{3}$, $-\dfrac{\pi}{12}$, $\dfrac{\pi}{6}$, $\dfrac{5\pi}{12}$, $\dfrac{2\pi}{3}$.

II. 54. Draw the graph of the functions:

a) $f(x) = \sin x + \cos x$;

b) $f(x) = 2\sin 2x - 4\sqrt{3}\cos^2 x + 2\sqrt{3} + 1$;

c) $f(x) = \sqrt{3}\sin 4x - 8\sin^2 x \cos^2 x - 1$;

d) $f(x) = 4\sin^2(\pi - x)\cos^2(\pi + x)$;

e) $f(x) = \sin^2\left(\dfrac{\pi}{8} + x\right) + \sin^2\left(\dfrac{\pi}{8} - x\right)$;

f) $f(x) = \cos\left(\dfrac{\pi}{6} - \dfrac{x}{2}\right)\sin\left(\dfrac{\pi}{3} - \dfrac{x}{2}\right)\sin\dfrac{x}{2} - 1$;

g) $f(x) = 2\cos^2\left(\dfrac{3\pi}{2} - \dfrac{x}{2}\right) + \sqrt{3}\cos\left(\dfrac{\pi}{2} - x\right) - 2$.

II. 55. Draw the graph of the functions:

a) $f(x) = \begin{cases} -1 & \text{if } x < -\dfrac{\pi}{2} \\ \sin x & \text{if } x \in \left[-\dfrac{\pi}{2}, \dfrac{\pi}{2} \right]; \\ 1 & \text{if } x > \dfrac{\pi}{2} \end{cases}$

b) $f(x) = \sin x - |\sin x|$.

II. 56. State the period and draw the graph of the function

$$f(x) = \frac{\sin x + \sin 3x}{\sqrt{1 + \cos 2x}} + \frac{\cos x - \cos 3x}{\sqrt{1 - \cos 2x}}, \quad x \in (0, 2\pi).$$

UNIT III

Identity Transformations of Trigonometric Expressions

III. 1. Calculate:

a) $\sin(-1920°)$;   b) $\cos(-2280°)$;   c) $\tan(-1830°)$;   d) $\cot(-1860°)$;

e) $\sin 7°30'$;   f) $\cos 7°30'$;   g) $\tan 7°30'$;   h) $\cot 7°30'$.

III. 2. Use the sum-product identities to find identities for each of the following:

a) $1 + \cos\dfrac{\pi}{4}$;   b) $1 - \cos\dfrac{\pi}{6}$;   c) $1 - \tan\dfrac{\pi}{3}$;   d) $1 + \tan\dfrac{\pi}{6}$;

e) $\sin\dfrac{\pi}{4} + \sin\dfrac{\pi}{6}$;   f) $\cos\dfrac{\pi}{6} - \cos\dfrac{\pi}{4}$;   g) $\sin\dfrac{\pi}{3} - \sin\dfrac{\pi}{6}$;

h) $\tan\dfrac{\pi}{4} + \tan\dfrac{\pi}{3}$;   i) $\sin\dfrac{\pi}{12} + \sin\dfrac{5\pi}{12}$;   j) $\cos\dfrac{\pi}{12} + \cos\dfrac{5\pi}{12}$.

III. 3. Calculate

a) $\sin\dfrac{5\pi}{7}\cos\dfrac{13\pi}{28} - \cos\dfrac{5\pi}{7}\sin\dfrac{13\pi}{28}$;   b) $\cos\dfrac{13\pi}{21}\cos\dfrac{2\pi}{7} + \sin\dfrac{13\pi}{21}\sin\dfrac{2\pi}{7}$;

c) $\dfrac{\tan\dfrac{\pi}{9} + \tan\dfrac{\pi}{18}}{1 - \tan\dfrac{\pi}{9}\cdot\tan\dfrac{\pi}{18}}$;   d) $4\cos\dfrac{\pi}{12}\sin\dfrac{5\pi}{48}\cos\dfrac{\pi}{48} - \left(\sin\dfrac{5\pi}{24} + \cos\dfrac{1}{24}\dfrac{\pi}{24} + \dfrac{1}{2}\right)$.

III. 4. Find a simpler form for the expressions:

a) $\dfrac{\sin^2 x - \sin^4 x}{\cos^2 x} + \cos^2 x$;   b) $\dfrac{\cos^2 \alpha - \sin^2 \alpha}{\cos^4 \alpha - \sin^4 \alpha}$;

c) $\dfrac{1+\sin^2 x}{2+\cot^2 x} + \dfrac{1+\cos^2 x}{2+\tan^2 x}$;   d) $\dfrac{(\sin x + \cos x)^2 - 1}{\cot x - \sin x \cos x}$;

e) $\dfrac{\cos^2 x - \cot^2 x + 1}{\sin^2 x + \tan^2 x - 1}$;   f) $\dfrac{\sin^2 2x - 4\sin^2 x \cos^2 x + 4\sin^4 x}{3 - 4\sin^2 x + \cos^2 2x}$;

g) $\dfrac{\cos^2 x + 2\sin^2 x}{\cos^3 x} + \dfrac{\cos^2 x + 4\sin x + \sin^2 x}{\cos x(4\sin x + 1)}$;   h) $\csc^4 x - \sec^4 x$;

i) $\dfrac{\cot x + \tan x}{1 + \tan 2x \tan x}$;   j) $\dfrac{2\tan^2 2x - (1 + \tan 2x)^2}{1 + \tan^2 x} + \sin 4x + 1$;

k) $\dfrac{\tan^3 x - \tan^5 x}{\cot^3 x - \cot x}$;   l) $\dfrac{\cos a \cdot (1 + \cos^2 a)}{\sin^2 a + 2\cos^2 a}$.

III. 5. Find the identities for each of the following:

a) $\dfrac{1 + \tan x + \cot x}{1 + \tan x + \tan^2 x} - \dfrac{\cot x}{1 + \tan^2 x}$;   b) $\dfrac{1 + 2\cot x \cot 2x}{1 + \cos 2x}$;

c) $\sin^2 x + \dfrac{\cot^2 x}{1 + \cot^2 x} - 2\sin^2 \dfrac{x}{2}$;   d) $\dfrac{1 + \sin^2 x \tan^2 x - 3\sin^2 x}{1 - \tan^2 x}$;

e) $\dfrac{\dfrac{1}{\sin 2x} - \dfrac{1}{\tan 2x}}{\dfrac{1}{\sin 2x} + \dfrac{1}{\tan 2x}}$;   f) $\dfrac{\csc 2x - \cot 2x}{\csc 2x + \cot 2x}$;   g) $\dfrac{\tan 2x}{\tan 4x - \tan 2x}$;

h) $\left[\dfrac{(1 + \cot x)^2 - (1 + \tan x)^2}{(1 + \cot x)^2 + (1 + \tan x)^2} \right]^2 + \left(\dfrac{\cot x}{1 + \cot^2 x} + \dfrac{\tan x}{1 + \tan^2 x} \right)^2$.

III. 6. Find the identities for each of the following:

a) $\sin a + \sin 3a$;   b) $\cos 3a - \cos 5a$;   c) $\sin(a + b) - \sin(a - b)$;

d) $\cos(a - b) + \cos(a + b)$;   e) $\sin\left(\dfrac{\pi}{4} + a\right) + \sin\left(\dfrac{\pi}{4} - a\right)$;

f) $\cos\left(\dfrac{a+b}{2}\right) + \cos\left(\dfrac{a-b}{2}\right)$;   g) $\tan(a + b) + \tan(a - b)$.

III. 7. Find the identities for each of the following:

a) $\cos(x+y+z)+\cos(x-y-z)$;

b) $\sin(x-y+z)+\sin(x+y+z)$;

c) $\sin(x+y-z)-\sin(x+y+z)$.

III. 8. Find the identities for each of the following:

a) $\cos x + \cos 2x + \cos 3x$;

b) $\cos x + \cos 2x + \cos 3x + \cos 4x$;

c) $\sin x + \sin 3x + \sin 9x + \sin 1\ x$;

d) $\cos\dfrac{2\pi}{9} + \cos\dfrac{4\pi}{9} + \cos\dfrac{6\pi}{9} + \cos\dfrac{8\pi}{9}$;

e) $\cos x + \cos 3x + \cos 9x + \cos 1\ x$;

f) $\cos\left(\dfrac{\pi}{2}+x\right) - \sin\left(\dfrac{31\pi}{2}-x\right) + \cos(7\pi+x) + \sin(15\pi-x)$;

g) $\left(\dfrac{\sqrt{2}}{2}+\sin x\right)\left(\dfrac{\sqrt{2}}{2}-\sin x\right) + \left(\dfrac{\sqrt{2}}{2}+\cos x\right)\left(\dfrac{\sqrt{2}}{2}-\cos x\right)$;

h) $\sin\left(x-\dfrac{3\pi}{2}\right) + \cos\left(x-\dfrac{2\pi}{3}\right) + \cos\left(x+\dfrac{2\pi}{3}\right)$.

III. 9. Find the identities for each of the following:

a) $\sin^2(x+2y)-\sin^2(x-2y)$; b) $\cos^2(x+y)+\sin^2(x-y)-1$;

c) $\sin^2(x+y)+\sin^2(x-y)-1$; d) $(\cos x-\cos 2y)^2 + (\sin x+\sin 2y)^2$;

e) $\cos^2 x+\cos^2 2y-\sin^2(x-2y)$; f) $\sin^2 2x+\sin^2 y-\sin^2(2x-y)$;

g) $\cos^2 \alpha + \cos^2 \beta - 2\cos\alpha\cos\beta\cos(\alpha-\beta)$;

h) $\cos^2 \alpha - \sin^2 \beta + 2\sin\alpha\sin\beta\cos(\alpha-\beta)$;

i) $\dfrac{\sin^2 x - \sin^2 y}{\sin(x-y)}$; j) $\dfrac{\sin^2 \alpha + \sin^2 \beta - \sin^2(\alpha+\beta)}{\cos^2 \alpha + \cos^2 \beta - \sin^2(\alpha+\beta)}$.

III. 10. Find the identities for each of the following:

a) $\tan\dfrac{x}{2} + \cot\dfrac{x}{2} + 2$; b) $\dfrac{\tan\dfrac{x}{2} - \cot\dfrac{x}{2}}{\tan\dfrac{x}{2} + \cot\dfrac{x}{2}}$; c) $\dfrac{1+\cot x\cot\dfrac{x}{2}}{\tan\dfrac{x}{2} + \cot\dfrac{x}{2}}$;

d) $\sin\left(\dfrac{\pi}{3}+\dfrac{x}{2}\right)\sin\left(\dfrac{\pi}{3}-\dfrac{x}{2}\right)\sin\dfrac{x}{2}$; **e)** $\dfrac{1+\sin 2x}{\cos 2x\cot\left(x-\dfrac{\pi}{4}\right)}+\cos^2 x$;

f) $\dfrac{(1-\cos x)\sin\left(\dfrac{\pi}{4}-x\right)}{2\sin^2 x-\sin 2x}$; **g)** $1-\dfrac{1}{1-\csc\left(\dfrac{\pi}{2}+2x\right)}$;

h) $\dfrac{-\tan\left(\dfrac{3\pi}{4}+x\right)(1+\sin 2x)}{\cos\left(\dfrac{3\pi}{2}+2x\right)}$; **i)** $\dfrac{\sin(\pi-6x)}{\sin 2x}+\dfrac{\cos(\pi-6x)}{\cos 2x}$.

III. 11. Find the identities for each of the following:

a) $\dfrac{\cot(x-270°)}{1-\tan^2(x-180°)}\cdot\dfrac{\cot^2(x-360°)-1}{\cot(x+180°)}$; **b)** $\dfrac{4\sin^2(\alpha+\pi)-\sin^2(2\alpha+\pi)}{\cos^2\left(2\alpha+\dfrac{3\pi}{2}\right)-4\cos^2(\alpha+\pi)}$;

c) $\dfrac{\sin^2(2x-\pi)-4\cos^2\left(\dfrac{3\pi}{2}-x\right)}{\sin^2\left(2x-\dfrac{5\pi}{2}\right)-4\sin^2\left(\dfrac{5\pi}{2}+x\right)+3}$; **d)** $\dfrac{4\cos^2(x-3\pi)-\cos^2(2x-3\pi)-3}{4\cos^2(x+3\pi)+\cos^2(2x+3\pi)-1}$;

e) $\dfrac{\cos^2(x-270°)}{\csc^2(x+90°)-1}+\dfrac{\sin^2(x+270°)}{\sec^2(x-90°)-1}$; **f)** $\dfrac{\left[1+\tan^2(x-90°)\right]\cdot\left[\csc^2(x-90°)-1\right]}{\left[1+\cot^2(x+90°)\right]\sec^2(x+90°)}$;

g) $\dfrac{\cos^2(x-90°)+\cot^2(x+90°)+1}{\sin^2(x+90°)+\tan^2(x+90°)+1}$; **h)** $\dfrac{4-\cos^2\left(\dfrac{\pi}{2}-2x\right)-4\cos^2\left(\dfrac{3\pi}{2}-x\right)}{8\sin^2(3\pi-x)-1-\cos(4x-3\pi)}$.

III. 12. Find the identities for each of the following:

a) $3\sin^2\left(\alpha+\dfrac{\pi}{2}\right)-\cos^2\left(\alpha-\dfrac{\pi}{2}\right)$;

b) $\cot(2x-7\pi)\left[\sin^4\left(\dfrac{7\pi}{4}-x\right)-\sin^4\left(\dfrac{7\pi}{4}+x\right)\right]$;

c) $\tan^2\left(\alpha-\dfrac{5\pi}{2}\right)-\cot^2\left(\alpha+\dfrac{5\pi}{2}\right)$;

d) $\sin^2\left(3x-\dfrac{3\pi}{2}\right)(1-\tan^2 3x)\tan\left(\dfrac{\pi}{4}+3x\right)\sec^2\left(\dfrac{\pi}{4}-3x\right)$;

e) $\dfrac{\sin^2\left(\dfrac{3\pi}{4}+\alpha\right)-\cos^2\left(\dfrac{3\pi}{4}+\alpha\right)}{\left(\cos\dfrac{\alpha}{4}+\sin\dfrac{\alpha}{4}\right)\left[\cos\left(4\pi-\dfrac{\alpha}{4}\right)+\cos\left(\dfrac{\pi}{2}+\dfrac{\alpha}{4}\right)\right]\cos\left(\dfrac{\pi}{2}+\dfrac{\alpha}{2}\right)}$;

f) $\dfrac{1}{2\tan\left(\dfrac{\pi}{2}-x\right)\cos^2\left(x-\dfrac{3\pi}{2}\right)}+\dfrac{1-\cos(4x+\pi)}{\cos^3\left(2x-\dfrac{\pi}{2}\right)}-\dfrac{1}{2\cot\left(x+\dfrac{\pi}{2}\right)\sin^2\left(x-\dfrac{3\pi}{2}\right)}$;

g) $\dfrac{\sec x-\cos\left(x-\dfrac{2\pi}{3}\right)\tan x}{1+\cos 2x}-\dfrac{1}{4\sin^2\left(\dfrac{x}{2}-\dfrac{\pi}{4}\right)\cot\left(\dfrac{x}{2}-\dfrac{\pi}{4}\right)}$.

III. 13. Find the identities for each of the following:

a) $1+\sin 2x-2\cos^2 x$; b) $1+\cos(2x+270°)-\sin(2x+270°)$;

c) $1-\cos(2x-270°)+\sin(2x+270°)$; d) $1+\cot\left(\dfrac{3\pi}{2}-2x\right)+\csc\left(\dfrac{3\pi}{2}+2x\right)$;

e) $\sin(2x-3\pi)\cos(x-3\pi)-\sin\left(2x-\dfrac{7\pi}{2}\right)\cos\left(x-\dfrac{7\pi}{2}\right)$;

f) $\sin(x+4\pi)\cos\left(2x-\dfrac{7\pi}{2}\right)-\sin\left(\dfrac{5\pi}{2}-x\right)\sin\left(2x-\dfrac{5\pi}{2}\right)$;

g) $\sin\left(x-\dfrac{3\pi}{2}\right)+\cos\left(x-\dfrac{2\pi}{3}\right)+\cos\left(x+\dfrac{2\pi}{3}\right)$;

h) $1+\sqrt{3}\cos\left(2x-\dfrac{3\pi}{2}\right)-2\cos^2\left(x-\dfrac{3\pi}{2}\right)$;

i) $1+\cos\left(x-\dfrac{3\pi}{2}\right)+\sin\left(x+\dfrac{3\pi}{2}\right)-\cot\left(x+\dfrac{3\pi}{2}\right)$;

j) $\dfrac{1+\cos\alpha+\cos 2\alpha+\cos 3\alpha}{1-\cos\alpha-2\cos^2\alpha}$; k) $\dfrac{1+\cos(2x-2\pi)-\cos\left(2x+\dfrac{3\pi}{2}\right)}{1-\cos(2x+2\pi)+\cos\left(2x+\dfrac{3\pi}{2}\right)}$.

III. 14. Find the identities for each of the following:

a) $\dfrac{\cos 3x-\cos 5x}{\sin 5x-\sin 3x}$; b) $\dfrac{\sin 7x+\sin 3x}{\cos 7x+\cos 3x}$; c) $\dfrac{2\cos x+\cos 5x+\cos 3x}{2\sin x+\sin 5x-\sin 3x}$;

d) $\dfrac{\cos 3x+\cos 4x+\cos 5x}{\sin 3x+\sin 4x+\sin 5x}$; e) $\dfrac{\sin 4x+\sin 5x+\sin 6x}{\cos 4x+\cos 5x+\cos 6x}$;

f) $\dfrac{\sin x - 2\cos 3x - \sin 5x}{\cos x - 2\sin 3x - \cos 5x}$; g) $\dfrac{\sin(2x+2\pi)+2\sin(4x-5\pi)+\sin(6x+6\pi)}{\cos(2x-2\pi)+2\cos(4x-5\pi)+\cos(6x-6\pi)}$;

h) $\dfrac{\sin 13x + \sin 14x + \sin 15x + \sin 16x}{\cos 13x + \cos 14x + \cos 15x + \cos 16x}$; i) $\dfrac{\cos 7x - \cos 8x - \cos 9x + \cos 10x}{\sin 7x - \sin 8x - \sin 9x + \sin 10x}$;

j) $\dfrac{\cos \alpha - \cos 3\alpha + \cos 5\alpha - \cos 7\alpha}{\sin \alpha + \sin 3\alpha + \sin 5\alpha + \sin 7\alpha}$; k) $\dfrac{\sin 2x + \cos 2x - \cos 6x - \sin 6x}{\sin 4x - \cos 4x}$.

III. 15. Find the identities for each of the following:

a) $\cos x(1 + \sec x + \tan x)(1 - \sec x + \tan x)$;

b) $\sin^2 x(1 + \csc x + \cot x)(1 - \csc x + \cot x)$;

c) $3 + 4\cos 2x + \cos 4x$; d) $1 + \sin 2x + \cos 2x + \tan 2x$;

e) $1 + \sin \dfrac{x}{2} - \cos^2 \dfrac{x}{4} + \sin^2 \dfrac{x}{4}$; f) $(\sin 4x - \cos 4x \tan 2x)(\sin 4x + \cos 4x \cot 2x)$;

g) $2\cos^2 x - \sqrt{3}\sin 2x - 1$; h) $\cos^2 \alpha - 3\sin^2 \alpha$;

i) $3 + 4\cos 2x + \cos 4x - 8\cos^4 x$; j) $\dfrac{\sin 4\alpha}{\sin^4 \alpha - \cos^4 \alpha} - 2$.

III. 16. Find the identities for each of the following:

a) $\dfrac{\cos(10° - 4x)}{4\cos(70° - x)\cos(20° + x)}$; b) $\dfrac{\cos\left(2x - \dfrac{\pi}{2}\right)\sin\left(\dfrac{\pi}{2} + x\right)}{(1 + \cos x)(1 + \cos 2x)}$;

c) $\dfrac{\sin^2\left(2\alpha - \dfrac{\pi}{2}\right)}{\tan\left(\alpha + \dfrac{3\pi}{2}\right) - \cot\left(\alpha - \dfrac{3\pi}{2}\right)}$; d) $\dfrac{4\sin\left(\dfrac{\pi}{2} + \alpha\right)}{\tan^2\left(\dfrac{3\pi}{2} - \dfrac{\alpha}{2}\right) - \cot^2\left(\dfrac{3\pi}{2} + \dfrac{\alpha}{2}\right)}$;

e) $\dfrac{\cos\left(\dfrac{5\pi}{2} - \alpha\right)\sin\left(\dfrac{5\pi}{2} + \dfrac{\alpha}{2}\right)}{\cos^2\left(\dfrac{\pi}{4} - \dfrac{\alpha}{4}\right)\left[2\sin\left(\dfrac{5\pi}{2} - \dfrac{\alpha}{2}\right) + \cos\left(\dfrac{3\pi}{2} - \alpha\right)\right]}$;

f) $\dfrac{\cos\left(2x + \dfrac{3\pi}{2}\right)\sin\left(\dfrac{3\pi}{2} - 3x\right) - \sin\left(2x - \dfrac{3\pi}{2}\right)\cos\left(3x + \dfrac{3\pi}{2}\right)}{\sin\left(\dfrac{5\pi}{2} - x\right)\cos 4x + \sin x \cos\left(\dfrac{5\pi}{2} + 4x\right)}$;

$$\text{g)}\ \dfrac{\sin\left(\dfrac{5\pi}{2}-\alpha\right)-\csc\left(\alpha+\dfrac{5\pi}{2}\right)}{\cos\left(\dfrac{5\pi}{2}-\alpha\right)+\sec\left(\alpha+\dfrac{5\pi}{2}\right)};\qquad \text{h)}\ \dfrac{1-\tan^2\left(x-\dfrac{5\pi}{4}\right)}{1+\tan^2\left(x-\dfrac{5\pi}{4}\right)}.$$

III. 17. Find the identities for each of the following:

a) $3-4\sin^2\left(\dfrac{\pi}{2}-\dfrac{x}{2}\right)$;

b) $6\sin^2 2x-1-\cos 4x$;

c) $2\cos^2 x+3\cos 2x-3$;

d) $\tan^4 x-4\tan^2 x+3$;

e) $\tan^3 x+\tan^2 x-3\tan x-3$.

III. 18. Find the identities for each of the following:

a) $4\cos\left(x-\dfrac{5\pi}{2}\right)\sin^3\left(\dfrac{3\pi}{2}+x\right)+4\sin\left(\dfrac{5\pi}{2}-x\right)\cos^3\left(\dfrac{3\pi}{2}+x\right)$;

b) $\sec^2 2x-\tan^2(\pi+2x)-2\cos^2\dfrac{x}{2}+\sqrt{3}\sin(x-\pi)$;

c) $\dfrac{1}{1+\cos 8x}\left[4\sin 2x\sin^2\left(\dfrac{3\pi}{2}+x\right)+\sin 3x\cos x-3\sin x\sin\left(\dfrac{3\pi}{2}-x\right)\right]$;

d) $4\cos^2\left(x-\dfrac{5\pi}{2}\right)+\cos(x-5\pi)+\sin\left(\dfrac{5\pi}{2}-3x\right)$;

e) $2\sin^2 x+\sqrt{3}\sin 2x-\dfrac{4\tan x\left(1-\tan^2 x\right)}{\sin 4x\left(1+\tan^2 x\right)^2}$;

f) $\sin^3\left(\dfrac{3\pi}{2}+\alpha\right)+\cos^3\left(\dfrac{3\pi}{2}+\alpha\right)+\cos\left(\dfrac{3\pi}{2}-\alpha\right)-\sin\left(\dfrac{3\pi}{2}+\alpha\right)$;

g) $\dfrac{\cos^2 x}{2\tan\left(\dfrac{5\pi}{4}+x\right)\cos^2\left(\dfrac{5\pi}{4}+x\right)}-\tan x-\sin\left(\dfrac{3\pi}{2}+x\right)-\cos\left(\dfrac{3\pi}{2}+x\right)$;

Radical expressions

III. 19. Simplify the expressions:

a) $\sqrt{\tan x + \sin x} - \sqrt{\tan x - \sin x}$, $0 < x < \dfrac{\pi}{2}$;

b) $\sqrt{\left(1 - \tan^2 x\right)\left(\cot^2 x - 1\right)}$;   c) $\sqrt{\dfrac{\cos 2x}{\cot^2 x - \tan^2 x}}$, $90° < x < 135°$;

d) $\sqrt{\left(1 + \sin\alpha\sin\beta\right)^2 - \cos^2\alpha\cos^2\beta}$;   e) $\sqrt{\csc^2\left(\dfrac{3\pi}{2} - x\right) + \sec^2\left(\dfrac{3\pi}{2} + x\right)}$;

f) $\sqrt[3]{\dfrac{\sin\left(x - \dfrac{\pi}{2}\right) - \csc\left(x - \dfrac{\pi}{2}\right)}{\cos\left(x - \dfrac{3\pi}{2}\right) + \sec\left(x + \dfrac{3\pi}{2}\right)}}$;   g) $\sqrt{\left(1 - \tan^2 x\right)\left(\cot^2 x - 1\right)}$;

h) $\dfrac{\sqrt{\tan x} + \sqrt{\cot x}}{\sqrt{\tan x} - \sqrt{\cot x}}$, $0 < x < \dfrac{\pi}{2}$, $x \neq \dfrac{\pi}{4}$;   i) $\sqrt{\dfrac{2\sin x - \sin 2x}{2\sin x + \sin 2x}}$;

j) $\sqrt{\dfrac{1 + \sin x}{1 - \sin x}} - \sqrt{\dfrac{1 - \sin x}{1 + \sin x}}$;   k) $\dfrac{\sqrt{1 + \sin\alpha} + \sqrt{1 - \sin\alpha}}{\sqrt{1 + \sin\alpha} - \sqrt{1 - \sin\alpha}}$;

l) $\sqrt{1 + \sin\dfrac{x}{2}} - \sqrt{1 - \sin\dfrac{x}{2}}$, $0° < x \leq 180°$.

III. 31. Solve the equation $x^2 \sin^2 b - 2x\left(1 - \cos a \cos b\right) + \sin^2 a = 0$. If x_1 and x_2 are the roots of the given equation, calculate the expression

$$E = \dfrac{\sqrt{x_1} - \sqrt{x_2}}{\sqrt{x_1} + \sqrt{x_2}} .$$

UNIT IV

Trigonometric Identities

IV. 1. Prove the identities:

a) $\cos x \tan x = \sin x$;   b) $\sin x \cot x = \cos x$;   c) $\csc x \cos x = \cot x$;

d) $\sin x \cot x \sec x = 1$;   e) $\cot x \tan x \sec x \cos x = 1$;

f) $\sin x + \tan x = \tan x (1 + \cos x)$;   g) $\sin x + \cos x \cot x = \csc x$;

h) $(\csc x - \cot x)(\sec x + 1) = \tan x$;   c) $\sec x (1 + \cos x) = \sec x + 1$;

i) $\tan x + \cot x = \sec x \csc x$;   j) $(\sin x + \cos x)(\tan x + \cot x) = \sec x + \csc x$;

k) $1 - \sin x \cos x \tan x = \cos^2 x$;   l) $\csc x (\csc x + \sin x) = \csc^2 x + 1$;

m) $2 \sin^2 x - 1 = \sin^2 x - \cos^2 x$;   n) $\cos^2 x = \sin^2 x + 2\cos^2 x - 1$.

IV. 2. Prove the identities:

a) $(\sin x + \cos x)^2 = 1 + 2 \sin x \cos x$;   b) $(\sin x + \cos x)^2 + (\sin x - \cos x)^2 = 2$;

c) $(1 - \cos^2 x)(1 + \cot^2 x) = 1$;   d) $\tan x = \tan^2 x \cdot \cot x$;   d) $\sec^2 x - \tan^2 x = 1$;

e) $\tan^2 x - \sin^2 x = \sin^2 x \cdot \tan^2 x$;   f) $\cot^2 x + \sec^2 x = \tan^2 x + \csc^2 x$;

g) $\sec^2 x - \sin^2 x = \cos^2 x + \tan^2 x$;   h) $\sec^2 x + \csc^2 x = \sec^2 x \csc^2 x$;

i) $(1 - \sec x)(1 + \sec x) = -\sin^2 x \sec^2 x$;   j) $(1 + \tan x)^2 = \sec^2 x + 2 \tan x$;

k) $(1 - \sin^2 x)(1 + \cot^2 x) = \cot^2 x$;   l) $(1 - \cos^2 x)(1 + \cos^2 x) = 2 \sin^2 x - \sin^4 x$;

m) $(a \cos x + b \sin x)^2 + (a \sin x - b \cos x)^2 = a^2 + b^2$ for any $a, b, x \in \mathbf{R}$.

IV. 3. Prove the identities:

a) $\dfrac{\csc x}{\sec x} = \cot x$;   b) $\dfrac{\cos x}{1-\sin x} = \sec x + \tan x$;   c) $\sin x + \cos x = \dfrac{1+\tan x}{\sec x}$;

d) $\dfrac{1}{\sin^2 x} + \dfrac{1}{\cos^2 x} = \dfrac{1}{\sin^2 x \cos^2 x}$;   e) $\dfrac{1-2\sin^2 x}{2\cos^2 x - 1} = 1$;

f) $\dfrac{1+\sin x}{\sin x} = 1 + \csc x$;   g) $\dfrac{\sin x - 1}{\cos x} = \tan x - \sec x$;   h) $\dfrac{\sin x + \tan x}{1+\cos x} = \tan x$;

i) $\cot^2 x = \dfrac{\cos^2 x}{1-\cos^2 x}$;   j) $\dfrac{1}{\sec^2 x} + \dfrac{1}{\csc^2 x} = 1$;

k) $\dfrac{1-\sin a}{1-\cos a} \cdot \dfrac{1+\csc a}{1+\sec a} = \cot^3 a$;   l) $\dfrac{\csc x}{\tan x + \cot x} = \cos x$;

m) $(1-\cos x)(1+\cos x) = \dfrac{1}{\csc^2 x}$;   n) $(1-\sin x)(1+\sin x) = \dfrac{1}{\sec^2 x}$;

o) $\dfrac{1}{1+\sin x} + \dfrac{1}{1-\sin x} = 2\sec^2 x$;   p) $\dfrac{1}{1-\cos x} + \dfrac{1}{1+\cos x} = 2\csc^2 x$;

q) $\dfrac{1}{1-\sec x} + \dfrac{1}{1+\sec x} = -2\cot^2 x$;   r) $\dfrac{\cos x}{1+\sin x} + \dfrac{\cos x}{1-\sin x} = 2\sec x$;

s) $\dfrac{\dfrac{1+\sin x}{\cos x} - \dfrac{\cos x}{1+\sin x}}{\dfrac{1+\cos x}{\sin x} - \dfrac{\sin x}{1+\cos x}} = \tan^2 x$;   s) $\dfrac{\dfrac{1}{(1-\sin x)^2} - \dfrac{1}{(1+\sin x)^2}}{\dfrac{1}{(1-\cos x)^2} - \dfrac{1}{(1+\cos x)^2}} = \tan^5 x$.

IV. 4. Prove the identities:

a) $\tan^2 x(1+\cot^2 x) = \dfrac{1}{1-\sin^2 x}$;   b) $\dfrac{1+\sin^2 x}{2+\cot^2 x} + \dfrac{1+\cos^2 x}{2+\tan^2 x} = 1$;

c) $\dfrac{\csc x + \cot x}{\csc x - \cot x} = \dfrac{1+2\cos x + \cos^2 x}{\sin^2 x}$;   d) $\dfrac{\cot x}{\tan x} = \dfrac{1-\sin^2 x}{1-\cos^2 x}$;

e) $\tan x + 1 = \dfrac{\sin x + \cos x}{\cos x}$;   f) $\dfrac{\sin x + \cos x \cot x}{\cot x} = \sec x$;

g) $\dfrac{\sec x + 1}{\sec x - 1} + \dfrac{\cos x + 1}{\cos x - 1} = 0$;   h) $\dfrac{\csc x}{\csc x - 1} + \dfrac{\csc x}{\csc x + 1} = 2\sec^2 x$;

i) $\dfrac{\cot x + \tan x}{\cot x} + \dfrac{\cot x - \tan x}{\tan x} = \tan^2 x + \cot^2 x$;   j) $\dfrac{\tan x \sin x}{\tan x + \sin x} = \dfrac{\tan x - \sin x}{\tan x \sin x}$;

k) $\dfrac{\tan x}{1+\tan^2 x} = \sin x \cos x$;   l) $\dfrac{(\tan x + \sec x)(\cos x - \cot x)}{(\cos x + \cot x)(\tan x - \sec x)} = 1$;

m) $\dfrac{\cos x}{\sec x}+\dfrac{\sin x}{\cot x}=\dfrac{\cos x\cot x+\tan x}{\csc x}$; **n)** $\dfrac{(\sin x+\cos x)^2}{(\sin x-\cos x)^2}=\dfrac{\sec^2 x+2\tan x}{\sec^2 x-2\tan x}$.

IV. 5. Prove the identities:

a) $\dfrac{\cos x-\sin x}{\cos^2 x-\sin^2 x}=\dfrac{\sec x}{\tan x+1}$; **b)** $\dfrac{1+2\sin x\cos x}{\sin x+\cos x}=\sin x+\cos x$;

c) $\tan^2 x+\cos^2 x+\dfrac{1}{\csc^2 x}=\sec^2 x$; **d)** $\dfrac{1+\tan^2 x}{1+\cot^2 x}=\tan^2 x$;

e) $\dfrac{1}{\cos^2 x-\sin^2 x}=\dfrac{1+\tan^2 x}{1-\tan^2 x}$; **f)** $\tan x\cot x+\dfrac{1-\sin^2 x}{1-\cos^2 x}=\dfrac{1}{\sin^2 x}$;

g) $\dfrac{\tan x-\sec x}{\cos x-\cot x}=\tan x\sec x$; **h)** $\dfrac{\tan^2 x-\sin^2 x}{\cot^2 x-\cos^2 x}-\tan^6 x=0$;

i) $\dfrac{\cos^2 x+3\sin x-1}{\cos^2 x+2\sin x+2}=\dfrac{1}{1+\csc x}$; **j)** $\dfrac{\sin^2 x+2\cos x-1}{2+\cos x-\cos^2 x}=\dfrac{1}{1+\sec x}$.

IV. 6. Prove the identities:

a) $\cos^3 x+\sin^3 x=(\cos x+\sin x)(1-\sin x\cos x)$;

b) $\cos^4 x+\sin^4 x=1-2\sin^2 x\cos^2 x$; **c)** $\sin^4 x-\cos^4 x=1-2\cos^2 x$;

d) $\cos^4 x+2\cos^2 x\sin^2 x+\sin^4 x=1$; **e)** $\sec^4 x-\tan^4 x=\sec^2 x+\tan^2 x$;

f) $\cos^3 x+\cos x\sin^2 x=\dfrac{1}{\sec x}$; **g)** $\dfrac{\sin^4 x+\cos^4 x}{\sin^2 x\cos^2 x}+2=\sec^2 x\csc^2 x$;

h) $\dfrac{\sin^4 x+\cos^4 x-1}{\sin^6 x+\cos^6 x-1}=\dfrac{2}{3}$; **i)** $\dfrac{\sin^3 x+\cos^3 x}{\sin x+\cos x}+\dfrac{\sin^3 x-\cos^3 x}{\sin x-\cos x}=2$;

j) $\left(\dfrac{\cos^3 x}{\sin x}+\dfrac{\tan x}{1+\tan^2 x}\right)\left(\dfrac{\sin^3 x}{\cos x}+\dfrac{\cot x}{1+\cot^2 x}\right)=1$.

IV. 7. Prove each of the following identities:

a) $\sin^6 x+\cos^6 x+3\sin^2 x\cos^2 x=1$;

b) $\sin^6 x-\cos^6 x=\dfrac{\sin^2 2x-4}{4}\cos 2x$;

c) $\cos^6 x+\sin^6 x=\dfrac{1+3\cos 4x}{8}$;

d) $\sec^6 x-\tan^6 x-3\tan^2 x\sec^2 x=1$;

e) $2\left(\sin^6 x + \cos^6 x\right) - 3\left(\sin^4 x + \cos^4 x\right) + 1 = 0$;

f) $\cos^8 x - \sin^8 x = \dfrac{\cos 2x(3 + \cos 4x)}{4}$.

IV. 8. Prove each of the following identities:

a) $\dfrac{1 - 2\sin^2 x}{1 + \sin 2x} = \dfrac{1 - \tan x}{1 + \tan x}$; **b)** $\dfrac{\sin 2x}{1 + \cos 2x} \dfrac{\cos x}{1 + \cos x} = \tan \dfrac{x}{2}$;

c) $\dfrac{\cos 4x + 1}{\cot x - \tan x} = \dfrac{\sin 4x}{2}$; **d)** $\cot\left(\dfrac{\pi}{4} + x\right) = \dfrac{\cos 2x}{1 + \sin 2x}$;

e) $\dfrac{1 - \cos 2x}{\sec^2 x - 1} + \dfrac{1 + \cos 2x}{\csc^2 x - 1} = 2$; **f)** $\dfrac{\tan 2x}{\tan^2 2x - 1} \cdot \dfrac{1 - \cot^2 2x}{\cot 2x} = 1$;

g) $\dfrac{1}{1 + \tan x} - \dfrac{1}{1 - \tan x} = -\tan 2x$; **h)** $\dfrac{\tan x}{\tan 2x - \tan x} = \cos 2x$;

i) $\dfrac{1 + \tan x + \cot x}{1 + \tan x + \tan^2 x} - \dfrac{\cot x}{1 + \tan^2 x} = \dfrac{\sin 2x}{2}$; **j)** $\dfrac{\tan x \tan 2x}{\tan 2x - \tan x} = \sin 2x$;

k) $\sin^2 2x + \dfrac{1}{1 + \tan^2 2x} - \cos 2x = 2\sin^2 x$; **l)** $\dfrac{\sin 2x}{\cos x + \cos^2 x} = 2\tan \dfrac{x}{2}$;

m) $\tan 2x - \sec 2x = \dfrac{\sin x - \cos x}{\sin x + \cos x}$; **n)** $\dfrac{\sin 3x}{\cos 3x + 2\cos x} = \tan x$;

o) $\dfrac{\sin 3a + \sin^3 a}{\cos 3a - \cos^3 a} = -\cot a$; **p)** $\dfrac{\sin 3a}{3} \cdot \cos^3 a + \dfrac{\cos 3a}{3} \cdot \sin^3 a = \dfrac{\sin 4a}{4}$;

q) $\dfrac{2\tan^2 x}{\tan^2 2x} = \dfrac{1 + \tan^4 x}{2 + \tan^2 2x}$; **r)** $\tan 3x = \dfrac{3\tan x - \tan^3 x}{1 - 3\tan^2 x}$;

s) $\dfrac{\sin^2 7x - \sin^2 4x}{\cos^2 5x - \cos^2 6x} = \dfrac{\sin 3x}{\sin x}$.

IV. 9. Prove the identities:

a) $\dfrac{\sin 7° \sin 65° - \sin 25° \cos 7°}{\sin 23° \cos 41° - \sin 41° \cos 23°} = 1$; **b)** $\dfrac{\sin 19° \cos 1° + \cos 161° \cos 101°}{\sin 21° \cos 9° + \cos 159° \cos 99°} = 1$;

c) $\dfrac{\cos 67° \cos 7° + \cos 83° \cos 23°}{\cos 68° \cos 8° + \cos 82° \cos 2°} = 1$; **d)** $\dfrac{\sin 20°. \cos 25° + \cos 20°. \sin 25°}{\cos 35°. \cos 10° - \sin 35°. \sin 10°} = 1$;

e) $\dfrac{\sin \dfrac{7\pi}{30} . \cos \dfrac{4\pi}{15} + \cos \dfrac{7\pi}{30} . \sin \dfrac{4\pi}{15}}{\sin \dfrac{7\pi}{12} . \cos \dfrac{5\pi}{12} - \cos \dfrac{7\pi}{12} . \sin \dfrac{5\pi}{12}} = 2$.

IV. 10. Prove the identities:

a) $\cos^4 x - \cos^2 y + \dfrac{1}{4}\sin^2 2x = -\sin(x+y)\sin(x-y);$

b) $\sin^2 x - \sin^2 y = \sin(x+y)\sin(x-y);$

c) $(\cos x - \cos y)^2 - (\sin x - \sin y)^2 = -4\sin^2 \dfrac{x-y}{2}\cos(x+y);$

d) $(\cos x + \cos y)^2 + (\sin x + \sin y)^2 = 4\cos^2 \dfrac{x-y}{2};$

e) $\sin^2(x+y) - \sin^2(x-y) = -4\sin^2 x \sin y \cos y;$

f) $\sin x + \sin y + \sin(x+y) = 4\sin \dfrac{x+y}{2}\cos\dfrac{x}{2}\cos\dfrac{y}{2}.$

IV. 11. Prove the identities:

a) $(\sin 2x + \sin 4x)^2 + (\cos 2x + \cos 4x)^2 = 4\cos^2 x;$

b) $\dfrac{\sin 4\alpha}{1+\cos 4\alpha}\dfrac{\cos 2\alpha}{1+\cos 2\alpha} = \tan\alpha;$

c) $\dfrac{1+\sin 2\alpha}{\sin\alpha+\cos\alpha} - \dfrac{1-\tan^2 \dfrac{\alpha}{2}}{1+\tan^2 \dfrac{\alpha}{2}} = \sin a;$

d) $\dfrac{\cos^2 \alpha}{\cot \dfrac{\alpha}{2} - \tan \dfrac{\alpha}{2}} = \dfrac{\sin 2\alpha}{4};$

e) $\dfrac{1+\tan x \tan \dfrac{x}{2}}{\cot x + \tan \dfrac{x}{2}} = \tan x;$

f) $\dfrac{\sin 2a}{1+\cos 2a}\cdot\dfrac{\cos a}{1+\cos a}\cdot\dfrac{\cos \dfrac{a}{2}}{1+\cos \dfrac{a}{2}}\cdot\dfrac{\cos \dfrac{a}{4}}{1+\cos \dfrac{a}{4}} = \tan\dfrac{a}{8};$

g) $\dfrac{3+4\cos 2x + \cos 4x}{3-4\cos 2x + \cos 4x} = \cot^4 x;$

h) $\dfrac{1}{\tan 3a - \tan a} - \dfrac{1}{\cot 3a - \cot a} = \cot 2a.$

IV. 12. Prove each of the following identities:

a) $\dfrac{\tan 2x + \cot 3y}{\cot 2x + \tan 3y} = \dfrac{\tan 2x}{\tan 3y};$

b) $\dfrac{\tan a + \tan b}{\tan a - \tan b} = \dfrac{\sin(a+b)}{\sin(a-b)};$

c) $\dfrac{\cot a - \cot b}{\cot a + \cot b} = \dfrac{\sin(a-b)}{\sin(a+b)};$

d) $\cot x\left(\dfrac{\tan x + \tan y}{\cot x + \cot y}\right) = \tan x;$

e) $\dfrac{\tan x + \tan y}{\tan(x+y)} + \dfrac{\tan x - \tan y}{\tan(x-y)} + 2\tan^2 x = 2\sec^2 x;$

52

f) $\tan x + \tan y = \tan x \tan y \left(\dfrac{\cos x}{\sin x} + \dfrac{\cos y}{\sin y} \right)$;

g) $\dfrac{\tan x \sec y - \tan y \sec x}{\sec x + \sec y} = \tan\left(\dfrac{x-y}{2} \right)$;   h) $\cot^2 x - \cot^2 y = \dfrac{\cos^2 x - \cos^2 y}{\sin^2 x \sin^2 y}$;

i) $\dfrac{\tan 2a + \tan 2b - \cot 2a - \cot 2b}{\tan a + \tan b - \cot a - \cot b} = \dfrac{1 - \tan 2a \cdot \tan 2b}{2}$;

j) $\dfrac{\tan(x+y) - \tan y}{1 + \tan y \tan(x+y)} + \dfrac{\tan y + \tan(x-y)}{1 - \tan y \tan(x-y)} = 2\tan x$;

k) $\dfrac{\sin(x-y)}{\cos x \cos y} + \dfrac{\sin(y-z)}{\cos y \cos z} + \dfrac{\sin(z-x)}{\cos z \cos x} = 0$;

l) $\dfrac{\cos(x+y) \cdot \cos(x-y)}{\cos^2 x \cdot \cos^2 y} = 1 - \tan^2 x \cdot \tan^2 y$;

m) $\dfrac{(\tan x + \tan y)(1 - \tan x \tan y)}{(1 + \tan^2 x)(1 + \tan^2 y)} = \dfrac{1}{2} \sin 2(x+y)$.

IV. 13. Prove the identities:

a) $1 + \sin x - \cos x = 2\sqrt{2} \sin\dfrac{x}{2} \cos\left(\dfrac{x}{2} - \dfrac{\pi}{4} \right)$;

b) $1 + \cos x + \cos 2x = 4\cos x \cos\left(\dfrac{\pi}{6} + \dfrac{x}{2} \right) \cos\left(\dfrac{\pi}{6} - \dfrac{x}{2} \right)$;

c) $1 + \sin x + \cos x - \tan x = \dfrac{2\sqrt{2} \sin^2 \dfrac{x}{2} \cos\left(\dfrac{\pi}{4} + x \right)}{\cos x}$;

d) $\tan x \tan\left(\dfrac{\pi}{3} - x \right) \tan\left(\dfrac{\pi}{3} + x \right) = \tan 3x$;

e) $\left[2\sin 3x + 3\left(\sin x + \sqrt{3} \cos x \right) \right] + \left[2\cos 3x + 3\left(\sqrt{3} \sin x - \cos x \right) \right] =$

$= 8\left[\sin^3\left(x + \dfrac{\pi}{3} \right) - \cos^3\left(x + \dfrac{\pi}{3} \right) \right]$;

f) $\cos 2x - \tan\left(x - \dfrac{\pi}{4} \right) + \dfrac{\sin^3 x - \cos^3 x}{\sin^3 x + \cos^3 x} = \cos 2x \cdot \tan\left(x - \dfrac{\pi}{4} \right) \cdot \dfrac{\sin^3 x - \cos^3 x}{\sin^3 x + \cos^3 x}$,

where $x \in \mathbf{R} \setminus \left\{ k\pi + \dfrac{3\pi}{4} \right\}, k \in \mathbf{Z}$.

IV. 14. Prove each of the following identities:

a) $\cos x + \cos(x + 120°) + \cos(x + 240°) = 0$;

b) $\sin x + \sin(x + 120°) + \sin(x + 240°) = 0$;

e) $\sin(90° - x) - \sin(90° + x) + 2\cos(180° - x) = 0$;

d) $\tan x + \tan(x + 60°) + \tan(x + 120°) = 3\tan 3x$;

e) $\tan 3x \tan(30° - x)\tan(30° + x) = \tan x$;

f) $\csc x \csc(60° - x)\csc(60° + x) = 4\csc 3x$.

IV. 15. Prove the identities:

a) $4\sin x \cos x \cos 2x = \sin 4x$;

b) $\cos 4x \tan 2x - \sin 4x = -\tan 2x$;

c) $\sin a + \sin 3a + \sin 5a + \sin 7a = 4\cos a \cos 2a \cos 4a$;

d) $\cos a + \cos 2a + \cos 6a + \cos 7a = 4\cos\dfrac{a}{2}\cos\dfrac{5a}{2}\cos 4a$;

e) $4\cos x \cos 2x \cos 3x = 1 + \cos 2x + \cos 4x + \cos 6x$;

f) $\sin 4x \cot 2x - \cos 4x = 2\sin^2 x - \cos 2x$;

g) $\tan x + 2\tan 2x + 4\tan 4x + 8\cot 8x = \cot x$;

h) $\tan 3x - \tan 2x - \tan x = \tan 3x \tan 2x \tan x$.

IV. 16. Prove the identities:

a) $\sin \alpha \sin\left(\dfrac{\pi}{3} - \alpha\right)\sin\left(\dfrac{\pi}{3} + \alpha\right) = \dfrac{1}{4}\sin 3\alpha$;

b) $\cos \alpha + 2\sin\left(\dfrac{\alpha}{2} + 15°\right)\sin\left(\dfrac{\alpha}{2} - 15°\right) = \dfrac{\sqrt{3}}{2}$;

c) $\tan 9° + \tan 15° - \tan 27° - \cot 27° + \cot 9° + \cot 15° = 8$;

d) $\tan 5° \cdot \tan 55° \cdot \tan 65° = 2 - \sqrt{3}$.

X. 17. Prove the identities:

a) $1 - 2\cos 40° = \dfrac{1}{2}\sec 160°$;

b) $\tan 20° + 4\sin 20° = \sqrt{3}$;

c) $\dfrac{(\tan 1° + 1)(\tan 2011° + 1) - 2}{(\tan 1° - 1)(\tan 2011° - 1) - 2} = \tan 13°$;

d) $\dfrac{\sqrt{1 + \sin 20°} + \sqrt{1 - \sin 20°}}{\sqrt{1 + \sin 20°} - \sqrt{1 - \sin 20°}} = \cot 20°$.

IV. 18. Prove each of the following identities:

a) $\csc x + \cot x = \cot \dfrac{x}{2}$;

b) $\tan x + \cot x + \tan 3x + \cot 3x = \dfrac{8\cos^2 2x}{\sin 6x}$;

c) $\sec 2x + \cot\left(\dfrac{3\pi}{2} + 2x\right) = \tan\left(\dfrac{5\pi}{4} - x\right)$;

d) $\sin 2x(1 - \sec 2x + \tan 2x \tan x) + \dfrac{1 + \sin x}{1 - \sin x} = \tan^2\left(\dfrac{\pi}{4} + \dfrac{x}{2}\right)$;

e) $4\sin x \sin\left(\dfrac{\pi}{3} + x\right) \sin\left(\dfrac{\pi}{3} - x\right) = \sin 3x$;

f) $\tan\left(\dfrac{\pi}{8} + x\right) - \tan\left(\dfrac{\pi}{8} - x\right) + \cot\left(\dfrac{\pi}{8} - x\right) - \cot\left(\dfrac{\pi}{8} + x\right) = 4\tan 4x$;

g) $(1 + \sec x + \tan x)(1 - \sec x + \tan x) = 2\tan x$.

IV. 19. Prove the identities:

a) $\dfrac{2\sin a - \sin 2a}{2\sin a + \sin 2a} = \tan^2 \dfrac{a}{2}$;   b) $\dfrac{1 + \sin 2a + \cos 2a}{1 + \sin 2a - \cos 2a} = \cot a$;

c) $\dfrac{\sin a + 2\sin 3a + \sin 5a}{\sin 3a + 2\sin 5a + \sin 7a} = \dfrac{\sin 3a}{\sin 5a}$;   d) $\dfrac{\sin 2x - \sin 3x + \sin 4x}{\cos 2x - \cos 3x + \cos 4x} = \tan 3x$;

e) $\dfrac{\cos 5x + \sin 3x - \cos x}{\sin 5x - \cos 3x - \sin x} = -\tan 3x$;   f) $\dfrac{\sin 2x - \cos 2x \tan x}{\sin 2x + \cos 2x \cot x} = \tan^2 x$;

g) $\dfrac{\cos 2a - \cos 6a + \cos 10a - \cos 14a}{\sin 2a + \sin 6a + \sin 10a + \sin 14a} = \tan 2a$;

h) $\dfrac{\sin 6x - \sin 7x - \sin 8x + \sin 9x}{\cos 6x - \cos 7x - \cos 8x + \cos 9x} = \tan \dfrac{15x}{2}$;

i) $\dfrac{\tan x - \tan 3x - \tan 5x + \tan 7x}{\cot x - \cot 3x - \cot 5x + \cot 7x} = \tan x \tan 3x \tan 5x \tan 7x$.

IV. 20. Prove the identities:

a) $1 - \tan\left(\dfrac{\pi}{4} - 2x\right)\cos 4x = \sin 4x$;

b) $\dfrac{1 - \tan(2\pi - 2x)\tan(\pi + x)}{\cot x + \tan x} = \dfrac{1}{2}\tan 2x$;

c) $\dfrac{1 - \tan(270° + x)}{1 - \cot(360° + x)} = \dfrac{\tan(360° + x) + 1}{\cot(270° - x) - 1}$;

d) $2\left[\csc 4x - \tan\left(\dfrac{5\pi}{2} + 4x\right)\right] + \tan(3\pi + 2x) = \cot 2x$;

e) $\tan x \tan\left(x + \dfrac{\pi}{3}\right)\tan\left(x + \dfrac{2\pi}{3}\right) = -\tan 3x$.

IV. 21. Prove the identities:

a) $2\cot\left(\dfrac{\pi}{4} + x\right)\cos^2\left(\dfrac{5\pi}{4} - x\right) = \cos 2x$;

b) $2\tan\left(\dfrac{\pi}{4} - x\right)\sin^2\left(\dfrac{\pi}{4} + x\right) = \cos 2x$;

c) $\sin^2 2x + \sin\left(\dfrac{\pi}{6} + 2x\right)\sin\left(\dfrac{\pi}{6} - 2x\right) = \dfrac{1}{4}$;

d) $\sin^2 2x + \cos\left(\dfrac{\pi}{3} + 2x\right)\cos\left(\dfrac{\pi}{3} - 2x\right) = \dfrac{1}{4}$;

e) $\cos^2\left(\dfrac{\pi}{8} + x\right) - \sin^2\left(\dfrac{\pi}{8} - x\right) = \dfrac{\cos 2x}{\sqrt{2}}$;

f) $\cos^2\left(\dfrac{5\pi}{8} + \dfrac{x}{2}\right) - \cos^2\left(\dfrac{3\pi}{8} + \dfrac{x}{2}\right) = \dfrac{\sin x}{\sqrt{2}}$;

g) $\sin^2\left(\dfrac{\pi}{8} + 2x\right) - \sin^2\left(\dfrac{\pi}{8} - 2x\right) = \dfrac{\sin 4x}{\sqrt{2}}$.

IV. 22. Prove the identities:

a) $\dfrac{\sin 2x + \sin 5x - \sin 3x}{\cos x + \cos 4x} = 2\sin x$;

b) $\sin 4x + \cos 4x \cot 2x + \dfrac{1}{2}\left(1 - \cot^2 x\right)\tan x = 0$;

c) $\dfrac{\sin 7x}{\sin x} - 2(\cos 2x + \cos 4x + \cos 6x) = 1$;

d) $\sin x + \sin 3x + \sin 5x + \sin 7x = \dfrac{1 - \cos 8x}{2\sin x}$;

e) $\dfrac{1 - \cos 2(n+1)x}{2\sin x} + \sin(2n+3)x = \dfrac{1 - \cos 2(n+2)x}{2\sin x}$;

f) $(\cos a + \cos b)(\cos 2a + \cos 2b)(\cos 4a + \cos 4b) = \dfrac{(\cos 8a - \cos 8b)}{8(\cos a - \cos b)}$.

IV. 24. Prove the identities:

a) $\dfrac{2\sqrt{3}\sin\left(\dfrac{\pi}{3}+2x\right)-3\sin\left(\dfrac{5\pi}{2}-2x\right)}{\cos\left(\dfrac{5\pi}{2}-2x\right)+2\cos\left(\dfrac{\pi}{6}+2x\right)}=\tan 2x$;

b) $\sqrt{2}(1+\cot x+\csc x)\sin\dfrac{x}{2}\cos\left(\dfrac{\pi}{4}+\dfrac{x}{2}\right)=\cos x$;

c) $\dfrac{(1+\sin x)\tan\left(\dfrac{\pi}{4}+\dfrac{x}{2}\right)}{2\sin\left(\dfrac{\pi}{4}+\dfrac{x}{2}\right)\cos\left(\dfrac{5\pi}{4}+\dfrac{x}{2}\right)}=-\tan^2\left(\dfrac{\pi}{4}+\dfrac{x}{2}\right)$;

d) $\dfrac{\tan^2\left(\dfrac{\pi}{2}+x\right)}{1+\tan^2\left(\dfrac{3\pi}{2}-x\right)}-3\cos^2\left(\dfrac{9\pi}{2}-x\right)=4\cos\left(\dfrac{\pi}{3}+x\right)\sin\left(\dfrac{\pi}{6}+x\right)$;

e) $\left(\tan^2 x-1\right)\sin^2\left(\dfrac{3\pi}{2}-x\right)\csc^2\left(\dfrac{5\pi}{4}+x\right)=2\cot\left(\dfrac{3\pi}{4}-x\right)$.

IV. 25. Prove the identities:

a) $\sin x(2\cos 2x+1)\cot\left(\dfrac{\pi}{6}-x\right)\cot\left(\dfrac{\pi}{6}+x\right)=\sin 3x\tan 3x\cot x$;

b) $4\sin(x-\pi)\sin\left(x-\dfrac{2\pi}{3}\right)\sin\left(x+\dfrac{2\pi}{3}\right)=\sin 3x$;

c) $4\sin\left(x-\dfrac{3\pi}{2}\right)\sin\left(\dfrac{\pi}{6}+x\right)\sin\left(\dfrac{\pi}{6}-x\right)=\cos 3x$;

d) $\dfrac{1-\cos(2x+90°)+\sin(2x-90°)}{1+\cos(2x-90°)+\sin(2x+90°)}=\tan x$;

e) $\dfrac{1+\cos(2x-2\pi)+\cos(4x+2\pi)-\cos(6x-\pi)}{\cos^2\left(2x+\dfrac{\pi}{2}\right)-\sin^2(x-\pi)}=4\cos 2x\cot x\cot 3x$.

IV. 26. Prove the identities:

a) $\dfrac{3+\cos 2x}{2+\tan^2 x}+\dfrac{3-\cos 2x}{2+\cot^2 x}=2$;

b) $\dfrac{\cos 2x+4\cos x+3}{1+\cos x}+\dfrac{\cos 2x-4\cos x+3}{1-\cos x}=4$;

c) $\dfrac{1+\sin x}{\sin x + \cos x} \cdot \cos\left(x - \dfrac{\pi}{4}\right) + \dfrac{1-\sin x}{\sin x - \cos x} \cdot \sin\left(x - \dfrac{\pi}{4}\right) = \sqrt{2}$;

d) $\dfrac{\sin 4x + 2\sin 2x + 3\cos x + \cos 3x}{2(2\sin x + 1)(2\cos x + \sin 2x)} = 1 - \sin x$;

e) $\dfrac{1 + \sin 2x - 2\sin^2 x}{1 + \cot^2 x} = \sqrt{2}\sin^2 x \cos\left(\dfrac{\pi}{4} - 2x\right)$.

IV. 27. Prove the identities:

a) $\dfrac{\cos\left(2x - \dfrac{5\pi}{2}\right)}{\cot\left(x + \dfrac{5\pi}{4}\right)\left[1 - \cos\left(2x + \dfrac{5\pi}{2}\right)\right]} = \tan 2x$;

b) $\dfrac{\cos\left(2x - \dfrac{\pi}{2}\right)}{\cot\left(x + \dfrac{\pi}{4}\right)\left[1 + \cos\left(2x + \dfrac{3\pi}{2}\right)\right]} = \tan 2x$;

c) $\dfrac{\cos\left(\dfrac{\pi}{2} - \dfrac{x}{2}\right) - \sin\left(\dfrac{\pi}{2} - \dfrac{x}{2}\right)\tan\dfrac{x}{4}}{\sin\left(\dfrac{3\pi}{2} - \dfrac{x}{2}\right) - \sin\left(3\pi - \dfrac{x}{2}\right)\tan\dfrac{x}{4}} = -\tan\dfrac{x}{4}$;

d) $\dfrac{\cot\left(\dfrac{\pi}{4} + \dfrac{x}{2}\right)[1 + \sin(5\pi - x)]\sec x - 2\cos 2x}{\cot\left(\dfrac{\pi}{4} + \dfrac{x}{2}\right)[1 + \sin(6\pi + x)]\sec x + 2\cos 2x} = \tan\left(\dfrac{\pi}{6} + x\right)\tan\left(x - \dfrac{\pi}{6}\right)$.

IV. 28. Prove the identities:

a) $8\cos^4 x + 4\cos(3\pi - 2x) + \cos(3\pi + 4x) = 3$;

b) $16\sin^5 x - 20\sin^3 x + 5\sin x = \sin 5x$;

c) $8\sin^4 x - 4\sin\left(2x + \dfrac{3\pi}{2}\right) - \sin\left(4x - \dfrac{3\pi}{2}\right) = 3$.

IV. 29. Prove the identities:

a) $\sin^4\dfrac{\pi}{8} + \cos^4\dfrac{3\pi}{8} + \sin^4\dfrac{5\pi}{8} + \cos^4\dfrac{7\pi}{8} = \dfrac{3}{2}$;

b) $\sin^4 \dfrac{\pi}{8} + \sin^4 \dfrac{3\pi}{8} + \sin^4 \dfrac{5\pi}{8} + \sin^4 \dfrac{7\pi}{8} = \dfrac{3}{2}$;

c) $\cos^4 \dfrac{\pi}{8} + \cos^4 \dfrac{3\pi}{8} + \cos^4 \dfrac{5\pi}{8} + \cos^4 \dfrac{7\pi}{8} = \dfrac{3}{2}$;

d) $\sin^4 \dfrac{\pi}{16} + \sin^4 \dfrac{3\pi}{16} + \sin^4 \dfrac{5\pi}{16} + \sin^4 \dfrac{7\pi}{16} = \dfrac{3}{2}$.

IV. 30. Prove the identities:

a) $\dfrac{1 - \tan 13^\circ}{1 + \tan 13^\circ} = \tan 32^\circ$;

b) $\dfrac{\left(1 + \sqrt{3}\right)\cos 7^\circ + \left(1 - \sqrt{3}\right)\sin 7^\circ}{\sqrt{3}\cos 8^\circ + \sin 8^\circ} = \sqrt{2}$;

c) $\tan^2 15^\circ \tan 75^\circ = 14$;

d) $\tan^2 29^\circ + 2\tan 29^\circ \tan 32^\circ = 1$;

e) $\tan^3 20^\circ - 3\sqrt{3}\tan^2 20^\circ - 3\tan 20^\circ + \sqrt{3} = 0$.

IV. 31. Prove the identities:

a) $\sin 20^\circ \sin 40^\circ \sin 80^\circ = \dfrac{\sqrt{3}}{8}$;

b) $\sin 10^\circ \sin 50^\circ \sin 70^\circ = \dfrac{1}{8}$;

c) $\cos^2 20^\circ \sin^2 50^\circ \cos^2 80^\circ = \dfrac{1}{64}$;

d) $\tan 20^\circ \tan 40^\circ \tan 80^\circ = \sqrt{3}$.

IV. 32. Prove the identities:

a) $\sin 10^\circ \sin 20^\circ \sin 30^\circ \sin 40^\circ \sin 50^\circ \sin 60^\circ \sin 70^\circ \sin 80^\circ = \dfrac{3}{256}$.

b) $\tan 10^\circ \tan 20^\circ + \tan 20^\circ \tan 60^\circ + \tan 60^\circ \tan 10^\circ = 1$;

c) $\sin 40^\circ + 8\cos 20^\circ \cos 40^\circ \sin 10^\circ = 2\cos^2 25^\circ$;

d $\cos 70^\circ + 8\cos 20^\circ \cos 40^\circ \cos 80^\circ = 2\cos^2 35^\circ$;

e) $\tan 9^\circ + \cot 9^\circ - \tan 27^\circ - \cot 27^\circ = 4$;

f) $\tan 30^\circ + \tan 40^\circ + \tan 50^\circ + \tan 60^\circ = \dfrac{8}{\sqrt{3}}\sin 70^\circ$.

IV. 33. Prove the identity $\cot 5° \cdot \tan 15° \cdot \tan 25° \cdot \tan 35° = 1$.

IV. 34. Prove the identities:

a) $\tan\left(\dfrac{\pi}{4} - x\right)(1 + \sin 2x)\sec 2x + 2\cos(4x - 4\pi) = \dfrac{\sin 6x}{\sin 2x}$;

b) $\cos^2\left(60° + x\right) + \cos^2\left(60° - x\right) + \cos\left(60° + x\right)\cos\left(60° - x\right) = \dfrac{3}{4}$;

c) $\sin(3\pi - 3x)\cos^3\left(\dfrac{\pi}{2} - x\right) - \cos(3\pi - 3x)\sin^3\left(\dfrac{\pi}{2} - x\right) = \cos^3 2x$;

d) $8\cos^4 2x + 4\cos^3 2x - 8\cos^2 2x - 3\cos 2x + 1 = 2\cos 7x \cos x$;

e) $8\cos^4 2x - 4\cos^3 2x - 8\cos^2 2x + 3\cos 2x + 1 = -2\sin 7x \sin x$.

IV. 35. Prove the identities:

a) $\sin\dfrac{3\pi}{10} - \sin\dfrac{\pi}{10} = \dfrac{1}{2}$; **b)** $\cos\dfrac{1}{5}\pi - \cos\dfrac{2\pi}{5} = \dfrac{1}{2}$;

c) $8\cos\dfrac{4\pi}{9}\cos\dfrac{2\pi}{9}\cos\dfrac{\pi}{9} = 1$; **d)** $\sin\dfrac{\pi}{14} + \cos\dfrac{2\pi}{14} - \sin\dfrac{3\pi}{14} = \dfrac{1}{2}$;

e) $\sin\dfrac{\pi}{14}\cos\dfrac{2\pi}{14}\sin\dfrac{3\pi}{14} = \dfrac{1}{8}$; **f)** $\sin\dfrac{\pi}{24}\sin\dfrac{5\pi}{24}\sin\dfrac{7\pi}{24}\sin\dfrac{1}{24}\pi = \dfrac{1}{16}$;

g) $\cos\dfrac{\pi}{5} + \cos\dfrac{2\pi}{5} + \cos\dfrac{4\pi}{5} + \cos\dfrac{6\pi}{5} = -\dfrac{1}{2}$;

h) $\cos\dfrac{\pi}{1} - \cos\dfrac{2\pi}{1} + \cos\dfrac{3\pi}{1} - \cos\dfrac{4\pi}{1} + \cos\dfrac{5\pi}{1} = \dfrac{1}{2}$.

IV. 36. Prove the identities:

a) $\cos\dfrac{\pi}{3}\cos\dfrac{2\pi}{3}\cos\dfrac{4\pi}{3}\cos\dfrac{8\pi}{3}\cos\dfrac{16\pi}{3} = \dfrac{1}{32}$;

b) $\cos\dfrac{2\pi}{31}\cos\dfrac{4\pi}{31}\cos\dfrac{8\pi}{31}\cos\dfrac{16\pi}{31}\cos\dfrac{32\pi}{31} = \dfrac{1}{32}$;

c) $\cos\dfrac{\pi}{15}\cos\dfrac{2\pi}{15}\cos\dfrac{3\pi}{15}...\cos\dfrac{12\pi}{15}\cos\dfrac{13\pi}{15}\cos\dfrac{14\pi}{15} = -\dfrac{1}{2^{14}}$;

d) $\cos\dfrac{\pi}{15}\cos\dfrac{2\pi}{15}\cos\dfrac{3\pi}{15}\cos\dfrac{4\pi}{15}\cos\dfrac{5\pi}{15}\cos\dfrac{6\pi}{15}\cos\dfrac{7\pi}{15} = \dfrac{1}{2^7}$.

IV. 37. Prove the identities:

a) $\sin^2 \alpha - \cos^2 \beta + 2\cos\alpha\cos\beta\cos(\alpha - \beta) = \cos^2(\alpha - \beta)$;

b) $\cos^2 \alpha - \sin^2 \beta - 2\cos\alpha\cos\beta\cos(\alpha - \beta) = -\cos^2(\alpha - \beta)$;

c) $(\sin\alpha - \sin\beta)(\sin\alpha + \sin\beta) = \sin(\alpha - \beta)\sin(\alpha + \beta)$;

d) $\sin\alpha\sin(\beta - \gamma) + \sin\beta\sin(\gamma - \alpha) + \sin\gamma\sin(\alpha - \beta) = 0$;

e) $(\sin\alpha + \sin\beta)^2 + (\cos\alpha + \cos\beta)^2 = 4\cos^2\dfrac{\alpha - \beta}{2}$.

IV. 38. Verify the identities:

a) $\sin x + \sin y + \sin z - \sin(x + y + z) = 4\sin\dfrac{x+y}{2}\sin\dfrac{y+z}{2}\sin\dfrac{z+x}{2}$;

b) $\cos x + \cos y + \cos z + \cos(x + y + z) = 4\cos\dfrac{x+y}{2}\cos\dfrac{y+z}{2}\cos\dfrac{z+x}{2}$;

c) $\tan x + \tan y + \tan z - \tan x \tan y \tan z = \dfrac{\sin(x + y + z)}{\cos x \cos y \cos z}$, where

$x, y, z \neq (2k + 1)\dfrac{\pi}{2}, k \in Z$;

d) $\cos(x + y)\cos(x - y) + \cos(y + z)\cos(y - z) + \cos(z + x)\cos(z - x) =$

$= \cos 2x + \cos 2y + \cos 2z$;

e) $\dfrac{\sin x + \sin y + \sin z - \sin(x + y + z)}{\cos x + \cos y + \cos z + \cos(x + y + z)} = \tan\dfrac{x+y}{2}\tan\dfrac{y+z}{2}\tan\dfrac{z+x}{2}$;

f) $\sin a + \sin b + \sin c - \sin(a + b + c) = 4\sin\dfrac{a+b}{2}\sin\dfrac{b+c}{2}\sin\dfrac{a+c}{2}$.

IV. 39. Prove the identities:

a) $\sqrt{\sin^4 x + 4\cos^2 x} + \sqrt{\cos^4 x + 4\sin^2 x} = 3$;

b) $\sqrt{2 + 2\cos x} \cdot \sqrt{2 + \sqrt{2 + 2\cos x}} \cdot \sqrt{2 + \sqrt{2 + \sqrt{2 + 2\cos x}}} = \sin x \csc\dfrac{x}{8}$;

c) $\left(\dfrac{\sin x + \cos x}{\sqrt{\tan x} + \sqrt{\cot x}}\right)^2 = \sin x \cos x$;   d) $\dfrac{\sqrt{\cot x} + \sqrt{\tan x}}{\sqrt{\cot x} - \sqrt{\tan x}} = \tan\left(\dfrac{\pi}{4} + x\right)$;

e) $\dfrac{\sqrt{1 + \sin x} - \sqrt{1 - \sin x}}{\sqrt{1 + \sin x} + \sqrt{1 - \sin x}} = \tan\dfrac{x}{2}$;   f) $\left(\sqrt{\dfrac{1 + \sin a}{1 - \sin a}} - \sqrt{\dfrac{1 - \sin a}{1 + \sin a}}\right)^2 = 4\tan^2 a$;

g) $\dfrac{\sqrt[4]{1 + \sin 2x} + \sqrt[4]{1 - \sin 2x}}{\sqrt[4]{1 + \sin 2x} - \sqrt[4]{1 - \sin 2x}} = \tan x + \sqrt{\tan^2 x - 1}$, $x \in \left(\dfrac{\pi}{4}, \dfrac{\pi}{2}\right)$;

h) $\sqrt[3]{\left[2\sin 3x + 3\left(\sin x + \sqrt{3}\cos x\right)\right]^2} + \sqrt[3]{\left[2\cos 3x + 3\left(\sqrt{3}\sin x - \cos x\right)\right]^2} = 4$.

Unit V

Trigonometric Equations

Elementary trigonometric equations

1. $\sin x = a$, $a \in [-1,1]$ with the solution $x = (-1)^k \arcsin a + k\pi$, $k \in \mathbf{Z}$.

2. $\cos x = a$, $a \in [-1,1]$ with the solution $x = \pm \arccos a + 2k\pi$, $k \in \mathbf{Z}$.

3. $\tan x = a$, $a \in \mathbf{R}$ with the solution $x = \arctan a + k\pi$, $k \in \mathbf{Z}$.

4. $\cot x = a$, $a \in \mathbf{R}$ with the solution $x = \operatorname{arc cot} a + k\pi$, $k \in \mathbf{Z}$.

Trigonometric Equations of the form

1. $\sin f(x) = \sin g(x)$, $f(x) = (-1)^k g(x) + k\pi$, $k \in \mathbf{Z}$.

2. $\cos f(x) = \cos g(x)$, $f(x) = \pm g(x) + 2k\pi$, $k \in \mathbf{Z}$.

3. $\tan f(x) = \tan g(x)$, $f(x) = g(x) + k\pi$, $k \in \mathbf{Z}$.

4. $\cot f(x) = \cot g(x)$, $f(x) = g(x) + k\pi$, $k \in \mathbf{Z}$.

Linear equations in $\sin x$ and $\cos x$ have the form

(1) $a\sin x + b\cos x = c$, or in general (1') $a\sin \varphi(x) + b\cos \varphi(x) = c$.

Method 1: It is equivalent to an elementary equation of the form

$$\sin(\varphi(x) + \theta) = \frac{c}{\sqrt{a^2 + b^2}}, \text{ where } \sin \theta = \frac{b}{\sqrt{a^2 + b^2}} \text{ and } \cos \theta = \frac{a}{\sqrt{a^2 + b^2}}.$$

Method 2: Since $\sin x = \dfrac{2\tan \dfrac{x}{2}}{1 + \tan^2 \dfrac{x}{2}}$ and $\cos x = \dfrac{1 - \tan^2 \dfrac{x}{2}}{1 + \tan^2 \dfrac{x}{2}}$, $\cos x \neq 0$, denoting $\tan\dfrac{x}{2} = t$ the equation (1) becomes $a\dfrac{2t}{1+t^2} + b\dfrac{1-t^2}{1+t^2} = c$. This equation could be solved as a quadratic equation.

Method 3: Denoting $\sin x = y$ and $\cos x = z$, we have to solve the system
$$\begin{cases} ay + bz = c \\ y^2 + z^2 = 1 \end{cases}$$

Method 4. (*Auxiliary angle*) Denoting $\tan\theta = \dfrac{b}{a}$ the equation

(1) becomes $a\sin x + \dfrac{\sin\theta}{\cos\theta}\cos x = c \Rightarrow \sin(x+\theta) = \dfrac{c}{a}\cos\theta \Rightarrow$

$\sin(x+\theta) = \pm\dfrac{c}{a}\dfrac{1}{\sqrt{1+\tan^2\theta}} \Rightarrow$

$\sin(x+\theta) = \pm\dfrac{1}{\sqrt{a^2+b^2}}$.

Symmetric equations in $\sin x$ **and** $\cos x$ **have the form**

(2) $a(\sin x + \cos x) + b\sin x \cos x = c$.

Method 1: Denote $\tan\dfrac{x}{2} = t$ and use the formulae $\sin x = \dfrac{2t}{1+t^2}$ and

$\cos x = \dfrac{1-t^2}{1+t^2}$.

Method 2: Denote $\sin x + \cos x = t$ then $\sin x \cos x = \dfrac{t^2-1}{2}$ and the equation (2) becomes $at^2 + 2at - b - 2c = 0$, where $t \in \left[-\sqrt{2}, \sqrt{2}\right]$.

Method 3: Denote $x = \dfrac{\pi}{4} + t$ then $\sin x = \dfrac{\sin t + \cos t}{\sqrt{2}}$, $\cos x = \dfrac{\cos t - \sin t}{\sqrt{2}}$ and the equation (2) becomes $b(\cos^2 t - \sin^2 t) + 2a\sqrt{2}\cos t - 2c = 0$.

Homogeneous equations in $\sin x$ **and** $\cos x$ **have the form**

(3) $a_0\sin^n x + a_1\sin^{n-1} x \cos x + a_2\sin^{n-2} x \cos^2 x + ... + a_n\cos^n x = 0$,

$a_1, a_2, a_3, ..., a_n \in \mathbf{R}$. Divide each term in the LHS by $\cos^n x$ and making the substitution $\tan x = t$, where $\cos x \neq 0$, the equation (3) becomes

$a_0 t^n + a_1 t^{n-1} + a_2 t^{n-2} + ... + a_n = 0$.

Equations containing sums of the form $\sin^{2n} x + \cos^{2n} x$, $n \geq 2$,

$$a_0\left(\sin^{2n} x + \cos^{2n} x\right) + a_1\left(\sin^{2(n-1)} x + \cos^{2(n-1)} x\right) + \ldots = 0.$$

In general, this kind of equations could be solved using the identity $\sin^2 x + \cos^2 x = 1$ and rising it at a convenient power and finally, we obtain an equation in terms of $\sin 2x$.

V. 1. Determine the solutions for each trigonometric equation, where $x \in \left[0, 360°\right]$:

a) $\sin x = 1$;   b) $\sin x = 0.5$;   c) $\cos x = -1$;

d) $\cos x = -\dfrac{\sqrt{2}}{2}$;   e) $3\sin x = -2$;   f) $-5\cos 4x + 3 = 2$;

g) $3\sin 2x = -\dfrac{1}{2}$;   h) $\cos 3x = -\dfrac{1}{2}$;   i) $\sin 6x = \dfrac{1}{2}$;

j) $4\tan x = 1$;   k) $3\tan 3x = -2$;   l) $4\cot x = -1$;

m) $5\cot 3x = -2$;   n) $\csc 4x = -3$;   o) $3\sec 3x = -4$.

V. 2. Solve the trigonometric equations:
a) $\sin x + \cos x = 1$;   b) $3\sin x - 2\cos x = 3$;
c) $3\sin 2x + 2\cos 2x = 3$;   d) $4\sin x + \cos x = 4$;
e) $12\sin 5x + \cos 5x = 9$;   f) $3\sin 7x - 2\cos 7x = 3$;

g) $\sin 2x - \sqrt{3}\cos 2x = \sqrt{3}$;   h) $\sin 4\alpha - \cos 4\alpha = \sqrt{\dfrac{3}{2}}$.

V. 3. Solve the trigonometric equations:

a) $\cos x - \sqrt{3}\sin x = 2\cos 3x$;   b) $\cos x - \sqrt{3}\sin x = \cos 3x$;
c) $2\sin 3\alpha + \sqrt{3}\sin 4\alpha + \cos 4\alpha = 0$;   d) $\sqrt{3}\sin 2x + \cos 5x - \cos 9x = 0$;
e) $\sin 2x + \cos 2x = \sqrt{2}\sin 3x$;   f) $\cos 3x - \sin 2x = \sqrt{3}\left(\cos 2x - \sin 3x\right)$;
g) $2\left(\cos 3\alpha - \sin \alpha \cos 3\alpha\right) = \sin 4\alpha + \sin 2\alpha$;   h) $\sin 3x - \sin 7x = \sqrt{3}\sin 2x$;

i) $\sqrt{3}\left(\sin\dfrac{3x}{2}\cos\dfrac{x}{2} - \sin\dfrac{x}{2}\cos\dfrac{3x}{2}\right) + \sin 2x = 0$.

V. 4. Solve the trigonometric equations:
a) $\sin 3x + \sin 5x = \sin 4x$;   b) $\cos \alpha - \cos 3\alpha = \sin 2\alpha$;
c) $\cos x - \cos 2x = \sin 3x$;   d) $\sin 3x - 2\cos 2x = -2$;
e) $\sin 9x = 2\sin 3x$;   f) $\cos 9x - \cos 7x + \cos 3x - \cos x = 0$;

g) $1 - \cos 6x = \tan 3x$; h) $1 + \cos x + \cos 2x + \cos 3x = 0$;

i) $1 + \sin 2x = \sin x + \cos x$; j) $1 + \sin x = \cos x + \sin 2x$;

k) $1 + \sin x + \cos x + \sin 2x + \cos 2x = 0$;

l) $\sin z + \sin 2z + \sin 3z = \cos z + \cos 2z + \cos 3z$.

V. 5. Solve the trigonometric equations:

a) $6\cos x + 4\sin x = 5\sec x$; b) $\tan 3x - \tan x = 4\sin x$;

c) $\tan x + \cot x = 2\sec 4x$; d) $\sec x + \cot 3x = \cot \dfrac{3x}{2}$;

e) $4\cos x \tan x = \sqrt{3} + \tan x$; f) $\sin 3x - 4\sin x \cos 2x = 0$;

g) $\sin 3\alpha \tan 3\alpha + 1 = \cos 6\alpha$; h) $2 + \cos x = 2\tan \dfrac{x}{2}$;

i) $\cos \dfrac{\alpha}{2} \cos \dfrac{3\alpha}{2} - \sin \alpha \sin 3\alpha - \sin 2\alpha \sin 3\alpha = 0$;

j) $1 + \tan 2x \tan 5x = \sqrt{2} \tan 2x \cos 3x \sec 5x$;

k) $2(\cos 4x - \sin x \cos 3x) = \sin 4x + \sin 2x$;

l) $\sqrt{3}(1 + \tan 2x \tan 3x) = \tan 2x \sec 3x$.

V. 6. Solve the trigonometric equations:

a) $\sin 3x \cos 3x = \sin 2x$; b) $\sin 2\alpha \sin 6\alpha = \cos \alpha \cos 3\alpha$;

c) $\sin x \sin 3x + \sin 4x \sin 8x = 0$; d) $\cos 3x \cos 6x = \cos 4x \cos 7x$;

e) $4\sin x \cos x \cos 2x \cos 8x = \sin 12x$; f) $\tan 3x \cos x = \sin x + 2\sin 2x$;

g) $4\cos x \cos 2x \sin 3x = \sin 2x$; h) $4\sin 2x \sin 6x \cos 4x + \cos 12x = 0$;

i) $4\cos x \cos 2x \cos 3x = \cos 6x$; j) $\cos x \cos 2x \cos 4x \cos 8x = 16^{-1}$;

k) $2\sin 2x + \sin 3x - \cos 3x \sec 5x \sin 5x = 0$.

V. 7. Solve the trigonometric equations:

a) $\sin 2x + 2\cot x = 3$; b) $2\sin 2x + 5\tan x = 7$;

c) $2\tan x - 2\cot x = 1$; d) $\tan 2x = \cot 3x - \cot 5x$;

e) $2\tan 2x = \tan 5x - \tan 3x$; f) $\tan \dfrac{3x}{2} - \tan \dfrac{x}{2} = \dfrac{2}{3}\sin x$;

g) $\csc x = \csc 2x + \csc 4x$; h) $\cot x - \tan x = \sin x + \cos x$;

i) $\sec x + \csc x + \sec x \csc x + \dfrac{5}{2} = 0$; j) $\tan x + \cot x - \cos 4x = 3$;

k) $\csc 6x - \cot x = \tan \dfrac{x}{2}$.

V. 8. Solve the trigonometric equations:

a) $\cos 2x = \sqrt{2}(\cos x - \sin x)$;   b) $\sin x + \cos x = \left(\sqrt{3} - 1\right)\cos 2x$;

c) $\sin 2x + 4(\sin x + \cos x) + 1 = 0$;   d) $\sin 2x + 4(\cos x - \sin x) = 4$;

e) $3\sin 2x - 7(\sin x + \cos x) + 5 = 0$;   f) $2\sin 2x - 9(\sin x + \cos x) + 6 = 0$;

g) $2\sin 2x + 5(\sin x - \cos x) - 4 = 0$;   h) $(1 - \tan x)(\sin x + \cos x) = \tan x$;

i) $\tan x - \sin x + \sqrt{3}(\cot x - \cos x) + 1 + \sqrt{3} = 0$;

j) $\dfrac{a + b\cos x}{b + a\sin x} = \dfrac{a + b\cos x}{b + a\sin x}$, $a, b \in \mathbf{R}$.

V. 9. Solve the trigonometric equations:

a) $1 + \sin x = (\sec x - \cos x)\tan\left(\dfrac{\pi}{4} + \dfrac{x}{2}\right)$;

b) $2(1 + \sin 2x) + \cot\left(x - \dfrac{\pi}{4}\right) = 0$;

c) $(1 + \sin 2x)(\cot x - 1)\tan x = 1 + \tan x$;

d) $\tan 2x + 4\cos x \cos 3x = \cot x$;

e) $\tan 4x - 2\cos 4x + \cot 2x = 2$;

f) $\sin x\left(1 + \tan x \tan\dfrac{x}{2}\right) = 4 - \cot x$;

g) $\sin x(\cos x - 2) + \tan x = \cos x + \sec x - 2$;

h) $2\tan x + \tan\dfrac{x}{2} = \cot 3x - 4\cot 2x$;

i) $\tan x - \sin 2x - \cos 2x(1 - 2\sec x) = 0$.

V. 10. Solve the trigonometric equations:

a) $1 - \sin 2x + \tan x = 2\sin^2 x$;   b) $1 + \sin 2x = (\cos 3x + \sin 3x)^2$;

c) $7\cos 2x - 3\cos^2 2x = 0$;   d) $2\cos^2 x + 7\sin x - 5 = 0$;

e) $25\sin^2 x + 90\cos x = 81$;   f) $\cos 2x + \sin x + 0.75 = \cos^2 x$;

g) $2\cos 2\alpha \cot^2 \alpha + 2 = 5\cot^2 \alpha$;   h) $2\cot^2 x \cos^2 x - \cot^2 x + 2 = 4\sin^2 x$;

i) $2\cot\left(\dfrac{\pi}{2} + x\right) - (1 - \cos 2x)\sec^2 x = \tan^2 x$;   j) $\tan^2 x - 2\cos^2 x = 0$;

k) $\tan\left(\dfrac{\pi}{2} - x\right) + (1 + \cos 2x)\csc^2 x = \cot^2 x$.

V. 11. Solve the trigonometric equations:

a) $\sin x + \sin 3x = 4\cos^3 x$;   b) $1 + \sin 3x = 2\cos^2 \dfrac{x}{2}$;

c) $4\cos x \sin^2 x + \sin x = \cos x$;   d) $\sin^2 2x + 9\cos 2x = 9$;

e) $\cos 9x = 4\cos^2 3x$;   f) $\sin^2 4x = 3\cos^2 4x$;

g) $\sin^2 x - \sin 2x = 3\cos^2 x$;   h) $5\cos^2 x = \sin^2 x + 2\sin 2x$;

i) $12\sin^2 x + \sin 2x - 2\cos^2 x = 4$;   j) $\sin 6x + \sin 2x = 4\sin^3 2x$;

k) $\sec^2 2\alpha - \csc^2 2\alpha = \dfrac{8}{3}$;   l) $4\tan^2 3x - \sec^2 3x = 2$.

V. 12. Solve the trigonometric equations:

a) $\sin^2 x - \sqrt{2}\sin x = 0$;   b) $\cos 4x + 2\cos^2 x = 1$;

c) $(1 + \cos 8x)\sin 4x = \cos^2 4x$;   d) $(\csc x + \sec x)(\sin x + \cos x) + 2 = 0$;

e) $1 + 2\cos^2 \dfrac{x}{2}(1 - \sin x) = \sin^2 x$;   f) $1 + \cos 5x = \left(\sin\dfrac{3x}{2} - \cos\dfrac{3x}{2}\right)^2$;

g) $\left(\cos\dfrac{x}{2} + \sin\dfrac{x}{2}\right)^2 = \cos 5x + \sin 7x + 2\cos^2\dfrac{3x}{2}$;

h) $\cos 2\alpha + \cos 6\alpha + 2\sin^2\alpha = 1$;

i) $\sin x + \cos x + \sin 2x + \sqrt{2}\sin 7x = \dfrac{2\tan x}{\tan^2 x + 1}$;

j) $\dfrac{1}{2}\sin 4x \sin x + \sin 2x \sin x + 2\sin^2 x = 2$;

k) $2\sin^2 3x + \sin^2 6x = (\sin 2x + \sin 4x)\sec x \csc 3x$;

l) $\sin 2x + 2\sin^2 x + 4(\sin x - \cos x + \tan x - 1) = 2$;

m) $2\sin 2x \sin^2 x - \sin^2 2x - 2\sin^2 x = \cot 2x$.

V. 13. Solve the trigonometric equations:

a) $11\tan 3x = \tan x$;   b) $\tan x - \tan 2x = \sin x$;

c) $\tan x + \tan 2x - \tan 3x = 0$;   d) $4\sin 2x = 3(\tan x - \cot x)$;

e) $\tan x + \tan 2x = \tan x \tan 2x$;   f) $\tan x - \cot x = \csc x - \sec x$;

g) $\cos 2x + 2\tan x = 2(\tan x \tan 2x + 1)\cos 2x$;

h) $\cot x - \tan x - 2\tan 2x - 4\tan 4x + 8 = 0$;

i) $3\tan x + \tan 2x + 3\tan x \tan 3x + \tan 2x \tan 3x = 0$;

j) $\tan(ax + b)\tan(ax - b) = 1$, $a \neq 0$, $a, b \in \mathbf{R}$;

k) $\tan(2x+1)\cot(x+3)=1$.

V. 14. Solve the trigonometric equations:

a) $\dfrac{\cos^2 x(1+\cot x)-3}{\sin x-\cos x}=3\cos x$;　b) $\cot^2 2x+\dfrac{4(\cos 3x-\cos x)}{\sin 3x-\sin x}+3=0$;

c) $1+\sin 2x=\dfrac{\cot x+1}{\cot x-1}$;　d) $\cot x-\tan x=\dfrac{2(\cos x-\sin x)}{\sin 2x}$;

e) $\dfrac{8\cos x}{\cot^2 \dfrac{x}{2}-\tan^2 \dfrac{x}{2}}=1-\sin 2x$;　f) $\dfrac{3(\cos 2x+\cot 2x)}{\cot 2x-\cos 2x}=2(\sin x+\cos x)^2$;

g) $\dfrac{\sin x-\tan x}{\sin x+\tan x}+\dfrac{3}{5}(1-\cos x)=0$;　h) $\tan 7x-2\tan 3x=\tan^2 3x\tan 7x$;

i) $(3-4\sin^2 x)\csc x=\cot x(1+2\cos 2x)$;　j) $\csc^2 2x+\tan x=\cot x+4$;

k) $4(\tan 3x-\tan 2x)=\tan 3x\sec^2 2x$.

V. 15. Solve the trigonometric equations:

a) $\dfrac{\sin^2 2x-4\sin^2 x}{\sin^2 2x-4\cos^2 x}+1=2\tan^2 x$;　b) $\dfrac{\cos^2 x-\cot^2 x}{\sin^2 x-\tan^2 x}+2\cot^3 x+1=0$;

c) $\dfrac{1-2\cos^2 x}{\sin x\cos x}+2\tan 2x+\cot^2 4x+3=0$;　d) $\dfrac{\sin 2x-\cos^2 x}{\cos^2 x-\sin 2x+1}=\tan x$;

e) $\dfrac{1+\sin^2 2x}{1+\sin^2 x}=2-\dfrac{4}{3}\sin^2 x$;　f) $\dfrac{4\cot x}{\cot^2 x+1}+2=\cos^2 2x$;

g) $\dfrac{3\cot^2 x-1}{\cot^2 x-3}=\sin 6x\cot x$;　h) $\tan^2 x-2\cot 2x=\dfrac{2\sin 2x+\sin 4x}{2\sin 2x-\sin 4x}$;

i) $\dfrac{3-\tan^2 x}{1-\tan^2 x}(\cos 3x+\cos x)=2\sin 5x\cot x$;

j) $\dfrac{\tan x}{2-\sec^2 x}(\sin 3x-\sin x)=\dfrac{2}{\cot^2 x-3}$;

k) $\dfrac{1}{4}\left(\tan\dfrac{x}{4}-\cot\dfrac{x}{4}\right)+\dfrac{1}{2}\tan\dfrac{x}{2}+\tan x=2\sqrt{3}$;

l) $\dfrac{3\tan x-\tan^3 x}{2-\sec^2 x}=\dfrac{4+2\cos\dfrac{6x}{5}}{\cos 3x+\cos x}$.

V. 16. Solve the trigonometric equations:

a) $\dfrac{\sin\alpha}{1+\cos\alpha}=2-\cot\alpha$; b) $\dfrac{1-\cos 2x}{\sin x}=1$; c) $\dfrac{\sin 2x}{1+\cos 2x}\cdot\dfrac{\sin x}{1+\cos x}+1=\csc x$;

d) $\dfrac{\cos x\cot\dfrac{x}{2}}{2\cos^2\dfrac{x}{2}}-\dfrac{\tan\dfrac{x}{2}\tan x+1}{\tan\dfrac{x}{2}+\cot x}=2\sqrt{3}$; e) $\dfrac{\cos\left(\dfrac{3\pi}{2}+x\right)\cos\left(\dfrac{3\pi}{2}-x\right)}{\sin\left(\dfrac{\pi}{4}+x\right)\sin\left(\dfrac{\pi}{4}-x\right)}=\dfrac{1}{3}$;

f) $\tan\left(\dfrac{7\pi}{2}-x\right)+\dfrac{\cos\left(\dfrac{7\pi}{2}+x\right)}{1+\cos x}=2$; g) $\dfrac{1}{\sqrt{3}-\tan x}-\dfrac{1}{\sqrt{3}+\tan x}=\sin 2x$;

h) $\dfrac{1}{\tan x}-\tan x+2\left(\dfrac{1}{\tan x+1}+\dfrac{1}{\tan x-1}\right)=4$; i) $\cot x+\tan 2x=1+\dfrac{\sin 3x}{\sin x}$;

j) $\left(3-\tan^2 x\right)\left(\cos 3x+\cos x\right)=\dfrac{4\cos 3x}{\tan 2x}$.

V. 17. Solve the trigonometric equations:

a) $\dfrac{\tan x}{\cos^2 5x}-\dfrac{\tan 5x}{\cos^2 x}=0$; b) $\dfrac{\cos^2 3x}{\tan x}+\dfrac{\cos^2 x}{\tan 3x}=0$;

c) $\dfrac{\cot 4x}{\sin^2 x}+\dfrac{\cot x}{\sin^2 4x}=0$; d) $\dfrac{\tan 2x}{\cos^2 x}-\dfrac{\tan x}{\cos^2 2x}=0$;

e) $\dfrac{\tan 4x}{\tan 2x}+\dfrac{\tan 2x}{\tan 4x}+\dfrac{10}{3}=0$; f) $\dfrac{\cot 2x}{\cot x}+\dfrac{\cot x}{\cot 2x}+2=0$;

g) $\dfrac{\sin^2 x}{\cos x(1+\tan x)}+\dfrac{\cos^2 x}{\sin x(1+\cot x)}=\sqrt{2}$;

h) $\dfrac{1}{\tan^2 x+\sec^2 x}+\dfrac{1}{\cot^2 x+\csc^2 x}=\dfrac{2}{3}$;

i) $\dfrac{1}{\tan 5x+\tan 2x}-\dfrac{1}{\cot 5x+\cot 2x}=\tan 3x$;

j) $\dfrac{1}{\tan 3x+\tan 4x}-\dfrac{1}{\cot 3x+\cot 4x}=-\cot^2 7x$;

k) $\dfrac{1}{\tan x+\cot x}+\dfrac{1}{\tan 2x+\cot 2x}=\dfrac{1}{\tan 3x+\cot 3x}+\dfrac{1}{\tan 4x+\cot 4x}$.

V. 18. Solve the trigonometric equations:

a) $\tan x+\cot x=\dfrac{2}{\cos 4x}$; b) $\dfrac{1}{1-\tan^2 x}=1+\cos 2x$;

c) $\dfrac{\sin x + \cos x}{\cos x - \sin x} + \dfrac{1 + \tan x \tan\dfrac{x}{2}}{\cot x + \tan\dfrac{x}{2}} = 1$; d) $\dfrac{1 + \sin 2x}{\sin 2x} = \sqrt{2}\cos 2x \tan\left(\dfrac{\pi}{4} + x\right)$;

e) $\dfrac{(1 - \cos 2x)^2}{(1 + \cos 2x)^2} - \dfrac{2}{\cos^2 x} = 1$; f) $\dfrac{\tan x + \cot x}{\cot x - \tan x} = 4\sin 2x + 6\cos 2x$;

g) $\cot^2 x - \dfrac{\sin 4x - 2\sin 2x}{\sin 4x + 2\sin 2x} = \dfrac{10}{3}$; h) $3\dfrac{1 + \tan x}{1 - \tan x} + \dfrac{1 + \sin 2x}{1 - \sin 2x} = 4$;

i) $\cot x - \tan x + \dfrac{4}{\cos 4x} = \dfrac{4\tan x}{\tan^2 x - 1}$;

j) $\dfrac{1 + \sin 2x - 2\sin^2 x}{1 + \cot^2 x} = \left[\cos\left(\dfrac{\pi}{4} + x\right) - \cos\left(\dfrac{\pi}{4} + 3x\right)\right]\cos\dfrac{\pi}{4}$.

V. 19. Solve the trigonometric equations:

a) $2\cos(3x - 13°)\cot 20° + 2\sin(3x - 13°) + \sec 70° = 0$;

b) $2\cos(3x - 50°) - 2\sin(3x - 50°)\cot 10° - \csc 170° = 0$;

c) $\sin x \sin(60° - x)\sin(60° + x) = \dfrac{1}{8}$;

d) $\tan(x - 15°)\cot(x + 15°) = \dfrac{1}{3}$;

e) $\tan 2x \tan(2x + 60°)\tan(2x + 120°) = \sqrt{3}$;

f) $\tan(270° + x) + \csc^2 x(1 + \cos 2x) = \cos^2 x \csc^2 x$;

g) $\sin(x + 20°) + \cos(x - 10°) - \sin(x - 40°) = \sqrt{3}$;

h) $\tan(3x - 60°) + \tan(x - 140°) = 2\cos(2x - 10°)$;

i) $\cot x + \cot 14° + \cot(x + 26°) = \cot x \cot 14° \cot(x + 26°)$;

j) $\tan 20°(\tan x + \tan 40°) + \tan 40° \tan x = 1$;

k) $\tan x + \tan 50° + \tan 70° = \tan x \tan 50° \tan 70°$;

l) $\sin^2(x + 50°) + \sin^2(x - 50°) - \cos 80° \cos 2x = \sin 2x$.

V. 20. Solve the trigonometric equations:

a) $(1 + \sin 2x)^2 + (1 - \sin 2x)^2 = 3 - \sin 4x$;

b) $\tan 2x \tan\left(2x + \dfrac{\pi}{3}\right)\tan\left(2x + \dfrac{2\pi}{3}\right) = \sqrt{3}$;

c) $1 + \sin \dfrac{x}{2} \sin x - \cos \dfrac{x}{2} \sin^2 x = 2 \sin^2 \left(\dfrac{\pi}{4} + \dfrac{x}{2} \right)$;

d) $\sin^2 \left(\dfrac{\pi}{8} + x \right) - \sin^2 \left(\dfrac{\pi}{8} - x \right) = \sin x$;

e) $\tan^2 x \cot^2 2x \cot 3x = \tan^2 x - \cot^2 2x + \cot 3x$;

f) $\tan 2x \tan^2 3x \tan^2 5x = \tan 2x + \tan^2 3x - \tan^2 5x$;

g) $\tan^2 x \tan^2 3x \tan 4x = \tan^2 x - \tan^2 3x + \tan 4x$.

V. 21. Solve the trigonometric equations:

a) $4 - 5 \cot 2x \tan 3x = \tan^2 3x$;

b) $\tan^2 x - 3 \cot^2 x - (\tan x - 3 \cot x) = 2$;

c) $\tan^4 3x + \cos^2 6x = 1$;

d) $8 \csc^3 2x - \tan^3 x = 6 \csc 2x + 3 \cot x$;

e) $2 + \sin x + \csc x + 2(\csc^2 x - \cos^2 x) = 4$;

f) $\tan^2 x + \cot^2 x + 3 \tan x + 3 \cot x + 4 = 0$;

g) $\tan^4 x + \cot^4 x + \tan^2 x + \cot^2 x = 4$;

h) $\tan^4 \dfrac{x}{2} + 2 \tan^2 \dfrac{x}{2} + \cot^4 \dfrac{x}{2} - 2 \cot^2 \dfrac{x}{2} = \dfrac{130}{9}$;

i) $12 \sec^2 x + \dfrac{1}{3}(\cos 2x \csc^2 x - 2) + 10\left(2 \tan x + \dfrac{1}{3} \cot x \right) = 0$;

j) $18 \cos^2 x + 5(3 \cos x + \sec x) + 2 \sec^2 x + 5 = 0$;

k) $\tan x + \cot x + \tan^2 x + \cot^2 x + \tan^3 x + \cot^3 x = 6$;

l) $\tan^3 x + \tan^2 x + \cot^2 x + \cot^3 x - 4 = 0$.

V. 22. Solve the trigonometric equations:

a) $\cos^2 x + \cos^2 2x = \cos^2 3x + \cos^2 4x$;

b) $\sin^2 2\alpha + \sin^2 3\alpha + \sin^2 4\alpha = 1 + \cos^2 5\alpha$;

c) $2 \sin^3 x + \sin^2 x - \sin x = \cos^2 x$;

d) $1 + \cos^3 x - 4 \cos^2 \dfrac{x}{2} = \sin^2 x$;

e) $\cos^3 x + \sin x - 3 \sin^2 x \cos x - 4 \sin^3 x = 0$;

f) $3\sin^2 x\cos\left(\dfrac{3\pi}{2}+x\right)+3\sin^2 x\cos x-\sin x\cos^2 x=\sin^2\left(\dfrac{3\pi}{2}+x\right)\cos x\,;$

g) $2\sin^5 2x-\sin^3 2x=3-6\cos^2 2x\,;$

h) $\dfrac{\sin^3 x-\cos^3 x}{2+\sin 2x}=\dfrac{1}{3}\cos 2x\,;$ **i)** $\dfrac{20\left(\sin^3\dfrac{x}{2}-\cos^3\dfrac{x}{2}\right)}{16\sin\dfrac{x}{2}-25\cos\dfrac{x}{2}}=\sin\dfrac{x}{2}\cos\dfrac{x}{2}\,;$

j) $\dfrac{\sin^3 x+\cos^3 x}{2\cos x-\sin x}=\cos 2x\,;$ **k)** $\dfrac{6\cos^3 x+2\sin^3 x}{3\cos x-\sin x}=\cos 2x\,.$

V. 23. Solve the trigonometric equations:

a) $\sin^3 x+\cos^3 x=1\,;$

b) $\sin^3 x\cos x-\sin x\cos^3 x=\dfrac{\sqrt{2}}{8}\,;$

c) $2\sin^3 x+2\sin^2 x\cos x-\sin x\cos^2 x-\cos^3 x=0\,;$

d) $\sin^3 x\sin 3x+\cos^3 x\cos 3x=\cos^3 4x\,;$

e) $4\sin^3 x\cos 3x+4\cos^3 x\sin 3x=3\sin 2x\,;$

f) $\sin^3 x\cos 3x+\cos^3 x\sin 3x+0.375=0\,;$

g) $\sin^3 x(1-\cot x)+\cos^3 x(1-\tan x)=\sqrt{2}\cos 2x\,;$

h) $2\cos 13x+3\cos 3x+3\cos 5x=8\cos x\cos^3 4x\,;$

i) $\cos 8x+3\cos 4x+3\cos 2x+0.5=8\cos x\cos^3 3x\,;$

j) $4\cot^3 x-12\cot x+\left(\cot\dfrac{x}{2}-\tan\dfrac{x}{2}\right)^2-12=0\,;$

k) $\csc^3 x\sec^3 x-6\sec 2x=\tan^3 x+\cot^3 x\,.$

V. 24. Solve the trigonometric equations:

a) $\sin^4 x+\cos^4 x=1\,25-\sin^2 2x\,;$ **b)** $\sin^4 x+\cos^4 x=\sin 2x-0.5\,;$

c) $\sin^4 x-\cos^4 x=3+5\cos x\,;$ **d)** $\cos^4\dfrac{x}{2}-\sin^4\dfrac{x}{2}=\sin 2x\,;$

e) $\cos^4 x-\sin^4 x=2-3\sin x\,;$ **f)** $2\sin^4 x-6\cos^4 x=\sin^2 2x\,;$

g) $2\sin^4 x-\cos^4 x=\cos 2x-\dfrac{5}{4}\sin^2 2x\,;$

h) $\sin^4 x-8\cos^4 x=8\sin x\cos^3 x-\sin^3 x\cos x\,;$

72

i) $\cos^4 x - \sin^4 x + 1 = 3(1 - \sin x)$;

j) $(2\cos 2x + 3)\cos^4 x - (2\cos 2x + 3)\sin^4 x = 2$;

k) $\sqrt{2}\left(\cos^4 2x - \sin^4 2x\right) = \cos 2x + \sin 2x$;

l) $2\sin x \cos x\left(\sin^4 x + \cos^4 x - 1\right) = \sin^2 2x$;

m) $\sin^4 x - \cos^4 x = 1 - 2\cos^2 x(\cos 3x - \tan x \sin 3x)$;

n) $2\left(\cos^4 x + \sin^4 x\right)\sec 2x = \sec 2x + \cos 4x + 1$.

V. 25. Solve the trigonometric equations:

a) $12\tan^4 3x - 7\tan^2 3x + 1 = 0$; b) $\tan^2 x + \cot^2 x - 2 = 4\tan 2x$;

c) $\cot^4 2x + \csc^4 2x = 25$; d) $\sin^4 2x + \sin^4\left(\dfrac{5\pi}{4} + 2x\right) = \dfrac{1}{4}$;

e) $\dfrac{\cos^4 x + \sin^4 x}{\cos^4 x - \sin^4 x} - 0.5\cos 2x = \dfrac{\sqrt{3}}{6}\csc 2x$;

f) $\tan^4 x + \cot^4 x = \dfrac{17}{4}(\tan x \tan 2x + 1)\cos 2x$;

g) $0.5\left(\tan^2\dfrac{x}{2} + \cot^2\dfrac{x}{2}\right) = \left(\tan\dfrac{x}{2}\tan x + 1\right)\cos x + \dfrac{2}{\sqrt{3}}\cot x$;

h) $4\left(\sin x \cos^5 x + \cos x \sin^5 x\right) + \sin^3 2x = 1$;

i) $\sin^5 x + \cos^5 x + \dfrac{1}{\sin x} + \dfrac{1}{\cos x} = 0$;

j) $\tan^6 x - \tan^4 x + \tan^2 x = \dfrac{\sin^2 x - \tan^2 x}{\cos^2 x - \cot^2 x}$.

V. 26. Solve the trigonometric equations:

a) $\sin^6 x + \cos^6 x = \dfrac{7}{16}$;

b) $16\left(\cos^6 x + \sin^6 x\right) = 16\cos^2 2x + 1$;

c) $2\left(\sin^6 2x + \cos^6 2x\right) - 3\left(\sin^4 2x + \cos^4 2x\right) = \sin x + \cos x$;

d) $2\left(\sin^6 x + \cos^6 x\right) - 3\left(\sin^4 x + \cos^4 x\right) = \sin 2x$;

e) $2\left(\sin^6 x + \cos^6 x\right) - 3\left(\sin^4 x + \cos^4 x\right) = \cos 2x$;

f) $\dfrac{1-\sin^6 x-\cos^6 x}{1-\sin^4 x-\cos^4 x}=2\cos^2 4x$;

g) $\tan^3 x+\cot^3 x-8\csc^3 2x=12$;

h) $\tan^3 x+\cot^3 x=\sec^3 x\csc^3 x-2\sqrt{3}\sec 2x$;

i) $\sin^{10} x+\cos^{10} x=\dfrac{1}{16}\sin^2 2x$.

V. 27. Solve the trigonometric equations containing radicals:

a) $\sin^3 x(1+\cot x)+\cos^3 x(1+\tan x)=\sqrt{2}\sin 2x$;

b) $\cos x+\sin x=\sqrt{2\sin^2 x-1}$;   c) $1-\cos x=\sqrt{1-\sqrt{4\cos^2 x-7\cos^4 x}}$;

d) $(1+\cos x)\sqrt{\tan\dfrac{x}{2}-2\cos x+\sin x}=2$;   e) $\sqrt{\cos^2 x+0.5}+\sqrt{\sin^2 x+0.5}=2$.

f) $\sqrt{\sin x-\sqrt{\sin x+\cos x}}=\cos x$;   g) $\sqrt{\dfrac{1-\sin x}{1+\sin x}}+\tan\dfrac{x}{2}=2$;

h) $\sqrt{\tan x+\sin x}+\sqrt{\tan x-\sin x}=2\cos x\sqrt{\tan x}$;

i) $\dfrac{\sqrt{1-\cos x}+\sqrt{1-\cos x}}{\cos x}=4\sin x$;   j) $\sqrt{\dfrac{1}{2}-\cos 2x}+\sqrt{\dfrac{1}{2}+\cos 2x}=1$;

k) $\sqrt[3]{\sin^2 x}-\sqrt[3]{\cos^2 x}=\sqrt[3]{\cos 2x}$;   l) $2\cos x=\sqrt{2+\sin 3x}$;

l) $2\tan 2x-4\cot 2x=2\cos x\sqrt{\tan^2 x+\cot^2 x-2}$;

m) $\sqrt{1-2\sin^2 x}+\sqrt{1+\sin 2x}=2\sqrt{\sin x+\cos x}$;

n) $\dfrac{\sqrt{1+\tan x}+\sqrt{2(1+\sec x)}}{\sqrt{1-\tan x}+\sqrt{2(1+\sec x)}}=\dfrac{\sqrt{1-\tan x}}{\sqrt{1+\tan x}}$.

V. 28. Solve the trigonometric equations containing absolute value:

a) $\sqrt{2}|\cos x|=1+\cot x$;   b) $|\sin x|+|\cos x|=1.2$;

c) $|\sin x+\cos x|=\sqrt{2}$;   d) $|\tan 2x+\cot 2x|=\dfrac{4\sqrt{3}}{3}$.

V. 29. Solve the trigonometric equations, where $m\in\mathbf{R}$:

a) $\cos x+\sqrt{3}\sin x=m$;

b) $\cos x-\sin x=m\cos 2x$;

c) $m \sin x + (2 - m)\cos x = m + 1$;

d) $\cos^2 x = \sin^2 x - \sin 2x + m$;

e) $2(m + 1)\sin^2 x - (5m + 1)\sin x + 2m = 0$;

f) $4\cos^2 x - 2\sqrt{3} \sin x \cos x + 2 \sin^2 x = m$;

g) $\cos 2x - (2m - 1)\sin x + m - 1 = 0$;

h) $m \cos 2x + 2(m^2 + 3)\sin x - 7m = 0$;

i) $m(\sin x - \cos x)^2 - \cos 4x - 1 = 0$;

j) $\sin^4 x + \cos^4 x = m$;

k) $\sin^4 x + \cos^4 x = m(\sin^6 x + \cos^6 x)$.

V. 30. The equation $\tan 3x = a$, where $a \in \mathbf{R}$, has three solutions in the interval $\left(-\dfrac{\pi}{2}, \dfrac{\pi}{2} \right)$. If α, β, γ are these roots, show that

$\sin \alpha \sin \beta \sin \gamma \cos(\alpha + \beta + \gamma) + \cos \alpha \cos \beta \cos \gamma \sin(\alpha + \beta + \gamma) = 0$.

V. 31. Find a $\hat{1}\left(0, \dfrac{\pi}{4} \right)$ such that $\sin x = \dfrac{\sqrt{2 - \sqrt{3}}}{2}$ and solve the equation $\sin x + \sin 3x + \sin 5x + \sin 7x = 2\sqrt{2 - \sqrt{3}} \cos x \cos 2x$.

V. 32. Find a $\hat{1}\left(0, \dfrac{\pi}{2} \right)$ such that $\sqrt{2 + \sqrt{2}} = 2\cos \alpha$ and $n \hat{1} \mathbf{N}$ such that $\left(\sqrt{2 + \sqrt{2 + \sqrt{2}}} + i\sqrt{2 - \sqrt{2 + \sqrt{2}}} \right)^n + 2^n = 0$.

Unit VI

Inverses of Trigonometric Functions

$\arcsin : [-1, 1] \rightarrow \left[-\dfrac{\pi}{2}, \dfrac{\pi}{2} \right]$, defined as $\arcsin x = \theta \Leftrightarrow x \in [-1, 1]$,

$\theta \in \left[-\dfrac{\pi}{2}, \dfrac{\pi}{2} \right]$ and $\sin \theta = x$.

$\sin(\arcsin x) = x \Leftrightarrow x \in [-1, 1]$

$\arcsin(\sin x) = x \Leftrightarrow x \in \left[-\dfrac{\pi}{2}, \dfrac{\pi}{2} \right]$

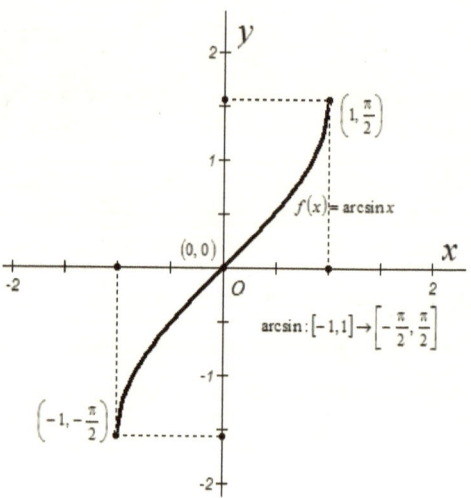

arccos : $[-1,1] \to [0, \pi]$, defined as $\arccos x = \theta \Leftrightarrow x \in [-1,1]$, $\theta \in [0, \pi]$ and $\cos \theta = x$.

$\cos(\arccos x) = x \Leftrightarrow x \in [-1,1]$

$\arccos(\cos x) = x \Leftrightarrow x \in [0, \pi]$

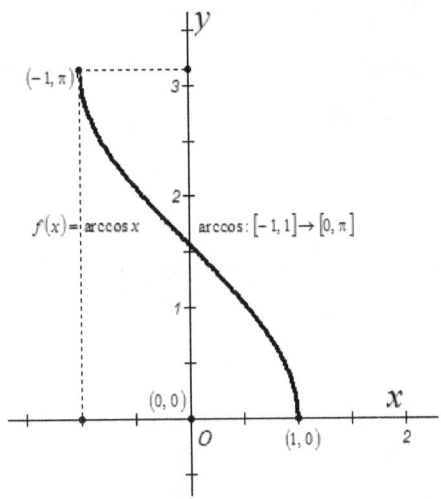

arctan : $\mathbf{R} \to \left(-\dfrac{\pi}{2}, \dfrac{\pi}{2}\right)$, defined as $\arctan x = \theta \Leftrightarrow x \in \mathbf{R}$, $\theta \in \left(-\dfrac{\pi}{2}, \dfrac{\pi}{2}\right)$ and $\tan \theta = x$.

$\tan(\arctan x) = x \Leftrightarrow x \in \mathbf{R}$

$\arctan(\tan x) = x \Leftrightarrow x \in \left(-\dfrac{\pi}{2}, \dfrac{\pi}{2}\right)$

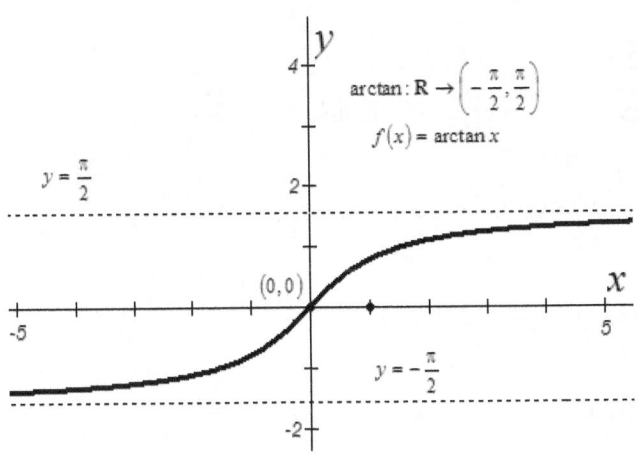

$\text{arc cot} : \mathbf{R} \to (0, \pi)$, defined as $\text{arc cot}\, x = \theta \Leftrightarrow x \in \mathbf{R}$, $\theta \in (0, \pi)$ and $\cos \theta = x$.

$\cot(\text{arc cot}\, x) = \theta \Leftrightarrow x \in \mathbf{R}$
$\text{arc cot}(\cot x) = \theta \Leftrightarrow x \in (0, \pi)$

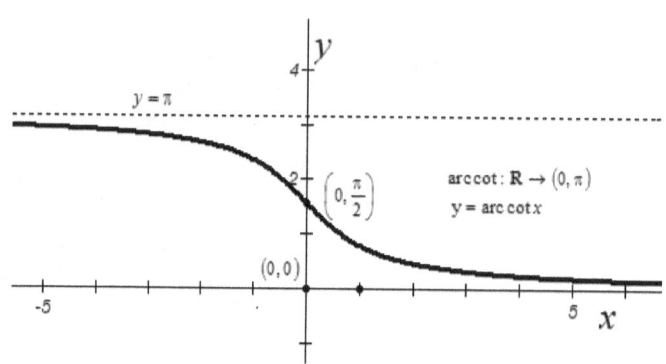

VI. 1. Prove the identities:

a) $\arcsin(-x) = -\arcsin x,\ x \in [-1, 1]$;

b) $\arccos(-x) = \pi - \arccos x,\ x \in [-1, 1]$;

c) $\arctan(-x) = -\arctan x,\ x \in \mathbf{R}$;

d) $\text{arc cot}(-x) = -\text{arc cot}\, x,\ x \in \mathbf{R}$;

e) $\arcsin x + \text{arc cos}\, x = \dfrac{\pi}{2}$; **f)** $\arctan x + \text{arc cot}\, x = \dfrac{\pi}{2}$.

VI. 2. Prove the identities:

a) $\sin(\arccos x) = \sqrt{1 - x^2}$; **b)** $\sin(\arctan x) = \dfrac{x}{\sqrt{1 + x^2}}$;

c) $\cos(\arcsin x) = \sqrt{1 - x^2}$; **d)** $\cos(\arctan x) = \dfrac{1}{\sqrt{1 + x^2}}$;

e) $\tan(\arcsin x) = \dfrac{x}{\sqrt{1 - x^2}}$; **f)** $\tan(\arccos x) = \dfrac{\sqrt{1 - x^2}}{x}$;

g) $\cot(\arctan x) = \dfrac{1}{x}$; **h)** $\tan(\text{arc cot}\, x) = \dfrac{1}{x}$;

i) $\cos\left(\dfrac{1}{2}\arccos x\right)=\sqrt{\dfrac{1+x}{2}}$; j) $\sin\left(\dfrac{1}{2}\arccos x\right)=\sqrt{\dfrac{1-x}{2}}$;

k) $\sin\left(\dfrac{1}{2}\arcsin x\right)=\dfrac{\sqrt{1+x}-\sqrt{1-x}}{2}$; l) $\cos\left(\dfrac{1}{2}\arcsin x\right)=\dfrac{\sqrt{1+x}+\sqrt{1-x}}{2}$;

m) $\cos(3\arccos x)=4x^3-3x$; n) $\sin(3\arccos x)=3x-4x^3$;

o) $\arctan x+\arctan y+\arctan z=\arctan\dfrac{x+y+z-xyz}{1-xy-yz-zx}$.

VI. 3. Calculate:

a) $\arcsin(\sin 1)$;

b) $\arcsin(\sin 3)$;

c) $\arcsin(\sin 10)$;

d) $\arccos(\cos 1)$;

e) $\arcsin(\cos 3)$;

f) $\arcsin(\cos 10)$;

g) $\arccos(\cos 10)$;

h) $\arcsin(\cos 2014)$.

VI. 4. Graph the function $f:[0,2\pi]\to \mathbf{R}$, $f(x)=\arcsin(\sin x)$.

VI. 5. Prove that the function $f:\left[0,\dfrac{\pi}{2}\right]\to[-1,1]$, $f(x)=\sin x-\cos x$ has an inverse and find the inverse function.

VI. 6. Find the inverses of the functions:

a) $f:\left[\dfrac{\pi}{2},\dfrac{3\pi}{2}\right]\to[-1,1]$, $f(x)=\sin x$;

b) $g:[\pi,2\pi]\to[-1,1]$, $g(x)=\cos x$;

c) $h:[3\pi,4\pi]\to[-1,1]$, $h(x)=\cos x$;

d) $f:\left[\dfrac{3\pi}{2},\dfrac{5\pi}{2}\right]\to \mathbf{R}$, $f(x)=\tan x$;

e) $f:\left[\dfrac{\pi}{6},\dfrac{7\pi}{6}\right]\to[-2,2]$, $f(x)=\sin x+\sqrt{3}\cos x$.

VI. 7. Find the domain of the functions:

a) $f(x) = \arcsin \dfrac{x}{3}$; b) $f(x) = \arccos \dfrac{3x+2}{5}$;

c) $f(x) = \arcsin \dfrac{1-x}{1+x}$; d) $f(x) = \arccos(4x^2 - 3)$;

e) $f(x) = \arcsin \dfrac{1}{x+1}$; f) $f(x) = \arccos \dfrac{x^2 - 6x + 8}{x^2 - 9}$.

VI. 8. Find the domain of the function

$f(x) = \arcsin \sqrt{x} + \arcsin \sqrt{1-x^2}$ and draw the graph.

VI. 9. Find the domain and range of the function
$f(x) = \cos(2 \arccos x)$ and draw the graph.

VI. 10. If $x \in [-1, 1]$, find the range of the function
$f(x) = \arcsin x \cdot \arccos x$.

VI. 11. a) Determine $\arctan \dfrac{1}{4}$ in terms of arcsin and arccos;
b) Determine $\operatorname{arccot} 3$ in terms of arcsin, arccos, and arctan.

VI. 12. If $x = 2 \arctan \dfrac{1}{3}$, evaluate $\sin x - \cos x$.

VI. 13. Write in an increasing order the numbers:

a) $\arcsin \dfrac{1}{3}$, $\arcsin \dfrac{1}{2}$, $\arcsin 1$; b) $\arccos \dfrac{1}{\sqrt{2}}$, $\arccos \dfrac{1}{\sqrt{3}}$, $\arccos \dfrac{2}{3}$;

c) $\arccos \dfrac{1}{3}$, $\arcsin \dfrac{3}{5}$, $\arctan \dfrac{11}{5}$.

VI. 14. Calculate the value of the expressions:

a) $\sin\left(\arcsin \dfrac{1}{2} + \arcsin \dfrac{1}{3} \right)$; b) $\cos\left(\arccos \dfrac{1}{\sqrt{2}} + \arccos \dfrac{1}{\sqrt{3}} \right)$;

c) $\sin\left(2 \arctan \dfrac{1}{2} + \arccos \dfrac{4}{5} \right)$; d) $\tan\left[-\dfrac{\pi}{4} - \dfrac{1}{4} \arcsin\left(-\dfrac{4}{5} \right) \right]$;

e) $\sin\left(\arccos \dfrac{3}{5} + \arccos \dfrac{2}{3} \right) + \cos\left(\arcsin \dfrac{4}{5} - \arcsin \dfrac{2}{3} \right)$;

f) $\sin\left(\arcsin \dfrac{2}{5} + \arcsin \dfrac{3}{5} + \arcsin \dfrac{4}{5} \right)$;

g) $\sin\left(\arccos\dfrac{2}{\sqrt{5}} - \arccos\dfrac{1}{\sqrt{5}} - \arcsin\dfrac{1}{\sqrt{5}}\right)$;

h) $\tan\left(2\arccos\dfrac{5}{\sqrt{26}} - \arcsin\dfrac{12}{13}\right)$.

VI. 15. Prove the identities:

a) $\sin\left(4\arctan\dfrac{1}{3}\right) = \cos\left(2\arctan\dfrac{1}{7}\right)$;

b) $\sin(4\operatorname{arccot}3) = \cos(2\operatorname{arccot}7)$;

c) $\sin\left(2\arctan\dfrac{1}{3}\right) + \tan\left(\dfrac{1}{2}\arcsin\dfrac{3}{5}\right) = \dfrac{14}{15}$;

d) $\sin\left(2\arctan\dfrac{1}{2}\right) - \tan\left(\arccos\dfrac{3}{5}\right) = -\dfrac{8}{15}$;

e) $\cos\left(2\arctan\dfrac{1}{2}\right) - \sin\left(3\arcsin\dfrac{1}{2}\right) = -\dfrac{2}{5}$;

f) $\cos(2\arctan 2) - \sin(4\arctan 3) = \dfrac{9}{25}$.

VI. 16. Calculate the value of the expressions:

a) $\cot\left(\dfrac{7\pi}{4} + \dfrac{1}{2}\arccos 2k\right) + \cot\left(\dfrac{7\pi}{4} - \dfrac{1}{2}\arccos 2k\right)$;

b) $\tan\left(\dfrac{7\pi}{4} + \dfrac{1}{2}\arccos\dfrac{2}{k}\right) + \tan\left(\dfrac{7\pi}{4} - \dfrac{1}{2}\arccos\dfrac{2}{k}\right)$;

c) $\cos^4\left(\dfrac{5\pi}{2} - \dfrac{1}{2}\arcsin\dfrac{3}{5}\right) + \cos^4\left(\dfrac{5\pi}{2} + \dfrac{1}{2}\arccos\dfrac{4}{5}\right)$;

d) $\cos^6\left(\dfrac{5\pi}{2} - \dfrac{1}{2}\arcsin\dfrac{3}{5}\right) - \cos^6\left(\dfrac{5\pi}{2} + \dfrac{1}{2}\arcsin\dfrac{4}{5}\right)$;

e) $\cos^6\left(\dfrac{5\pi}{2} - \dfrac{1}{2}\arcsin\dfrac{4}{5}\right) + \cos^6\left(\dfrac{5\pi}{2} + \dfrac{1}{2}\arcsin\dfrac{3}{5}\right)$.

VI. 17. Calculate the value of the expressions:

a) $\arccos\left(\cos\left(2\arctan\left(\sqrt{2} - 1\right)\right)\right)$;

b) $\arccos\left(\cos\left(2\operatorname{arc}\cot\left(\sqrt{2} - 1\right)\right)\right)$;

c) $\arccos\left(\cos\left(2\arctan\left(\sqrt{2} - 1\right)\right)\right)$;

d) $\arccos\left(\cos\left(2\operatorname{arc\,cot}\left(\sqrt{2}-1\right)\right)\right)$;

e) $\arcsin\left\{\cos\left[2\operatorname{arccot}\left(\sqrt{2}-1\right)\right]\right\}$.

VI. 18. Calculate the value of the expressions:

a) $\sin^2\left(\operatorname{arccot}\dfrac{1}{2}+\arctan\dfrac{1}{3}\right)$;   b) $\sin^2\left[2\arctan 2-\dfrac{1}{2}\arcsin\left(-\dfrac{4}{5}\right)\right]$;

c) $\cot\left[\dfrac{1}{2}\arccos\dfrac{3}{5}-2\operatorname{arccot}\left(-\dfrac{1}{2}\right)\right]$;   d) $\tan\left[\dfrac{1}{2}\arccos\dfrac{3}{5}-3\operatorname{arccot}(-2)\right]$;

e) $\cos\left[\dfrac{1}{2}\arcsin\dfrac{4}{5}-2\operatorname{arccot}\left(-\dfrac{1}{2}\right)\right]$;   f) $\tan\left[\dfrac{1}{2}\arccos\dfrac{3}{5}-2\arctan(-2)\right]$;

g) $\cos\left[\dfrac{1}{2}\arccos\dfrac{3}{5}-2\arctan(-2)\right]$.

VI. 19. Calculate the value of the expressions:

a) $\dfrac{1}{2}\sin\dfrac{\pi}{6}-\cos^4\left(\dfrac{3\pi}{2}-\dfrac{1}{2}\arcsin\dfrac{3}{5}\right)$;   b) $\cot\dfrac{\pi}{4}-2\sin^2\left(\dfrac{\pi}{2}+\dfrac{1}{2}\arcsin\dfrac{2\sqrt{2}-1}{3}\right)$;

c) $\tan\left(\dfrac{5\pi}{4}+\dfrac{1}{2}\arccos\dfrac{b}{a}\right)+\tan\left(\dfrac{5\pi}{4}-\dfrac{1}{2}\arccos\dfrac{b}{a}\right)$;

d) $\dfrac{x^3}{2\sin^2\left(\dfrac{1}{2}\arctan\dfrac{x}{y}\right)}+\dfrac{y^3}{2\cos^2\left(\dfrac{1}{2}\arctan\dfrac{y}{x}\right)}$.

VI. 20. If $\arccos\alpha+\arccos\beta+\arccos\gamma=\pi$, $\alpha,\beta,\gamma\in[-1,1]$, prove that $\alpha^2+\beta^2+\gamma^2+2\alpha\beta\gamma=1$.

VI. 21. Prove the identities:

a) $\arcsin\dfrac{1}{\sqrt{5}}+\arccos\dfrac{3}{\sqrt{10}}=\dfrac{\pi}{4}$;   b) $\arccos\sqrt{\dfrac{2}{3}}-\arccos x\dfrac{\sqrt{6}+1}{2\sqrt{3}}=\dfrac{\pi}{6}$;

c) $\arctan\dfrac{1}{4}+2\arctan\dfrac{1}{5}=\arctan\dfrac{32}{43}$;

d) $\arccos\dfrac{\sqrt{3}}{2}+\arccos\dfrac{\sqrt{2}}{2}=\arccos\dfrac{\sqrt{6}-\sqrt{2}}{4}$;

e) $\arcsin\dfrac{7}{25}+\arccos\dfrac{24}{25}=\arcsin\dfrac{336}{625}$;   f) $\cos(2\operatorname{arccot}7)=\sin(4\operatorname{arccot}3)$;

g) $\arccos\dfrac{36}{85} + \arcsin\dfrac{4}{5} = \dfrac{\pi}{2} + \arccos\dfrac{15}{17};$

h) $\arccos\left(-\dfrac{1}{2}\right) + 2\arcsin\dfrac{1}{2} = 2\arctan 1 + 3\arctan\sqrt{3};$

i) $\arcsin\dfrac{2}{5} + \arcsin\dfrac{3}{5} + \arcsin\dfrac{4}{5} = \arcsin\dfrac{\sqrt{21}}{5};$

j) $\arctan\dfrac{1}{3} + \arctan\dfrac{1}{5} + \arctan\dfrac{1}{7} + \arctan\dfrac{1}{8} = \dfrac{\pi}{4};$

k) $\dfrac{1}{3}\arctan 1 + \dfrac{1}{4}\arccos\dfrac{1}{2} = \dfrac{1}{2}\arctan\sqrt{3};$

l) $\arctan\left(\dfrac{2}{\sqrt{3}}\dfrac{\alpha}{\beta} - \dfrac{1}{\sqrt{3}}\right) + \arctan\left(\dfrac{2}{\sqrt{3}}\dfrac{\beta}{\alpha} - \dfrac{1}{\sqrt{3}}\right) = \dfrac{\pi}{4}.$

VI. 22. Prove the identities:

a) $2\arctan\sqrt{\dfrac{1-x}{1+x}} = \arccos x, \ x \in [-1,1];$

b) $\tan\left[2\arctan\left(\dfrac{1-\cos x}{\sin x}\right)\right] = \tan x;$

c) $\tan\left(\arccos\dfrac{1}{\sqrt{1+x^2}} + \arccos\dfrac{x}{\sqrt{1+x^2}}\right) = \dfrac{1-x^2}{2x}, \ x < 0;$

d) $\arctan x^2 + \arctan\dfrac{1}{x^2} = \dfrac{\pi}{2};$

e) $\arctan\dfrac{1+x^2}{2+x^2} - \arctan(3+2x^2) = -\dfrac{\pi}{4};$

f) $2\arctan x + \arcsin\dfrac{2x}{1+x^2} = \begin{cases} \pi & if \ x \in [1,\infty) \\ -\pi & if \ x \in (-\infty,-1) \end{cases};$

g) $\arctan x + \arcsin\dfrac{x-1}{\sqrt{2(1+x^2)}} = -\dfrac{3\pi}{4}, \ \text{for } x \in (-\infty,-1);$

h) $\arctan x - \arcsin\dfrac{x-1}{\sqrt{2(1+x^2)}} = \dfrac{\pi}{4}, \ \text{for } x \in [-1,\infty);$

i) $\arccos x + \arccos\dfrac{x+\sqrt{1-x^2}}{\sqrt{2}} = \begin{cases} 2\arccos x - \dfrac{\pi}{4} & if \ x \in \left[-1, \dfrac{1}{\sqrt{2}}\right] \\ \dfrac{\pi}{4} & if \ x \in \left(\dfrac{1}{\sqrt{2}},1\right] \end{cases};$

j) $\arctan x + \arctan y = \dfrac{1}{2}\arcsin \dfrac{2(x+y)(1-2xy))}{\left(1+x^2\right)\left(1+y^2\right)}$;

k) $\arctan[\cos(\arcsin x)] + \operatorname{arccot}[\sin(\arccos x)] = \dfrac{\pi}{2}$;

l) $\arctan \dfrac{4x+4}{2-3x} - \arctan \dfrac{5x+2}{4} = \arctan \dfrac{3}{4}, \ x < \dfrac{2}{3}$.

Equations Containing
Inverses of Trigonometric Functions

VI. 23. Solve the equations:

a) $\cos(3\arccos x) = \cos(2\arccos x) + 1$;

b) $\sin(3\arcsin x) = \cos(2\arccos x) - 1$;

c) $2\cos(4\arccos x) - 8\sin^2(\arccos x) + 7 = 0$;

d) $2\cos(4\arcsin x) - 8\cos^2(\arcsin x) + 7 = 0$.

VI. 24. Solve the equations:

a) $6\arcsin^2 x - 5\arcsin x + 1 = 0$; **b)** $2\arctan(2x-1) = \arccos(x-1)$;

c) $2\arccos\left(\dfrac{2-x}{3}\right) = 3\arccos(x+2)$; **d)** $\arcsin 2x = 3\arcsin x$;

e) $2\arcsin x = \arcsin(x\sqrt{2})$; **f)** $\arcsin x = \arccos\sqrt{1-x^2}$;

g) $\sin(4\arctan x) = \dfrac{\sqrt{3}}{2}$; **h)** $\arccos x = \arctan \dfrac{\sqrt{1-x^2}}{x}$;

i) $\arctan^2(2x-1) - 3\arctan(2x-1) = 0$; **j)** $2\arccos\left(-\dfrac{x}{2}\right) = \arccos(x+3)$;

k) $\arccos 2x = 2\arccos\sqrt{\dfrac{x+2}{2}}$.

VI. 25. Solve the equations:

a) $\arcsin x + \arcsin 2x = \dfrac{\pi}{3}$; **b)** $\arcsin 2x + \arcsin(2x\sqrt{3}) = -\dfrac{\pi}{2}$;

c) $\arccos x - \arcsin \dfrac{4x}{3} = \pi$; **d)** $\arccos(x\sqrt{3}) + \arccos x = \dfrac{\pi}{2}$;

e) $\arcsin(x+1) + \arccos x = \pi$; **f)** $\arccos x + \arcsin\sqrt{1-x^2} = \pi$;

84

g) $\arcsin \dfrac{1}{\sqrt{x}} - \arcsin \sqrt{1-x} = \dfrac{\pi}{2}$; **h)** $\arcsin^3 x + \arccos^3 x = \dfrac{\pi^3}{32}$.

VI. 26. Solve the equations:

a) $\arctan(x+1) - \arctan x = \dfrac{\pi}{4}$; **b)** $\arctan\left(x+\dfrac{1}{2}\right) + \arctan\left(x-\dfrac{1}{2}\right) = \dfrac{\pi}{4}$;

c) $\arctan 2\sqrt{x} + \arctan 3\sqrt{x} = \dfrac{3\pi}{4}$; **d)** $\arctan 3x - \text{arccot} 3x = \dfrac{\pi}{4}$;

e) $\text{arccot}\, x + 2\arctan x = \dfrac{\pi}{4}$; **f)** $\arctan(x+1) + \arctan 2x - \arctan 3x = \dfrac{\pi}{4}$;

g) $\arctan x + \dfrac{1}{2}\text{arcsec}\, 5x = \dfrac{\pi}{4}$; **h)** $\arctan^2 x + \text{arccot}^2 x = \dfrac{5\pi^2}{8}$;

i) $\arctan x + \arctan(x^2 - x - 1) = \dfrac{5\pi}{12}$;

j) $\arctan \sqrt{2-\sqrt{4-x^2}} + \arctan \sqrt{2+\sqrt{4-x^2}} = \dfrac{\pi}{2}$;

k) $\arctan x + \dfrac{1}{2}\text{arc sec}\, 5x = \dfrac{\pi}{4}$.

VI. 27. Solve the equations:

a) $\arccos x - \arcsin x = \arccos(x\sqrt{3})$;

b) $\arcsin x - \arccos x = \arcsin(5x - 3)$;

c) $\arccos x + \arccos(1-x) = \arccos(-x)$;

d) $\arccos(2x-1) - \arcsin(2x-1) = \arcsin\dfrac{3}{4}$;

e) $\arcsin x + \arccos \sqrt{x} = \arctan\left(\sqrt{x+1} + \sqrt{x}\right)^2$, $x \in (0,1)$;

f) $\arcsin(x^2 - 1) + 2\arctan(x^2 - 1) = \arccos(1 - x^2)$.

VI. 28. Solve the equations:

a) $\arctan x + \arctan 2x = \arctan 3x$;

b) $\arctan x - \arctan 2x = \arctan 3x$;

c) $\arctan(x-1) + \arctan(x+1) = \arctan 3x$;

d) $\arctan(x-1) + \arctan x + \arctan(x+1) = \arctan 3x$;

e) $\arctan\dfrac{\sqrt{x}}{3} + \arctan\dfrac{\sqrt{x}}{2} = \arctan \sqrt{x}$;

f) $\arcsin x + \arctan x = \arctan 2x$.

VI. 29. Solve the equations:

a) $\arcsin x \cdot \arccos x = \dfrac{\pi^2}{18}$;

b) $\arccos x \cdot \arccos(-x) = \dfrac{5\pi^2}{36}$;

c) $\arctan x \cdot \mathrm{arccot}\, = \dfrac{\pi^2}{16}$;

d) $(\arcsin x)^2 + (\arccos x)^2 = (\arctan x)^2 + (\mathrm{arccot})^2$.

VI. 30. Find the roots x_1, x_2, x_3 of the equation $x^3 - ax + b = 0$, if $\arctan x_1 + \arctan x_2 = \arctan x_3$, where a, b are real numbers.

UNIT VII

Trigonometric Ratios in a Triangle

Notations in a Triangle

$\triangle ABC$

Length of Sides $AB = c, BC = a, AC = b$

Length of Median from A $AM = m_a$, $BM = MC$

Length of Hight from A $AD = h_a$, $AD \perp BC$

Length of Bisector from A $AE = l_a$, $\angle BAE = \angle EAC$

Semiperimeter $p = \dfrac{a+b+c}{2}$

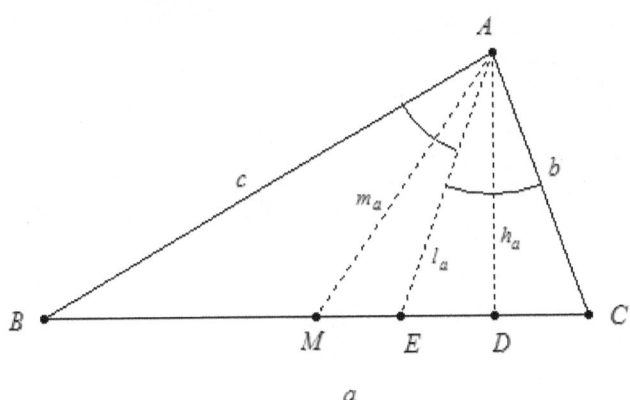

Length of Radius of the Circumscribed Circle

$R = BO = AO = CO$,
where O is the
point of intersection of the
perpendicular
bisectors ON, OM, OP
of the sides of the $\triangle ABC$.

S - area of the $\triangle ABC$.

$$S = \frac{ab\sin C}{2} = \frac{b\,\sin A}{2} = \frac{c a \sin B}{2}$$

Heron's formula $S = \sqrt{p(p-a)(p-b)(p-c)}$

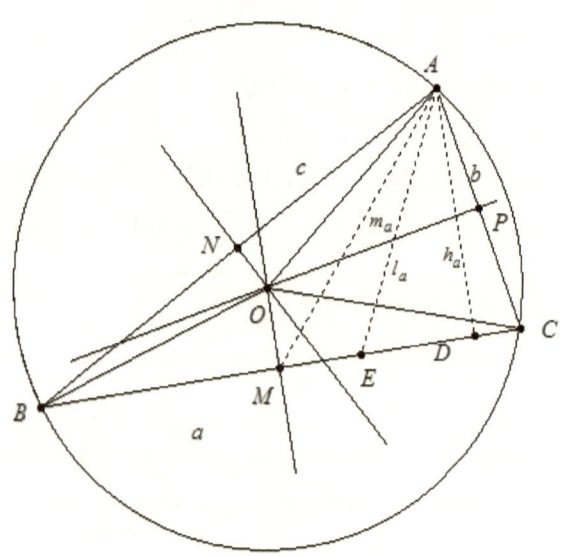

Length of Radius of the Inscribed Circle and Exinscribed Circle

$$r = \frac{S}{p};$$

$$r_a = \frac{S}{p-a};$$

$$r_a = p\tan\frac{A}{2}$$

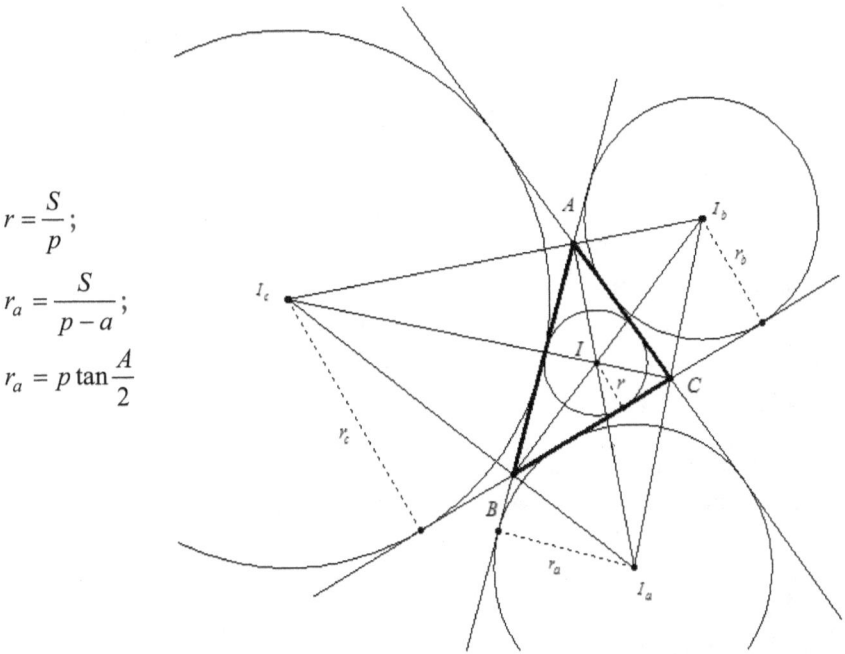

$$\sin\frac{A}{2} = \sqrt{\frac{(p-b)(p-c)}{b}}, \quad \cos\frac{A}{2} = \sqrt{\frac{p(p-a)}{b}}, \quad \tan\frac{A}{2} = \sqrt{\frac{(p-b)(p-c)}{p(p-a)}}.$$

Relations in a Triangle

VII. 1. **a)** Prove that $\dfrac{165°42'}{27°37'} \in N$.

b) The difference of two complimentary angles is $13°14'16''$. Find the measure of each angle.

c) The difference of two supplementary angles is $63°24'12''$. Find each angle.

d) Consider two concurrent lines. The sum of the three angles formed by the two lines is $240°$. Find the measure of the four angles.

e) An angle is $\dfrac{7}{5}$ of its complementary angle. What are the measurements of each angle?

VII. 2. In a triangle a = 4.7 cm, b = 2.4 cm, $A = 60°$, decide if the triangle exists. If it does exist, determine the angle B to the nearest degree.

VII. 3. Determine the number of possible triangles that could be drawn with the given measures and find the measures of the other angles in each possible triangle. Round to the nearest tenth of a degree, if necessary.

a) $\triangle ABC$, a = 30 cm, b = 25 cm, and m(A) = 44°;

b) $\triangle ABC$, c = 30 cm, b = 25 cm, and m(B) = 27°;

c) $\triangle ABC$, a = 24 cm, b = 48 cm, and m(A) = 30°;

d) $\triangle PQR$, r = 85 cm, q = 90 cm, and m(R) = 37.3°;

e) $\triangle PQR$, r = 8.5 cm, q = 24 cm, and m(R) = 37.6°.

VII. 4. An equilateral triangle has a side length of 6 cm. If a perpendicular is dropped from a vertex to the opposite side, what is the length of the perpendicular?

VII. 5. The hypotenuse, c, of the right $\triangle ABC$ is 5.0 cm long. Given the trigonometric ratio $\cos A = 0.75$ for angle A, what is the area of the triangle to the nearest tenth of a cm²?

VII. 6. A ladder is leaning against a wall. If the angle between the ground and the ladder is 60° and the height of the wall is 3.6 m, how long is the ladder?

VII. 7. A triangular garden has side lengths of 10 m, 12 m, and 5 m. Determine the angle between the side lengths of 10 m and 12 m to the nearest degree.

VII. 8. The angles of a triangle are in the ratio $\dfrac{1}{2} : \dfrac{1}{3} : \dfrac{1}{6}$, and the perimeter is $30\left(\sqrt{3}+1\right)$. Find the length of the sides.

VII. 9. In $\triangle ABC$, AB = 6, AC = 9, and m(B) = 60°. Calculate $\sin A$.

VII. 10. In $\triangle ABC$, BC = 5, AC = 4, and $C = \dfrac{\pi}{4}$. Calculate $\sin B$.

VII. 11. In the acute $\triangle ABC$, BC = 8 cm, $\sin B = \dfrac{1}{3}$, and $\sin C = \dfrac{1}{9}$. Calculate the length of AB.

VII. 12. Solve the triangle in which:

a) $a = 12$ cm, $B = \dfrac{\pi}{4}$, $C = \dfrac{5\pi}{12}$;

b) $a = 2$ cm, $b = \sqrt{2}$ cm, $c = 1 + \sqrt{3}$;

c) $c = 16$ cm, $A = \dfrac{5\pi}{12}$, $C = \dfrac{\pi}{4}$;

d) $b = 4$ cm, $B = \dfrac{\pi}{3}$, $C = \dfrac{\pi}{4}$;

e) $b = 2$ cm, $c = 3$ cm, $A = \dfrac{\pi}{3}$.

VII. 13. Solve the triangle ABC in the following cases:

a) $a = 15$, $\sin B = \dfrac{24}{25}$, and $C = 30°$;

b) $b = 7\sqrt{2}$, $A = 45°$, and $\sin C = \dfrac{12}{13}$;

c) $a = 12$, $b = 7$, and $c = 9$;

d) $a = \sqrt{3}$, $b = \dfrac{\sqrt{2}}{2}$, and $c = \dfrac{\sqrt{2} + \sqrt{6}}{4}$;

e) $b = 5$, $A = 18°$, and $c = 4$;

f) $a = 2\sqrt{7}$, $c = 6$, $\cos 2B = \dfrac{1}{7}$.

VII. 14. In $\triangle ABC$, $A = 60°$, $a = \sqrt{6}$, and $b + c = 3 + \sqrt{3}$. Find the area of the triangle.

VII. 15. In a triangle ABC the radius for the circumscribed circle is $R = 3$ and $a = \sqrt{2}$. Calculate $\sin A$.

VII. 16. In $\triangle ABC$, $AC = 6$, $BA = 8$, and the radius for the circumscribed circle is $R = 5$. Find the sum $\sin A + \sin B$.

VII. 17. In $\triangle ABC$ $B = C = \dfrac{\pi}{12}$ and $a + b + c = 8 + 2\sqrt{2} + 2\sqrt{6}$. Find a, b, c, and the area of $\triangle ABC$.

VII. 18. Prove the Law of Sines using analytical coordinates.

Conditional Relations in a Triangle

VII. 19. In the triangle ABC, $A = 45°$, $AB = a$, and $AC = \dfrac{2\sqrt{2}}{3}a$. Prove that $\tan B = 2$.

VII. 20. In $\triangle ABC$, $A = 60°$ and $\dfrac{c}{b} = 2 - \sqrt{3}$. Calculate $\tan\dfrac{B-C}{2}$ and the angles B and C.

VII. 21. Show that in a triangle ABC $\cot B + \cot C = 2\cot A$, then $b^2 + c^2 = 2a^2$.

VII. 22. Ii in a triangle ABC $\cos 2B + \cos 2C = 2\cos 2A$, then $b^2 + c^2 = 2a^2$.

VII. 23. In the triangle ABC $b^3 + c^3 - a^3 = a^2(b + c - a)$. Show that $A = 60°$.

Right Triangles

VII. 24. Solve the right triangle ABC $\left(A = \dfrac{\pi}{2}\right)$, in cases:

a) $BC = 15$, $C = 30°$; b) $AB = 8\sqrt{3}$, $B = 60°$; c) $BC = 17$, $AC = 8$;

d) $AB = 12$, $AC = 5$; e) $a = 30$, $B = \dfrac{\pi}{3}$; f) $a = 13$, $c = 12$;

g) $C = \dfrac{\pi}{6}$, $b = \dfrac{\sqrt{3}}{2}$; h) $b = 12$, $c = 5$; i) $b = 3$, $\sin C = \dfrac{1}{2}$,

j) $BC = 4$, $C = 15°$; k) $m_a = 3$ and $m_b = 4$, where m_a and m_b are the medians of the triangle from A and C, respectively.

VII. 25. If in a triangle ABC $\sin^2 A + \sin A = 2$, then the triangle is right.

VII. 26. In $\triangle ABC$ $(A = 90°)$, $a + b = 3c$. Calculate $\sin 2B$ and $\tan\dfrac{C}{2}$.

VII. 27. Show that if in a triangle ABC,
$b\cos B + c\cos C = 2a\sin B\sin C$, then $A = \dfrac{\pi}{2}$.

VII. 28. Show that if in a triangle ABC, $2a \sin B \cos B = \dfrac{bc}{R}$, then the triangle is right.

VII. 29. If in a triangle ABC $\sin C = \cos A + \cos B$ then the triangle is right.

VII. 30. If in a triangle ABC $\cos B + \cos C = \sin B + \sin C$ then the triangle is right.

VII. 31. Show that if in a triangle ABC

a) $\dfrac{1}{\sin B} + \dfrac{1}{\tan B} = \dfrac{a+c}{b}$ or

b) $\dfrac{1}{\cos 2B} + \dfrac{1}{\cot 2B} = \dfrac{c+b}{c-b}$, then the triangle is right.

VII. 32. If in a triangle ABC $\cos^2 A + \cos^2 B + \cos^2 C = 1$, then the triangle is right.

VII. 33. Prove that if in a triangle ABC $\sin^2 A + \sin^2 B + \sin^2 C = 2$, then the triangle is right.

VII. 34. Prove that if in a triangle ABC $\sin A = \dfrac{\sin B + \sin C}{\cos B + \cos C}$, then the triangle is right.

VII. 35. Prove that if in a triangle ABC

a) $\tan B = \dfrac{\sin B + \cos C}{\cos B + \sin C}$;

b) $\tan B = \dfrac{\cos(B - C)}{\sin A - \sin(B - C)}$;

c) $\sin 4A + \sin 4B + \sin 4C = 0$, then the triangle is right.

VII. 36. Prove that if in a triangle ABC $\tan(45° + B) = \dfrac{1 + \cot C}{1 - \cot C}$, then the triangle is right.

VII. 37. Prove that in a right triangle ABC $(A = 90°)$ the following identities hold:

a) $b \cos C + c \cos B = a$;

b) $b \cos B + c \cos C = a \cos(B - C)$;

c) $b(a^2 - c^2) \sin B - c(a^2 - b^2) \sin C = a(b^2 - c^2)$.

VII. 38. Prove that in a right triangle $ABC(A = 90°)$, the following identity holds:

a) $b^2 + c^2 = 4R^2$;

b) $a^2 + b^2 + c^2 = 8R^2$;

c) $\sin^2 B + \sin^2 C = 1 + 2\cos A \sin B \sin C$.

VII. 39. Prove that if in a triangle ABC, $\dfrac{a+c}{b} = \cot\dfrac{B}{2}$, then the triangle is right.

VII. 40. Prove that if the triangle ABC is a right triangle then
$$\tan 2C = \frac{2bc}{b^2 - c^2}.$$

VII. 41. Prove that in a right triangle ABC $(A = 90°)$ the following identity holds:
$$(1 + \cos B)(1 + \cos C) = \frac{2p^2}{a^2}, \text{ where } p = \frac{a+b+c}{2}.$$

VII. 42. Prove that if in a triangle ABC
$$\sqrt{p(p-a)} + \sqrt{(p-b)(p-c)} = \sqrt{2bc},$$ then the triangle is right.

VII. 43. Find the relation between the angles x, y, and z such that
$$\tan\frac{x}{2}\tan\frac{y}{2} + \tan\frac{x}{2}\tan\frac{z}{2} + \tan\frac{y}{2}\tan\frac{z}{2} = 1.$$

VII. 44. Prove that if in a triangle ABC $\dfrac{1}{2}\left(\dfrac{1}{r} + \dfrac{1}{r_a}\right) = \dfrac{1}{b} + \dfrac{1}{c}$, then the triangle is right.

Equilateral Triangle

VII. 45. If in a $\triangle ABC$ $p = b\cos A + c\cos B + a\cos C$ then the triangle is equilateral.

VII. 46. Let m_a, m_b, m_c; h_a, h_b, h_c be the lengths of the medians and the heights of $\triangle ABC$.

a) If $a \le b \le c$, prove that $h_a \ge h_b \ge h_c$ and that $m_a \ge m_b \ge m_c$;

b) If $\left(\dfrac{h_a^2}{h_b \cdot h_c}\right)^{m_a} \cdot \left(\dfrac{h_b^2}{h_c \cdot h_a}\right)^{m_b} \cdot \left(\dfrac{h_c^2}{h_a \cdot h_b}\right)^{m_c} = 1$, prove that $\triangle ABC$ is equilateral.

VII. 47. If in a $\triangle ABC$ $\cos A + \cos B + \cos C = \dfrac{3}{2}$ then the triangle is equilateral.

VII. 48. If in a $\triangle ABC$ $S = \dfrac{3}{4} \dfrac{abc}{\sqrt{a^2 + b^2 + c^2}}$, then the triangle is equilateral.

VII. 49. If in a $\triangle ABC$, one of the relations holds
a) $8\cos A\cos B\cos C - 1 = 0$;
b) $8\sin A\sin B\sin C + 1 = 0$, then the triangle is equilateral.

VII. 50. Show that if in a triangle ABC $a^2 = \dfrac{b^3 + c^3 - a^3}{b + c - a}$ and $\sin B\sin C = 0.75$, then the triangle is equilateral.

VII. 51. If in a $\triangle ABC$ $\tan\dfrac{A}{2} + \tan\dfrac{B}{2} + \tan\dfrac{C}{2} = \dfrac{1}{4S}\left(a^2 + b^2 + c^2\right)$, then the triangle is equilateral.

VII. 52. If in a $\triangle ABC$, one of the relations holds
a) $\sqrt{r_a} + \sqrt{r_b} + \sqrt{r_c} = \dfrac{\sqrt{r_a r_b r_c}}{r}$;
b) $r_a + r_b + r_c = p\sqrt{3}$, then the triangle is equilateral.

Isosceles Triangle

VII. 53. If in a $\triangle ABC$ $b + c = 4R\cos\dfrac{A}{2}$ then the triangle is isosceles.

VII. 54. If in a $\triangle ABC$ $2a = b + c$ and $2A = B + C$ then the triangle is isosceles.

VII. 55. If in a $\triangle ABC$ $a^2 \sin B\cos B = \dfrac{abc}{4R}$, then the triangle is isosceles.

VII. 56. In a triangle $2a = b + c$ and $2A = B + C$. Show that the triangle is isosceles.

VII. 57. If in a $\triangle ABC$, one of the relations holds

a) $\sin A = 2\sin B\cos C$; b) $a = 2b\cos C$;

c) $a = 2b\sin\dfrac{A}{2}$; d) $a = 2R\cos\dfrac{C}{2}$;

e) $a = 2r\cot\dfrac{B}{2}$; f) $2h_a = a\tan B$;

g) $a\cos B = b\cos A$; h) $r_a^2 + 2r_a \cdot r_c = p^2$,

then the triangle is isosceles.

VII. 58. If in a $\triangle ABC$, one of the relations holds

a) $\sin\dfrac{A}{2}\cos\dfrac{B}{2} = \sin\dfrac{B}{2}\cos\dfrac{A}{2}$;

b) $\sin\dfrac{A}{2}\cos^3\dfrac{B}{2} = \sin\dfrac{B}{2}\cos^3\dfrac{A}{2}$,

then the triangle is isosceles.

VII. 59. If in a $\triangle ABC$, one of the relations holds

a) $a\tan A + b\tan B = (a+b)\cot\dfrac{C}{2}$;

b) $\dfrac{\cot A + \cot B}{\cot C + \cot B} = \dfrac{a}{c}$,

then the triangle is isosceles.

VII. 60. If in a $\triangle ABC$, one of the relations holds

a) $(p-c)\cot\dfrac{A}{2} = p\tan\dfrac{C}{2}$;

b) $(b-c)\cot\dfrac{A}{2} + (c-a)\cot\dfrac{B}{2} = 0$;

c) $\dfrac{\cos B - \cos C}{p-a} + \dfrac{\cos C - \cos A}{p-b} = 0$,

then the triangle is isosceles.

VII. 61. If in a triangle $a\tan A + b\tan B = (a+b)\tan\dfrac{A+B}{2}$ then the triangle is isosceles.

VII. 62. If the relation $h_a = \dfrac{p\cos\dfrac{A}{2}}{1+\sin\dfrac{A}{2}}$ holds in a triangle ABC, then

ABC is an isosceles triangle.

Right or Isosceles Triangle

VII. 63. Determine the nature of the triangle ABC, such that $\sin B + \cos B = \sin C + \cos C$.

VII. 64. If in a $\triangle ABC$ $b\left(1+\cot\dfrac{B}{2}\right) = c\left(1+\cot\dfrac{C}{2}\right)$, then the triangle is isosceles or right.

VII. 65. If in a $\triangle ABC$, one of the relations holds

a) $a\cos A = b\cos B$;

b) $\dfrac{\tan B}{\tan C} = \dfrac{\sin^2 B}{\sin^2 C}$;

c) $\dfrac{b^2 - c^2}{b^2 + c^2} = \dfrac{\sin(B-C)}{\sin(B+C)}$, then the triangle is isosceles or right.

Trigonometric Relations in a Scalene Triangle

VII. 66. Show that in any triangle

a) $b\cos C - c\cos B = \dfrac{b^2 - c^2}{a}$;

b) $a^2 + b^2 + c^2 = 2(b\,\cos A + ac\cos B + ab\cos C)$;

c) $a^2 + b^2 + c^2 = 4S\left(\dfrac{1}{\tan A} + \dfrac{1}{\tan B} + \dfrac{1}{\tan C}\right)$;

d) $\dfrac{\tan A}{\tan B} = \dfrac{a^2 + c^2 - b^2}{b^2 + c^2 - a^2}$;

e) $\dfrac{bc\cos A + ca\cos B + ab\cos C}{a\sin A + b\sin B + c\sin C} = R.$

VII. 67. Show that in any triangle $S = 2R^2 \sin A \sin B \sin C$.

VII. 68. Show that in any triangle

a) $S = \dfrac{abc}{4R}$;   b) $S = \dfrac{1}{2}\sqrt[3]{(abc)^2 \sin A \sin B \sin C}$.

VII. 69. $\dfrac{\tan\dfrac{A-B}{2}}{\tan\dfrac{A+B}{2}} = \dfrac{a-b}{a+b}$ (*Law of Tangents*).

VII. 70. Show that in any triangle

a) $$\dfrac{\sin\dfrac{A-B}{2}\sin\dfrac{C}{2}}{a-b}=\dfrac{\sin\dfrac{B-C}{2}\sin\dfrac{A}{2}}{b-c}=\dfrac{\sin\dfrac{C-A}{2}\sin\dfrac{B}{2}}{c-a};$$

b) $$\dfrac{\cos\dfrac{A-B}{2}\sin\dfrac{C}{2}}{a+b}=\dfrac{\cos\dfrac{B-C}{2}\cos\dfrac{A}{2}}{b+c}=\dfrac{\cos\dfrac{C-A}{2}\cos\dfrac{B}{2}}{c+a}.$$

VII. 71. If A, B, C are the angle of a triangle, then:

a) $\sin\dfrac{A}{2}+\sin\dfrac{B}{2}+\sin\dfrac{C}{2}=1+4\sin\dfrac{\pi-A}{4}\sin\dfrac{\pi-B}{4}\sin\dfrac{\pi-C}{4}$;

b) $\cos\dfrac{A}{2}+\cos\dfrac{B}{2}+\cos\dfrac{C}{2}=4\cos\dfrac{\pi-A}{4}\cos\dfrac{\pi-B}{4}\cos\dfrac{\pi-C}{4}$;

c) $\sin A+\sin B+\sin C=4\cos\dfrac{A}{2}\cos\dfrac{B}{2}\cos\dfrac{C}{2}$;

d) $\sin A-\sin B+\sin C=4\sin\dfrac{A}{2}\cos\dfrac{B}{2}\sin\dfrac{C}{2}$;

e) $\cos A-\cos B-\cos C=1-4\sin\dfrac{A}{2}\cos\dfrac{B}{2}\cos\dfrac{C}{2}$;

f) $\cos A+\cos B+\cos C=1+4\sin\dfrac{A}{2}\sin\dfrac{B}{2}\sin\dfrac{C}{2}$;

g) $\sin 2A+\sin 2B+\sin 2C=4\sin A\sin B\sin C$;

h) $\cos 2A+\cos 2B-\cos 2C-1=-4\sin A\sin B\cos C$;

i) $\sin 3A+\sin 3B+\sin 3C+1=-4\cos\dfrac{3A}{2}\cos\dfrac{3B}{2}\cos\dfrac{3C}{2}$;

j) $\cos 4A+\cos 4B+\cos 4C=4\cos 2A\cos 2B\cos 2C-1$;

k) $\dfrac{\cos A}{\sin B\sin C}+\dfrac{\cos B}{\sin C\sin A}+\dfrac{\cos C}{\sin A\sin B}=2$;

l) $\dfrac{\sin A}{\sin B\sin C}+\dfrac{\sin B}{\sin C\sin A}+\dfrac{\sin C}{\sin A\sin B}=2\left(\cot A+\cot B+\cot C\right)$;

m) $\sin^2\dfrac{A}{2}+\sin^2\dfrac{B}{2}+\sin^2\dfrac{C}{2}=1-2\sin\dfrac{A}{2}\sin\dfrac{B}{2}\sin\dfrac{C}{2}$;

n) $\cos^2\dfrac{A}{2}+\cos^2\dfrac{B}{2}+\cos^2\dfrac{C}{2}=2\left(1+\sin\dfrac{A}{2}\sin\dfrac{B}{2}\sin\dfrac{C}{2}\right)$;

o) $\sin^2 A+\sin^2 B+\sin^2 C=2\left(1+\cos A\cos B\cos C\right)$;

p) $\sin^2 2A + \sin^2 2B + \sin^2 2C = 2\left(1 - \cos 2A \cos 2B \cos 2C\right)$;

q) $\cos^2 A + \cos^2 B - \cos^2 C = 2 \sin A \sin B \cos C$;

r) $\cos^2 A + \cos^2 B + \cos^2 C = 1 - 2 \cos A \cos B \cos C$;

s) $\cot \dfrac{A}{2} + \cot \dfrac{B}{2} + \cot \dfrac{C}{2} = \cot \dfrac{A}{2} \cot g \dfrac{B}{2} \cot \dfrac{C}{2}$;

t) $\cot \dfrac{A}{2} - \tan \dfrac{B}{2} - \tan \dfrac{C}{2} = \cot \dfrac{A}{2} \tan \dfrac{B}{2} \tan \dfrac{C}{2}$;

u) $\tan \dfrac{A}{2} \tan \dfrac{B}{2} + \tan \dfrac{B}{2} \tan \dfrac{C}{2} + \tan \dfrac{C}{2} \tan \dfrac{A}{2} = 1$;

v) $\tan A + \tan B + \tan C = \tan A \tan B \tan C$;

w) $\tan A \tan B + \tan B \tan C + \tan C \tan A = 1 + \dfrac{1}{\cos A \cos B \cos C}$;

y) $\cot A + \cot B + \cot C = \cot A \cot B \cot C + \dfrac{1}{\sin A \sin B \sin C}$;

z) $\cot A \cot B + \cot B \cot C + \cot g C \cot A = 1$.

VII. 72. Show that in an acute triangle
$\tan A + \tan B + \tan C = \tan A \tan B \tan C$.

VII. 73. Prove that in any triangle ABC, the following identities hold:

a) $1 - \tan \dfrac{A}{2} \tan \dfrac{B}{2} = \dfrac{c}{p}$;

b) $\tan \dfrac{A}{2} \tan \dfrac{B}{2} \tan \dfrac{C}{2} = \dfrac{S}{p^2}$;

c) $\dfrac{\cos^2 \dfrac{A}{2}}{a} + \dfrac{\cos^2 \dfrac{B}{2}}{b} + \dfrac{\cos^2 \dfrac{C}{2}}{c} = \dfrac{p^2}{abc}$;

d) $a \cos^2 \dfrac{A}{2} + b \cos^2 \dfrac{B}{2} + c \cos^2 \dfrac{C}{2} = p + \dfrac{S}{R}$;

e) $a \sin(B - C) + b \sin(C - A) + c \sin(A - B) = 0$;

f) $\dfrac{\cos A - \cos B}{p - c} + \dfrac{\cos B - \cos C}{p - a} + \dfrac{\cos C - \cos A}{p - b} = 0$;

c) $\dfrac{a^2 \sin(B - C)}{\sin B + \sin C} + \dfrac{b^2 \sin(C - A)}{\sin C + \sin A} + \dfrac{c^2 \sin(A - B)}{\sin A + \sin B} = 0$.

VII. 74. Prove that in any triangle ABC, the following identities hold:

a) $\dfrac{a\cos C - b\cos B}{a\cos B - b\cos A} + \cos C = 0, \quad c \neq b$;

b) $\dfrac{\sin(A-B)\sin C}{1+\cos(A-B)\cos C} = \dfrac{a^2 - b^2}{a^2 + b^2}$;

c) $(a+c)\cos\dfrac{B}{4} + b\cos\left(A + \dfrac{3B}{4}\right) = 2c\cos\dfrac{B}{2}\cos\dfrac{B}{4}$.

VII. 75. Prove that in any triangle $b\cos B + c\cos C = a\cos(B - C)$.

VII. 76. Prove that in any triangle ABC, the following identities hold:

a) $1 + \cos A\cos(B - C) = \dfrac{b^2 + c^2}{4R^2}$;

b) $4S = \left(b^2 + c^2 - a^2\right)\tan A$;

c) $\dfrac{b+c}{2c\cos\dfrac{A}{2}} = \dfrac{\sin\left(\dfrac{A}{2} + C\right)}{\sin(A+B)}$;

d) $p = r\left(\cot\dfrac{A}{2} + \cot\dfrac{B}{2} + \cot\dfrac{C}{2}\right) = r\cot\dfrac{A}{2}\cot\dfrac{B}{2}\cot\dfrac{C}{2}$.

VII. 77. In a triangle ABC $m_a^2 = \dfrac{2\left(b^2 + c^2\right) - a^2}{4}$, where m_a is the length of the median from A of the triangle.

VII. 78. Prove that in any triangle ABC, the following identities hold:

a) $r = (p - a)\tan\dfrac{A}{2}$;

b) $S = p(p - a)\tan\dfrac{A}{2}$;

c) $p = 4R\cos\dfrac{A}{2}\cos\dfrac{B}{2}\cos\dfrac{C}{2}$;

d) $r = 4R\sin\dfrac{A}{2}\sin\dfrac{B}{2}\sin\dfrac{C}{2}$;

e) $p - a = 4R\cos\dfrac{A}{2}\sin\dfrac{B}{2}\sin\dfrac{C}{2}$;

f) $m_a^2 = R^2\left(\sin^2 A + 4\cos A\sin B\sin C\right)$;

g) $8S = \left(4m_a^2 - a^2\right)\tan A$.

VII. 79. Prove that in a triangle ABC $l_a = \dfrac{2bc}{b+c}\cos\dfrac{A}{2}$, where l_a is the length of the bisector from A of the triangle.

VII. 80. Prove that in any triangle ABC, the following identities hold:

a) $r_a \cdot r_b \cdot r_c = p^2 r$;

b) $r \cdot r_a = bc\sin^2\dfrac{A}{2}$;

c) $r_b \cdot r_c - r \cdot r_a = bc\cos A$;

d) $\dfrac{r_a}{\tan\dfrac{A}{2}} = \dfrac{r_b}{\tan\dfrac{B}{2}} = \dfrac{r_c}{\tan\dfrac{C}{2}} = p$;

e) $r_a + r_b + r_c - r = 4R$;

f) $\dfrac{1}{r_a} + \dfrac{1}{r_b} + \dfrac{1}{r_c} = \dfrac{1}{r}$;

g) $\sqrt{r \cdot r_a \cdot r_b \cdot r_c} = S$;

h) $r_a r_b + r_b r_c + r_a r_c = p^2$.

VII. 81. Prove that in any triangle ABC, the following identities hold:

a) $\dfrac{h_a h_b}{r r_c} + 2\dfrac{h_a - h_b}{r_a - r_b} = 4\cos C$;

b) $\dfrac{h_b + h_c}{r_a} + \dfrac{h_c + h_a}{r_b} + \dfrac{h_a + h_b}{r_c} = 6$;

c) $\dfrac{1}{l_a \sin\dfrac{A}{2}} + \dfrac{1}{l_b \sin\dfrac{B}{2}} + \dfrac{1}{l_c \sin\dfrac{C}{2}} = \dfrac{2p}{S}$;

d) $\dfrac{(b+c)^2}{bc}l_a^2 + \dfrac{(c+a)^2}{ca}l_b^2 + \dfrac{(a+b)^2}{ab}l_c^2 = 4p^2$.

Regular Polygons

VII. 82. If S_n is the area of a regular polygon with n sides, find S_3, S_4, S_6, S_8, S_{12}, and S_{20} in terms of R, the radius of the circumcircle of the polygon.

VII. 83. Find the number of sides and area of a regular polygon $ABCDEF\ldots$ inscribed in a circle such that $\dfrac{1}{AB} = \dfrac{1}{AC} + \dfrac{1}{AD}$.

Inscribed Circle

VII. 84. If I is the incenter of the triangle ABC, prove that $AI = 4R \cdot \sin\dfrac{B}{2} \cdot \sin\dfrac{C}{2}$.

VII. 85. The inscribed circle in $\triangle ABC$ touches the triangle in the points A', B', C'. Show that $\dfrac{S_{\triangle A'B'C'}}{S_{\triangle ABC}} = \dfrac{r}{2R}$.

VII. 86. Prove that in any triangle ABC, the following inequality holds $\sin\dfrac{A}{2} \le \dfrac{a}{2\sqrt{bc}}$.

VII. 87. In $\triangle ABC$ $B - C = \dfrac{2\pi}{3}$ and $R = 8r$. Find the angles of the triangle.

Orthocenter of a Triangle

VII. 88. Let ABC be a triangle, H the orthocenter and h_a the height from the point A. Show that:

a) $h_a = 2R\sin B \sin C$;

b) $AH = 2R\cos A$;

c) $aAH + bBH + cCH = 4S$.

VII. 89. Let ABC be a triangle. If O is the center of the circumscribed triangle and I is the center of the inner circle (inscribed circle), then show that $OI^2 = R(R - 2r)$.

VII. 90. Show that in any triangle ABC $\cos^2\dfrac{B-C}{2} \ge \dfrac{2r}{R}$.

VII. 91. Prove that, in any triangle ABC,

$$\frac{\sin^n A + \sin^n B + \sin^n C}{3} \geq \left(\frac{p}{3r}\right)^n \quad \text{where } n \in (-\infty, 0] \cup [1, \infty).$$

VII. 92. Prove that, in any triangle ABC,

$$\frac{\sin^n A + \sin^n B + \sin^n C}{3} \geq \left(\frac{4}{3}\right)^n \cos^n \frac{A}{2} \cos^n \frac{B}{2} \cos^n \frac{C}{2}$$

for any $n \in (-\infty, 0] \cup [1, \infty)$.

UNIT VIII

Complex Numbers

$z = a + bi$, where $a \in \mathbf{R}$, $b \in \mathbf{R}$, $i^2 = -1$ is the algebraic representation of a complex number.

The set of the complex number is $\mathbf{C} = \{a + bi \mid a \in \mathbf{R}, \ b \in \mathbf{R}, \ i^2 = -1\}$, where a is the real part of the complex number z or $a = \mathrm{Re}(z)$,

$\quad$ b is the imaginary part of the complex number z or $b = \mathrm{Im}(z)$.
$\bar{z} = a - bi$ is the *conjugate* of the complex number z.

$|z| = \sqrt{a^2 + b^2}$ is the *modulus* (*absolute value*) of the complex number z. If $z = a + bi$ is a complex number, the point $M(x, y)$ is called the geometric image of z, therefore the distance from the origin to M is $|z|$.

Properties:

a) $z + \bar{z} = 2a$; $\quad$ **b)** $z \cdot \bar{z} = a^2 + b^2$; $\quad$ **c)** $\overline{\sum_{k=1}^{n} z_k} = \sum_{k=1}^{n} \bar{z}_k$; $\quad$ **d)** $\overline{\prod_{k=1}^{n} z_k} = \prod_{k=1}^{n} \bar{z}_k$;

e) $\overline{z^n} = (\bar{z})^n$; $\quad$ **f)** $\overline{\left(\dfrac{z_1}{z_2}\right)} = \dfrac{\bar{z}_1}{\bar{z}_2}$; $\quad$ **g)** $z \in \mathbf{R} \Leftrightarrow \bar{z} = z$.

$\quad$ If $z = a + bi$, where $a \in \mathbf{R}$, $b \in \mathbf{R}$, $i^2 = -1$ is the algebraic representation of a complex number, then the point $M(a, b)$ is called the *geometric image* of the complex number, where x-axis is the real axis and

y-axis is the imaginary axis (complex plane). Instead of $M(a,b)$ can be used $M(z)$.

The oriented angle $t^* \in [0, 2\pi)$ between the positive x-axis and OM is called the *argument* of the complex number z, denoted by $t^* = \arg z$.

$\arg \bar{z} = 2\pi - \arg z$

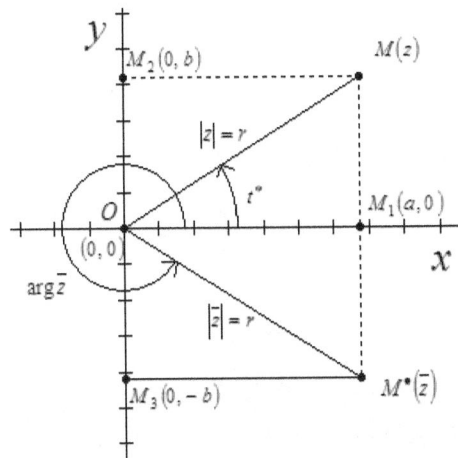

Using the definition of sine and cosine in ΔOM_1M, $a = r\cos t^*$ and $b = r\sin t^*$. Therefore $z = r(\cos t^* + i\sin t^*)$, which is called the *trigonometric form* (*polar form*) of the complex number z.

$M(a, b)$ represents the Cartesian coordinate of the point M,

$M(r, t^*)$ represents the polar coordinate of the point M.

$t = t^* + 2k\pi, k \in \mathbf{Z}$ is called the *extended argument* of the complex number z, denoted by $\text{Arg } z$.

If $z = r(\cos t^* + i\sin t^*)$, from $\tan t^* = \dfrac{b}{a}$, we obtain

$t^* = \arctan \dfrac{b}{a}$, if t^* is in the first quadrant;

$t^* = \pi - \arctan \left| \dfrac{b}{a} \right|$, if t^* is in the second quadrant;

$t^* = \pi + \arctan \left| \dfrac{b}{a} \right|$, if t^* is in the third quadrant;

$$t^* = 2\pi - \arctan\left|\frac{b}{a}\right|, \text{ if } t^* \text{ is in the fourth quadrant.}$$

Let $z_1 = r_1(\cos t_1 + i\sin t_1)$ and $z_2 = r_2(\cos t_2 + i\sin t_2)$,

$$z_1 z_2 = r_1 r_2\left[\cos(t_1 + t_2) + i\sin(t_1 + t_2)\right], \quad \frac{z_1}{z_2} = \frac{r_1}{r_2}\left[\cos(t_1 - t_2) + i\sin(t_1 - t_2)\right]$$

and, in general $\prod_{k=1}^{n} z_k = \sum_{k=1}^{n} r_k(\cos t_k + i\sin t_k)$.

If $z = r(\cos t^* + i\sin t^*)$,

$z^n = r^n(\cos nt^* + i\sin nt^*)$, $n \in \mathbf{Z}$ (De Moivre's Formula).

$$\sqrt[n]{r(\cos t^* + i\sin t^*)} = \sqrt[n]{r}\left(\cos\frac{t^* + 2k\pi}{n} + i\sin\frac{t^* + 2k\pi}{n}\right), \quad k = 0, 1, 2, \ldots, n-1.$$

Simple Geometric Notations and Properties

Let $z_1 = r_1(\cos t_1 + i\sin t_1)$ and $z_2 = r_2(\cos t_2 + i\sin t_2)$ be two complex numbers having the geometric images M_1 and M_2. Then the *distance* between the points M_1 and M_2 is given by $M_1 M_2 = |z_1 - z_2|$.

The *equation of the line* passing through the points M_1 and M_2 is $z = (1-\lambda)z_1 + \lambda z_2, 1 \hat{\mathbf{I}} \mathbf{R}$.

The complex angular coefficient (*slope*) of the line determined by the points with coordinates z_1 and z_2 is $m = \dfrac{z_2 - z_1}{\overline{z}_2 - \overline{z}_1}$ and the equation of the line is $z - z_1 = \dfrac{z_2 - z_1}{\overline{z}_2 - \overline{z}_1}(\overline{z} - \overline{z}_1)$.

The *circle* with the center $C(z_0)$ has the equation $(z - z_0)(\overline{z} - \overline{z}_0) = r^2$, where r is the radius of the circle.

The *ratio* in which a line segment $M_1 M_2$ is divided by an internally point P such that $\overrightarrow{M_1 P} = \lambda \overrightarrow{PM_2}$ is

$$z_P = \frac{1}{\lambda + 1}z_1 + \frac{\lambda}{\lambda + 1}z_2, 1 \hat{\mathbf{I}} \mathbf{R}, 1^1 -1,$$ where P has the geometric image z_P.

The *midpoint* P splits a segment $M_1 M_2$ into two congruent parts. The ratio of the lengths of these two parts is 1:1. Therefore

$$z_P = \frac{z_1 + z_2}{2}.$$

The *centroid* of a triangle (*centre of mass*) is the point where the three medians of the triangle intersect and it divides each median in the ratio 2:1. Let $A(z_A)$, $B(z_B)$, $C(z_C)$ be the geometric images of the vertices of $\triangle ABC$ and the centroid $G(z_G)$, then

$$z_G = \frac{z_A + z_B + z_C}{3}.$$

The *incenter* of a triangle is the point where the three angle bisectors of a triangle meet. Let $A(z_A)$, $B(z_B)$, $C(z_C)$ be the geometric images of the vertices of $\triangle ABC$ and the incenter $I(z_I)$, then

$$z_I = \frac{az_A + bz_B + cz_C}{a + b + c}, \text{ where } a, b, c \text{ are the length sides of the}$$

triangle.

VI. 1. Write the following complex numbers in standard algebraic form:

a) $z_1 = (3 + 3i) + (2 - i)$;   b) $z_2 = (3 + i) - (2 + 4i)$;

c) $z_3 = (3 + 2i)(1 + 4i)$;   d) $z_4 = (3 + i) \div (2 + i)$;

e) $z_5 = (1 - i) \div (1 + 2i)$;   f) $z_6 = \dfrac{1 + i}{3 - 4i} + \dfrac{4}{4 + 3i}$.

VI. 2. Compute:

a) $(1 + i)(1 - i)^2 (-1 - i)^3 (-1 + i)^4$;

b) $(1 + i) + (1 - i)^2 + (-1 - i)^3 + (-1 + i)^4$;

c) $i + i^3 + i^5 + i^7$;   d) $\dfrac{1}{i} + \dfrac{1}{i^3} + \dfrac{1}{i^5} + \dfrac{1}{i^7}$;   e) $\dfrac{i + i^2 + i^3 + i^4}{1 + i}$;

f) $i^{2010} + i^{2011} + i^{2012} + i^{2013}$;   g) $(1 + i)^{2013} - (1 - i)^{2013}$.

VI. 3. Consider the complex numbers:

a) $z_1 = -4 + 3i$;   b) $z_2 = 3 - 2i$;   c) $z_3 = -2 + i$;   d) $z_4 = 1 - 2i$.

Compute $\operatorname{Re}(z)$, $\operatorname{Im}(z)$, and $|z|$ in each case.

VI. 4. If $z \in \mathbf{C}$ show that:

a) $z \cdot \bar{z} \in \mathbf{R}$, $z + \bar{z} \in \mathbf{R}$, $\dfrac{z}{\bar{z}} + \dfrac{\bar{z}}{z} \in \mathbf{R}$;

b) $z - \bar{z} \in \mathbf{C} - \mathbf{R}$, $\dfrac{z}{\bar{z}} - \dfrac{\bar{z}}{z} \in \mathbf{C} - \mathbf{R}$.

VI. 5. Write each complex expression in standard form:

a) $\dfrac{1+i}{2-5i}$; b) $\dfrac{-3+i}{2+3i}$; c) $\dfrac{-3+2i}{2+i}$; d) $\dfrac{(2+i)^3 - (2-i)^3}{(2+i)^3 + (2-i)^3}$;

e) $\dfrac{(1+i)^4 + (1-i)^4}{(1+2i)^4 + (1-2i)^4}$; f) $\left(-2(1+i)^3 + \dfrac{17+31i}{3+4i} \right) \dfrac{-1+i}{6i} - 1$;

g) $\dfrac{i^{13} - i^{10}}{1+i^1} + i^{14}$; h) $\dfrac{3}{(1+i)^4} + \dfrac{31-17i}{4-28i} - \dfrac{1}{i}$; i) $\dfrac{1+i}{3-i} + \dfrac{i}{3+i}$;

j) $\dfrac{(1-i)(1-2i)(1-3i)(1-4i)}{(1+i)(1+2i)(1+3i)(1+4i)}$.

VI. 6. Find the real numbers x and y such that:

a) $(1+3i)x + (2-i)y = 3 + 2i$; b) $(x+i)x + (1+3yi)y = 2 + 4i$;

c) $3\sqrt{x^2 - 2y} + (1-i)x^2 = 2(1+2i)y - 12i$;

d) $4x + 5 + (y-2)i = -19 + 5i$; e) $\overline{3x - 5i} + 2x + y = \overline{3 + i} + 2yi$;

f) $\dfrac{x - 3 + (y-3)i}{x + 2 + (y+4)i} = i$; g) $\dfrac{x - 2 + (y-1)i}{y - 3} = -1 + 3i$

h) $\dfrac{2+5i}{x-y} + \dfrac{-1+3i}{x+y} = \dfrac{7x - 12i}{x^2 - y^2}$.

VI. 7. Find the real numbers a and b such that:

a) $\dfrac{1+i}{1-1i} = a + bi$; b) $\dfrac{5+2i}{2-5i} = a + bi$; c) $\dfrac{5-3i}{2+3i} = a + bi$;

d) $\dfrac{2-i}{2+i} + \dfrac{2+i}{2-i} = a + bi$; e) $\dfrac{(1+i)^3}{(1-i)^5} = a + bi$.

VI. 8. Show that the number $z = \left(1 + i\sqrt{3} \right)^3$ is an integer number.

VI. 9. Solve in $\mathbf{C}$ the equations:

a) $9x^2 + 6x + 10 = 0$;   b) $x^2 + 5x + 7 = 0$;

c) $2x^2 - 3x + 2 = 0$;   d) $-x^2 + 3x - 8 = 0$.

VI. 10. Prove that the complex numbers $1 + 3i$ and $1 - 3i$ are solutions of the equation $x^3 - 4x^2 + 14x - 20 = 0$.

VI. 11. Prove that the complex numbers $-\dfrac{1}{2} + i\dfrac{\sqrt{3}}{2}$ and $-\dfrac{1}{2} - i\dfrac{\sqrt{3}}{2}$ are roots of the equation $x^4 + x^2 + 1 = 0$.

VI. 12. If x_1 and x_2 are roots of the equation $x^2 + x + 1 = 0$, then calculate:

a) $\left(x_1^3 + x_1^2 - 1\right)^{2002} + \left(x_2^3 + x_2^2 - 1\right)^{2002}$;

b) $\left(x_1^4 + x_1^3\right)^{2002} + \left(x_2^4 + x_2^3\right)^{2002}$;

c) $\left(x_1^4 + x_1^3 + x_1^2 + 1\right)^{2013} + \left(x_2^4 + x_2^3 + x_2^2 + 1\right)^{2013}$.

VI. 13. If ε is one of the roots of the unity, $x^3 - 1 = 0$ and $\varepsilon \neq 1$, calculate the following complex numbers:

a) $z_1 = \varepsilon^4 + \varepsilon^2 + 1$;   b) $z_2 = \dfrac{1+\varepsilon}{(1-\varepsilon)^2} + \dfrac{1-\varepsilon}{(1+\varepsilon)^2}$;   c) $z_3 = \dfrac{\varepsilon^3 - 1}{1 - 2\varepsilon^2}$;

d) $z_4 = (a + b + c)(a + b\varepsilon + c\varepsilon^2)(a + b\varepsilon^2 + c\varepsilon)$;

e) $z_5 = (1+\varepsilon)(1+\varepsilon^2)(1+\varepsilon^3)(1+\varepsilon^4)(1+\varepsilon^5)(1+\varepsilon^6)$;

f) $z_6 = (1-\varepsilon)(1-\varepsilon^2)(1-\varepsilon^4)(1-\varepsilon^5)$.

VI. 14. Prove the identities:

a) $\left(\dfrac{-\sqrt{3}+i}{2}\right)^5 - \left(\dfrac{\sqrt{3}+i}{2}\right)^5 = \sqrt{3}$;   b) $\left(\dfrac{-1+i\sqrt{3}}{2}\right)^5 + \left(\dfrac{-1-i\sqrt{3}}{2}\right)^5 = -1$;

c) $\left(\dfrac{-1+i\sqrt{3}}{2}\right)^6 + \left(\dfrac{-1-i\sqrt{3}}{2}\right)^6 = 2$;   d) $\left(\dfrac{\sqrt{3}-i}{2}\right)^6 + \left(\dfrac{\sqrt{3}+i}{2}\right)^6 = -2$.

VI. 15. Calculate $z^{13} + \dfrac{1}{z^{13}}$ if z is a root of the equation $z + \dfrac{1}{z} = 1$.

VI. 16. Calculate the complex number:

$$z = \left(\frac{1+i}{1-i}\right)^{2010} + \left(\frac{1+i}{1-i}\right)^{2011} + \left(\frac{1+i}{1-i}\right)^{2012} + \left(\frac{1+i}{1-i}\right)^{2013}.$$

VI. 17. If x_1 and x_2 are the roots of the equation $x^2 + x + 1 = 0$ then calculate the value of the expression

$$E = \frac{(x_1+1)^{2014}}{x_1^{2}+1} + \frac{(x_2+1)^{2014}}{x_2^{2}+1}.$$

VI. 18. Let z_1, z_2 be two complex numbers such that $|z_1| = |z_2| = 1$.

Show that $\dfrac{z_1 + z_2}{1 + z_1 z_2}$ is a real number.

VI. 19. Let z be a complex number such that $|z| = 1$. Show that

$$\frac{1 + z^{2n}}{z^n}$$ is a real number.

VI. 20. Let $z \in \mathbf{C}$ be a complex number such that $z \neq \bar{z}$ and

$$\frac{1 - z + z^2}{1 + z + z^2} \in \mathbf{R}.$$ Prove that $|z| = 1$.

VI. 21. Show that the number $z = (3 + 2i)^{4n} + (2 + 3i)^{4n}$ is a real number for any n integer number.

Prove that in general $z = (a + bi)^{4n} + (b + ai)^{4n} \in \mathbf{R}$, where $a, b \in \mathbf{R}$.

VI. 22. Let n be a positive integer number and $z_n = (1 + i)^n + (1 - i)^n$. Prove that for any n:

a) z_n is a real number;

b) $z_{n+2} = 2z_{n+1} - 2z_n$; **c)** $z_n = 2(\sqrt{2})^n \cos\dfrac{n\pi}{4}$.

VI. 23. The geometric image of the complex numbers $z_1 = 1 + 3i$ and $z_2 = 3 - i$ in the complex plane is represented by the points A and B, respectively. Prove that $\triangle OAB$ is a right triangle.

VI. 24. Show that the geometric images of the complex numbers $z_1 = -1 - 2i$, $z_2 = 1 + 4i$, and $z_3 = 2 + 7i$ are three collinear points.

VI. 25. Let $z_1 = 1+i$, $z_2 = 4-2i$, and $z_3 = -2+4i$ be three complex numbers. Show that the geometric image in the complex plane is an isosceles triangle.

VI. 26. The points A, B, C, D are the geometric image in the complex plane of the numbers $z_1 = 1-2i$, $z_2 = 1+2i$, $z_3 = -2+i$, and $z_4 = -2-i$. Show that they lie on a circle and find the equation of this circle;   moreover A, B, C, D is an isosceles trapezoid.

VI. 27. Show that the midpoints of the sides of a quadrilateral form a parallelogram.

VI. 28. Consider the four points A, B, C, D in a plane and their geometric image in the complex plane $z_1 = 1-3i$, $z_2 = 7-2i$, $z_3 = 8+4i$, and $z_4 = 2+3i$, respectively. Prove that $ABCD$ is a parallelogram and find the geometric image of the point of intersection of the diagonals.

VI. 29. Let $A(z_A)$, $B(z_B)$, $C(z_C)$ be the geometric images of the vertices of $\triangle ABC$ and the centroid (*point of intersection of medians*) $G(z_G)$, where $z_A = 1+i$, $z_B = 3+2i$, and $z_G = 1-i$. Find the geometric image z_C of the vertex C.

VI. 30. Consider the complex linear function
$f : C \to C$, $f(z) = az + b$, $a, b, c \in C$, $a \neq 0$. If the points A, B and M, N have the geometric images z_1, z_2 and $f(z_1)$, $f(z_2)$, respectively, show that $AB = |a|MN$.

VI. 31. Solve for the complex variable z and express z in the form $a + bi$:

a) $|z| = |z + 1 - i| = |\overline{z} + 2 - 4i|$;   b) $|z + i| = |z + 1 - i| = |z - 3 - 2i|$;

c) $\left|\dfrac{z + 2i}{z + 4i}\right| = 1$ and $\left|\dfrac{z + 2i}{z - 1}\right| = \dfrac{1}{\sqrt{2}}$;   d) $\left|\dfrac{z + 1}{z}\right| = 2$ and $\left|\dfrac{iz + 1}{1 + i}\right| = 1$;

e) $\left|\dfrac{z - 1}{z - i}\right| = \dfrac{1}{2}$ and $\left|\dfrac{z + 1}{z}\right| = \sqrt{2}$.

VI. 32. Find the set of points in the complex plane such that:

a) $|z - i| = 2$; b) $|z - 1| + |z + 1| = 4$; c) $|z - 3| = |z + 3|$;

d) $2|z| = |z - i| + |z + i|$; e) $|z + i| < 2$; f) $|z - 2| - |z + 2| < 2$;

g) $1 < |2z + 3 - 2i| \le 2$; h) $|i - z| > 4$; i) $|z + i - 2| \le 2$.

VI. 33. Solve the equations:

a) $x^3 - 1 = 0$; b) $x^3 - 8 = 0$; c) $27x^3 + 1 = 0$;

d) $64x^3 - 27 = 0$; e) $x^4 - 16 = 0$; f) $x^4 - 81 = 0$;

g) $x^6 - x^3 - 2 = 0$; h) $x^6 + 3x^3 + 2 = 0$.

VI. 34. Let $z \in \mathbf{C}$ be a complex number. Show that:

a) $\left| iz \right| + \left| \left(1 - i\sqrt{3} \right) z \right| = 3|z|$; b) if $|z| < 1$ then $|2 - i\bar{z}| < 3$;

c) if $|z| < \dfrac{1}{2}$ then $\left\| (1 + 2i)z + 2i|z| \right\| < 3$;

d) if $|z| < \dfrac{1}{3}$ then $\left| \left(\sqrt{2} - i \right) z^3 - iz \right| < \dfrac{3}{4}$;

e) if $|z| < \dfrac{1}{2}$ then $\left\| (1 + i)z^3 + i|z| \right\| < 3$;

f) if $\left| z^3 + \dfrac{1}{z^3} \right| \le 2$ then $\left| z + \dfrac{1}{z} \right| \le 2$.

VIII. 35. If $\varepsilon \ne 1$ is one of the roots of the equation $x^3 - 1 = 0$, prove that:

a) $(1 + \varepsilon)^{2013} + \left(1 + \varepsilon^2 \right)^{2013} = -2$;

b) $(1 + \varepsilon)^n + \left(1 + \varepsilon^2 \right)^n + \left(\varepsilon + \varepsilon^2 \right)^n = \begin{cases} -3 \ if \ n = 3k \\ 0 \ if \ n \ne 3k \end{cases}, k \in \mathbf{N}$;

c) $|z - 1|^2 + |z - \varepsilon|^2 + \left| z - \varepsilon^2 \right|^2 = 3\left(1 + |z|^2 \right), \ \forall z \in \mathbf{C}$.

VIII. 36. Prove that the number

$z = \left(\sqrt[4]{a} + i \cdot \sqrt[4]{b} \right)\left(\sqrt[4]{a} + i^2 \cdot \sqrt[4]{b} \right) \ldots \left(\sqrt[4]{a} + i^{4k} \cdot \sqrt[4]{b} \right)$ is real,

where a, b are positive real numbers.

VIII. 37. Find $a \in \mathbf{R}$ such that:

a) $\left| \dfrac{2+4i}{1+ai} \right| = 2$; b) $\dfrac{a+2i}{3+ai} \in \mathbf{R}$.

VIII. 38. Find the sets of points in the complex plane such that:

a) $A = \left\{ (x,y) \in \mathbf{R} \times \mathbf{R} \middle| \ \mathrm{Re}\left(\dfrac{z-2}{z-6} \right) = \mathrm{Im}\left(\dfrac{z-2}{z-6} \right), \ z = x+iy \right\}$;

b) $B = \left\{ (x,y) \in \mathbf{R} \times \mathbf{R} \middle| \ \mathrm{Re}\left(\dfrac{z-2}{z-1} \right) = 0, \ z = x+iy \right\}$;

c) $C = \left\{ (x,y) \in \mathbf{R} \times \mathbf{R} \middle| \ \dfrac{i+\bar{z}}{i-\bar{z}} \in \mathbf{R}, \ z = x+iy \right\}$;

d) $D = \left\{ (x,y) \in \mathbf{R} \times \mathbf{R} \middle| \ \left| \dfrac{z^2-i}{z^2-2i} \right| = 1, \ z = x+iy, \ z^2 \neq 2i \right\}$;

e) $E = \left\{ (x,y) \in \mathbf{R} \times \mathbf{R} \middle| \ \mathrm{Im}\left(\dfrac{z+1+i}{2-iz} \right) = 0, \ z = x+iy \right\}$.

VIII. 39. Let x_1 and x_2 be the roots of the equation
$2x^2 - 2mx + m = 0$.
a) Calculate, in terms of the parameter $m \in \mathbf{R}$, the expressions
$x_1^2 + x_2^2$, $x_1^3 + x_2^3$, $\dfrac{1}{x_1} + \dfrac{1}{x_2}$;
b) Find m such that the equation does not have real roots;
c) Find m when $|x_1| + |x_2| = 4$.

VIII. 40. If $z = 1 + 2i$ is a root of the equation
$z^2 - (1-2i)z + 2a - i = 0$ find a C.

VIII. 41. If $z = 1 + i$ is a root of the equation
$z^3 + z^2 + (a-1)z + b = 0$ find a, b R.

VIII. 42. Factor each polynomial completely over the set of complex numbers:
a) $x^4 - x$; b) $x^3 + 8$; c) $x^4 + x^2$; d) $x^5 + 27x^2$.

VIII. 43. Find the complex numbers z such that:

a) $z^2 - \bar{z} = 0$; b) $z^2 - i\bar{z} = 0$; c) $z^2 + |z| = 0$;

d) $z^2 + 3|z| = 0$; e) $z^2 + z|z| + |z|^2 = 0$.

VIII. 44. Find $z = x + iy$, $x, y \in \mathbf{R}$ where:

a) $z^2 = -3 - 4i$; b) $z^2 = 1 - 4\sqrt{3}i$;

c) $z^2 = 3 + 4i$; d) $z^2 = 7 + 24i$.

VIII. 45. Solve for the complex variable z:

a) $z^2 - 4iz - 3 = 0$; b) $z^2 - (1 - i)z + 2 + i = 0$;

c) $(2 - i)z^2 - (4 - 12i)z + 14 - 12i = 0$; d) $z^2 - 5iz - 7 + i = 0$;

e) $z^2 + 4iz + 5 = 0$; f) $z^2 - (5 - 2i)z + 6 - 4i = 0$;

g) $iz^2 + (1 - 2i)z - 1 + 3i = 0$; h) $z^2 - (5 - 2i)z + 5 - 5i = 0$;

i) $z^2 - (2 + i)z - 1 + 7i = 0$; j) $z^2 - (5 + 2i)z + 9 + 7i = 0$;

k) $z^2 - (8 + 3i)z + 13 + 13i = 0.$

VIII. 46. Find $z = x + i y$, x, y $\mathbf{R}$ from the equations:

a) $(1 + i)z + (5 - 3i)\bar{z} = 20 - 4i$;

b) $(2 + 2i)z - 3\,\mathrm{Re}\,(z) = -18 + 20i$;

c) $(1 + 2i)(z + i) + (3 - 4i)(1 + zi) = 1 + 7i$;

d) $(-2 + i)(i - z) + (3 - 4i)(i + z) = 1 + 7i$;

e) $(1 + 3i)z - 1 \ i = (3 - 2i)z + 13$;

f) $2z + i \cdot \bar{z} = 9 + 3i$.

VIII. 47. Solve for the complex variable z and express z in the form $z = a + bi$, $a, b \in \mathbf{R}$:

a) $iz^4 + (4 + 2i)z^2 + (1 - i)^6 = 0$; b) $z^4 - 3(1 - 2i)z^2 - 8 - 6i = 0$;

c) $z^4 - 2z^3 + 4z^2 - 2z + 3 = 0$; d) $z^4 - (5 - 2i)z^2 + 8 + 10i = 0$;

e) $\left(\dfrac{5z - 2i}{2z - 5i}\right)^3 + \left(\dfrac{5z - 2i}{2z - 5i}\right)^2 + \dfrac{5z - 2i}{2z - 5i} + 1 = 0$.

VIII. 48. Find $z = a + bi$, $a, b \in \mathbf{R}$ such that:

a) $z^3 = -11 + 2i$; b) $z^3 = -9 - 46i$; c) $z^3 = -8i$;

d) $z^3 = -6 + 6i\sqrt{3}$; e) $z^4 = -8 - 8i\sqrt{3}$; f) $z^6 = -64$.

VI. 49. Simplify the complex fractions:

a) $\dfrac{z^2 - 8(1-i)z + 63 - 16i}{z^2 - 6z + 25}$; b) $\dfrac{z^2 - iz - 1 - i}{z^2 - 2z + 2}$;

c) $\dfrac{z^3 - (3+i)z^2 + (2+3i)z - 2i}{z^3 - (3-i)z^2 + (2-3i)z - 2i}$.

VIII. 50. Solve the system of equations, where $x, y \in \mathbf{C}$:

a) $\begin{cases} (1+3i)x - (1-2i)y + 1 - i = 0 \\ 2(2+i)x - (2+i)y - (7-4i) = 0 \end{cases}$; b) $\begin{cases} (2+3i)x + (2-3i)y = 4 - 2i \\ (1+i)x + (1-i)y = 2 \end{cases}$;

c) $\begin{cases} 4x + (3-2i)y = 5 - i \\ (3+2i)x + 4y = 5 + i \end{cases}$; d) $\begin{cases} (2+i)x + (2-i)y = 6 + 3i \\ (1+i)x + (1-i)y = 3 + i \end{cases}$.

VIII. 51. Solve in $\mathbf{C}$ the systems of equations:

a) $\begin{cases} x + y = 3 \\ y = 4 \end{cases}$; b) $\begin{cases} x + y = 2i \\ x^2 + y^2 = -1 \end{cases}$;

c) $\begin{cases} x^2 - y^2 = -3 \\ 2(x^2 + y^2) + xy = -4 \end{cases}$; d) $\begin{cases} 2x + y = 2 \\ x^2 + y^2 + y + 3x = 0 \end{cases}$.

VIII. 52. Consider the complex function

$f : \mathbf{C} \to \mathbf{C}$, $f(z) = z^2 + 3z - \bar{z}$.

a) Compute: $f(-1)$, $f(i)$, $f(2-i)$, $f(\bar{z})$, $f(|z|)$;

b) Verify the eguality $f(z) - f(\bar{z}) = 4(\bar{z} - z)$.

VIII. 53. Consider the complex function $f : \mathbf{C} \setminus \{1\}\ \mathbf{C}$,

$f(z) = \dfrac{z^2 - z - 2}{z - 1}$. Find the points in the plane such that f(z) R.

VIII. 54. Consider the complex function f having the property
$f(x) + f(\varepsilon x) = 4 - \varepsilon x$, $\forall x \in \mathbf{C}$, where ε is a complex root of the equation

$x^3 = 1$. Prove that $f(x) + f(\varepsilon^2 x) = 4 - x$, $\forall x \in \mathbf{C}$ and find $f(x)$.

VIII. 55. Consider the complex functions $f, g : \mathbf{C} \to \mathbf{C}$ having the properties $f(x)f(ix) = x^2$ and $g(x) + g(\varepsilon x) = x$, $\forall x \in \mathbf{C}$, where ε is a complex root of the equation $x^3 = 1$. Prove that $f(-x) = -f(x)$, $\forall x \in \mathbf{C}$ and find $g(x)$.

VIII. 56. Let ε be one of the roots of the equation $z^n - 1 = 0$, $n \in \mathbf{N}$, $n \geq 2$ and the function $f : \mathbf{R} \to \mathbf{R}$, such that
(1) $f(x + a) + \varepsilon \cdot f(x - a) = b$, $\forall x \in \mathbf{R}$,
where $a \in \mathbf{R}^*$, $b \in \mathbf{C}$. Show that f is a periodic function.

VIII. 57. If α is one of the roots of the equation $z^n - 1 = 0$, $n \in \mathbf{N}$ and $n \geq 2$, then calculate $E = \alpha + 2\alpha^2 + 3\alpha^3 + \dots + n\alpha^n$;

VIII. 58. Let be the complex number $z = \cos t + i \sin t$.
Prove that $\cos nt = \dfrac{z^{2n} + 1}{2z^n}$ and $\sin nt = \dfrac{z^{2n} - 1}{2iz^n}$, $n \in N^*$.
Calculate the product $A = \sin 10° \sin 50° \sin 70°$.

VIII. 59. Prove that if $z_1, z_2, \dots, z_n \in \mathbf{C} \setminus \{0\}$ and $|z_1| = |z_2| = \dots = |z_n|$ then the expression $\left(1 + \dfrac{z_2}{z_1}\right)\left(1 + \dfrac{z_3}{z_2}\right) \dots \left(1 + \dfrac{z_n}{z_{n-1}}\right)\left(1 + \dfrac{z_1}{z_n}\right)$ is a real number.

VIII. 60. Solve the equation $(2 - i)z^3 - 3 - i = 0$.

VIII. 61. Let z_1, z_2, z_3 be three complex numbers such that $z_1 \neq z_2 \neq z_3$. Show that the geometric images of the numbers $A(z_1)$, $B(z_2)$, $C(z_3)$ are collinear if and only if $\dfrac{z_3 - z_1}{z_2 - z_1} \in \mathbf{R}^*$.

VIII. 62. Consider the complex function $f : \mathbf{C} \to \mathbf{C}$, $f(z) = az + b$, where $a, b \in \mathbf{R}$, $a \neq 2$ and the complex numbers z_1, z_2, z_3.
Show that if the geometric images of the numbers z_1, z_2, z_3 are collinear,

then the geometric images of the numbers $f(z_1), f(z_2)$, and $f(z_3)$ are also collinear.

VIII. 63. Consider the function $f : \mathbf{C} \to \mathbf{C}$, $f(z) = 3z + |z|$. Prove that f is a bijection and find its inverse.

VIII. 64. Consider the complex number $z = x + yi$, $x, y \in \mathbf{R}$. Find the geometric image (*locus*) of the complex number z which satisfies the relation:

a) $z^2 = a(1 - 2i) + 1$;

b) $z^2 = 2aiz + 1$, $a \in \mathbf{R}$.

VIII. 65. For each of the polynomials prove that:

a) $f(x) = (x^2 + x + 1)^{8n+1} - x$, $n \in \mathbf{N}^*$ is divisible by $x^2 + 1$;

b) $f(x) = (x^2 - x + 1)^{2n+2} + (x^4 + x - 1)^{2n+5} - x$, $n \in \mathbf{N}^*$ is divisible by $x^2 + 1$;

c) $f(x) = (2x^2 + x + 2)^{4n+1} - x$, $n \in \mathbf{N}^*$ is divisible by $x^2 + 1$;

d) $f(x) = (x - 1)^{2n+1} - x^{n+2}$, $n \in \mathbf{N}^*$ is divisible by $x^2 - x + 1$;

e) $f(x) = (x + 1)^{6n+1} + (x + 1)^{6n+5} + (x^2 + 1)^{6n+3}$, $n \in \mathbf{N}^*$ is divisible by $x^2 + x + 1$;

f) $f(x) = x^{6n-1} + x + 1$, $n \in \mathbf{N}^*$ is divisible by $x^2 + x + 1$.

VIII. 66. Find roots of the polynomials in the set **C** (*complex numbers*), knowing that each of them has a real root.

a) $f(z) = z^3 - (4 + i)z^2 + (7 + 3i)z - 2i - 6$;

b) $f(x) = x^3 - 4x^2 + (6 - i)x - 3 + i$.

VIII. 67. Consider the polynomial $f(x) = x^3 + ax^2 + 4x + b$, where $a, b \in \mathbf{R}$. Find the numeric value for a and b such that

$x_1 = \dfrac{1 - i\sqrt{3}}{2}$ is a root of the equation $f(x) = 0$.

VIII. 68. The equation $x^3 - 3x^2 + ax + b = 0$ has rational coefficients, and one of its roots is $-1 + i\sqrt{3}$; determine the value of a and b.

VIII. 69. One of the roots of the polynomial function
$f(x) = x^4 - mx^3 + 13x^2 - 14x + n$, $m, n \in R$ is $x_1 = 1 - i$.
Find m and n and solve the equation $f(x) = 0$.

VIII. 70. A *quartic* (or *biquadratic*) polynomial with real coefficients has four zeroes, two of them are:
a) $-1 + i\sqrt{3}$ and $-1 + i$;
b) $3 + 2i$ and $1 - i$.
What are the other two roots? Find the polynomial (*a minimum degree polynomial*).

VIII. 71. Let $z = r(\cos t + i \sin t)$ be a nonzero complex number. Show that $-z = r[\cos(t + \pi) + i \sin(t + \pi)]$.

VIII. 72. Find the polar coordinates of the points: $M_1(1, \sqrt{3})$, $M_2(1, -1)$, $M_3(-1, -2)$, $M_4(0, -1)$, $M_5(1, 0)$, $M_6(0, 3)$.

VIII. 73. Find the polar representation of the complex numbers:
a) $z_1 = -1 - i$; **b)** $z_2 = -1 + 2i$; **c)** $z_3 = -1 + i\sqrt{3}$; **d)** $z_4 = \sqrt{2} - i\sqrt{2}$;
e) $z_5 = -i\sqrt{2}$; **f)** $z_6 = 1 + \cos\theta + i \sin\theta$, $\theta \in [0, 2\pi)$.

VIII. 74. Write the algebraic form of the complex numbers:
a) $z = \sqrt{3}(\cos 0 + i \sin 0)$; **b)** $z = \sqrt{2}\left(\cos\frac{\pi}{2} + i\sin\frac{\pi}{2}\right)$;
c) $z = \sqrt{3}(\cos\pi + i\sin\pi)$; **d)** $z = \cos\frac{\pi}{3} + i\sin\frac{\pi}{3}$;
e) $z = \cos\frac{3\pi}{4} + i\sin\frac{3\pi}{4}$; **f)** $z = \cos\frac{3\pi}{2} + i\sin\frac{3\pi}{2}$;
g) $z = \cos\frac{4\pi}{3} + i\sin\frac{4\pi}{3}$; **h)** $z = 2\left(\cos\frac{1}{6}\pi + i\sin\frac{1}{6}\pi\right)$;

i) $z = 2\left(\cos\dfrac{\pi}{12} + i\sin\dfrac{\pi}{12}\right)$;   j) $z = \cos\dfrac{\pi}{8} + i\sin\dfrac{\pi}{8}$;

k) $z = 2\left(\cos\dfrac{15\pi}{2} + i\sin\dfrac{15\pi}{2}\right)\left(\cos\dfrac{27\pi}{4} + i\sin\dfrac{27\pi}{4}\right)\cdot\sqrt{2}\left(\cos\dfrac{43\pi}{12} + i\sin\dfrac{43\pi}{12}\right)$.

VIII. 75. Calculate the following algebraic complex expressions using the polar representation of a complex number:

a) $(1-i)(\sqrt{3}+i)$;   b) $(2-2i)(\sqrt{3}+i)(-1-i\sqrt{3})$;   c) $\dfrac{(1-i)^8(\sqrt{3}+i)^5}{(-1-i\sqrt{3})^5}$;

d) $\left(\dfrac{\sqrt{3}+i}{1+i}\right)^D$;   e) $\left(\dfrac{3-i}{1-2i}\right)^8 + \left(\dfrac{3+i}{2-i}\right)^8$;   f) $\left(\dfrac{1}{2}+i\dfrac{\sqrt{3}}{2}\right)^n + \left(\dfrac{1}{2}-i\dfrac{\sqrt{3}}{2}\right)^n$;

g) $\dfrac{2(1+\tan^2\theta)}{1+\tan\theta+i(1-\tan\theta)}$;   h) $\dfrac{1-\cos\theta-i\sin\theta}{1+\cos\theta+i\sin\theta}$, $\theta\in[0,2\pi)$;

i) $\dfrac{\cos2\theta+\cos4\theta+i(\cos4\theta-\sin2\theta)}{\cos2\theta+\cos4\theta-i(\cos4\theta-\sin2\theta)}$;

j) $(1-\cos x+i\sin x)^n$.

VIII. 76. Consider the sequence $(a_n)_{n\geq1}$, $a_n = (1+i)^n + (1-i)^n$, $n\in\mathbf{N}^*$.

a) show that $a_n\in\mathbf{R}$, $\forall n\in\mathbf{N}^*$;

b) prove that $a_{n+2} = 2a_{n+1} - 2a_n$, $\forall n\in\mathbf{N}^*$;

c) show that $a_n = 2(\sqrt{2})^n \cos\dfrac{n\pi}{4}$, $\forall n\in\mathbf{N}^*$.

VIII. 77. Solve the equations:

a) $z^3 - 8 = 0$;   b) $(3-2i)z^6 = 1 - 5i$;   c) $z^6 = i$;   d) $z^6 - iz^3 + 2 = 0$;

e) $z^8 + z^4 - 1 + 3i = 0$;   f) $z^D - 3iz^5 - 3 + i = 0$;   g) $z^4 + z^3 + z^2 + z + 1 = 0$;

h) $z^4 = \dfrac{\sqrt{3}+3i}{3-i\sqrt{3}}$;   i) $z^8 = \dfrac{1+i}{\sqrt{3}-i}$;   j) $z^5 - iz^4 - z^3 + iz^2 + z - i = 0$.

VIII. 78. Let $x,y,z\in\mathbf{R}$ be three numbers such that

$\sin x + \sin y + \sin z = 0$ and $\cos x + \cos y + \cos z = 0$.

Show that:

a) $\sin2x + \sin2y + \sin2z = 0$ and $\cos2x + \cos2y + \cos2z = 0$;

b) $\sin 2^n x + \sin 2^n y + \sin 2^n z = 0$ and $\cos 2^n x + \cos 2^n y + \cos 2^n z = 0$;

c) $\sin(x + y + z) = \dfrac{\sin 3x + \sin 3y + \sin 3z}{3}$ and

$\cos(x + y + z) = \dfrac{\cos 3x + \cos 3y + \cos 3z}{3}$.

VIII. 79. Calculate $z^n + \dfrac{1}{z^n}$, $n \in \mathbf{N}^*$ if:

a) $z + \dfrac{1}{z} = 2\cos a$; b) $z + \dfrac{1}{z} = 2\sin a$, where $a \in \mathbf{R}$.

VIII. 80. Calculate $\tan 5x$ in terms of $\tan x$.

VIII. 81. Let $z_1 = \cos\alpha + i\sin\alpha$ and $z_2 = \cos\beta + i\sin\beta$ be

two complex numbers. Show that $\cos(\alpha + \beta) = \dfrac{z_1^2 z_2^2 + 1}{2 z_1 z_2}$ and

$\sin(\alpha + \beta) = \dfrac{z_1^2 z_2^2 - 1}{2 z_1 z_2}$.

VIII. 82. Show that if
$\dfrac{\sin(\alpha + \beta)}{\sin(\alpha - \beta)} = \dfrac{\sin(\theta + \omega)}{\sin(\theta - \omega)}$ then $\dfrac{\cos(\alpha + \omega)}{\cos(\alpha - \omega)} = \dfrac{\cos(\beta + \theta)}{\cos(\beta - \theta)}$.

VIII. 83. Find the geometric image of the complex numbers z such that $|z + 3 - 2i| = |z + 1 - i|$.

VIII. 84. Consider the points: $A(5,1)$, $B(4, 1 - \sqrt{5})$, $C(2 - 2\sqrt{2}, 2)$, and $I(2,1)$ in the complex plane. Show that the points A, B, C lie on the same circle with the center I.

VIII. 85. Solve the equations:

a) $\left(\dfrac{1 + iz}{1 - iz}\right)^n = -1$; b) $\left(\dfrac{1 + iz}{1 - iz}\right)^n = \dfrac{1 + ia}{1 - ia}$, $a \in \mathbf{R}$;

c) $\left(z+i\sqrt{1-z^2}\right)^n +\left(z-i\sqrt{1-z^2}\right)^n = 0$, $n \in \mathbf{N}^{\bullet}$;

d) $\left(1+\dfrac{iz}{n}\right)^n -\left(1-\dfrac{iz}{n}\right)^n = 0$, $n \in \mathbf{N}$;

e) $\left(a\,x-i\sqrt{1-x^2}\right)^n +\left(ax+i\sqrt{1-x^2}\right)^n = 0$, $a \in \mathbf{R}$.

VIII. 86. Let $z = r\left(\cos t + i\sin t\right)$ be a nonzero complex number. Calculate the product $P = \left(z+\bar z\right)\left(z^2 +\bar z^2\right)..\left(z^n +\bar z^n\right)$.

VIII. 87. Prove that $\left(\dfrac{1+i\tan t}{1-i\tan t}\right)^n = \dfrac{1+i\tan nt}{1-i\tan nt}$, where $t \in \mathbf{R}\setminus\left\{(2k+1)\dfrac{\pi}{2}, k \in \mathbf{Z}\right\}$, $n \in \mathbf{N}^{*}$.

VIII. 88. Solve the equation
$$\left(\frac{x-i}{x+i}\right)^{n-1} +\left(\frac{x-i}{x+i}\right)^{n-2} +...+\left(\frac{x-i}{x+i}\right)^{1} +1 = 0, \ n \in \mathbf{N}.$$

VIII. 89. Consider the set of complex numbers:

$A = \left\{\left(\dfrac{1+i\sqrt3}{2}\right)^n \mid n \in \mathbf{N}\right\}$,

$B = \left\{\left(\dfrac{1-i\sqrt3}{2}\right)^n \mid n \in \mathbf{N}\right\}$, $C = \left\{\left(\dfrac{-1+i\sqrt3}{2}\right)^n \mid n \in \mathbf{N}\right\}$, and

$D = \left\{\left(\dfrac{-1-i\sqrt3}{2}\right)^n \mid n \in \mathbf{N}\right\}$.

Prove that $C = D \subset A = B$.

UNIT IX

Trigonometric Sums

IX. 1. Show that

$$\cos 1° + \cos 3° + ... + \cos 89° = \sin 91° + \sin 93° + ... + \sin 179°.$$

IX. 2. Find the value of the sum $S = \cos 1° + \cos 2° + ... + \cos 359°$.

IX. 3. Calculate the sum $\sin 1° + \sin 2° + ... + \sin 359°$.

IX. 4. Calculate the sum

$$\sec 0° \sec 1° + \sec 1° \sec 2° + ... + \sec n° \sec(n+1)° = \sum_{k=0}^{n} \sec k° \sec(k+1)°.$$

IX. 5. Verify the identity

$$\sin \frac{h}{2} \cos(x+h) = \frac{1}{2}\left(\sin\left(x + \frac{3h}{2} \right) - \sin\left(x + \frac{h}{2} \right) \right);$$

and prove the next identities:

a) $\cos x + \cos(x+h) + ... + \cos(x+nh) = \dfrac{\sin \dfrac{n+1}{2} h \cos\left(x + \dfrac{nh}{2} \right)}{\sin \dfrac{h}{2}};$

b) $\sin x + \sin(x+h) + ... + \sin(x+nh) = \dfrac{\sin \dfrac{n+1}{2} h \sin\left(x + \dfrac{nh}{2} \right)}{\sin \dfrac{h}{2}};$

c) $\sin 1° + \sin 3° + \sin 5° + ... + \sin 99° = \dfrac{\sin^2 50°}{\sin 1°}$;

d) $\dfrac{\sin x + \sin 3x + ... + \sin(2n-1)x}{\cos x + \cos 3x + ... + \cos(2n-1)x} = \tan nx,\ n \in \mathbf{N}^*$.

IX. 6. Calculate the sums

$$S_1 = \cos^2 x + \cos^2 2x + ... + \cos^2 nx = \sum_{k=1}^{n} \cos^2 kx \text{ and}$$

$$S_2 = \sin^2 x + \sin^2 2x + ... + \sin^2 nx = \sum_{k=1}^{n} \sin^2 kx.$$

IX. 7. Calculate the sums

$$S_1 = \sin^3 x + \sin^3 2x + ... + \sin^3 nx = \sum_{k=1}^{n} \sin^3 kx \text{ and}$$

$$S_2 = \cos^3 x + \cos^3 2x + ... + \cos^3 nx = \sum_{k=1}^{n} \cos^3 kx.$$

IX. 8. Calculate the sum

$$S_n = \sin^3 \frac{x}{3} + 3^1 \sin^3 \frac{x}{3^2} + ... + 3^{n-1} \sin^3 \frac{x}{3^n} = \sum_{k=1}^{n} 3^{k-1} \sin^3 \frac{x}{3^k}.$$

IX. 9. Evaluate the sums:

a) $S_n = \dfrac{1}{\cos x - \cos 3x} + \dfrac{1}{\cos x - \cos 5x} + ... + \dfrac{1}{\cos x - \cos(2n+1)x} =$

$$= \sum_{k=1}^{n} \frac{1}{\cos x - \cos(2k+1)x};$$

b) $S_n' = \dfrac{1}{\cos x + \cos 3x} + \dfrac{1}{\cos x + \cos 5x} + ... + \dfrac{1}{\cos x + \cos(2n+1)x} =$

$$= \sum_{k=1}^{n} \frac{1}{\cos x + \cos(2k+1)x};$$

c) $S_n'' = \dfrac{1}{\cos x \cos 2x} + \dfrac{1}{\cos 2x \cos 3x} + ... + \dfrac{1}{\cos nx \cos(n+1)x} =$

$$= \sum_{k=1}^{n} \frac{1}{\cos kx \cos(k+1)x}.$$

IX. 10. Evaluate the sums:

a) $S_n = \displaystyle\sum_{k=1}^{n} \frac{1}{\cos(a + (k-1)b)\cos(a + kb)}$;   **b)** $S_n = \displaystyle\sum_{k=1}^{n} \frac{1}{\cos ka \cos(k+1)a}$.

IX. 11. Evaluate the sums

a) $S_n = \tan x + \dfrac{1}{2}\tan\dfrac{x}{2} + \dfrac{1}{2^2}\tan\dfrac{x}{2^2} + \dots + \dfrac{1}{2^n}\tan\dfrac{x}{2^n} = \displaystyle\sum_{k=0}^{n} \dfrac{1}{2^k}\tan\dfrac{x}{2^k}$;

b) $S'_n = \tan x + 2\tan 2x + \dots + 2^n \tan 2^n x = \displaystyle\sum_{k=0}^{n} 2^k \tan 2^k x$.

IX. 12. Evaluate the sum

$S_n = \dfrac{\tan x}{\cos 2x} + \dfrac{\tan 2x}{\cos 4x} + \dfrac{\tan 4x}{\cos 8x} + \dots + \dfrac{\tan 2^n x}{\cos 2^{n+1} x} = \displaystyle\sum_{k=0}^{n} \dfrac{\tan 2^k x}{\cos 2^{k+1} x}$.

IX. 13. Evaluate the sum

$S_n = \dfrac{\sin x}{\cos x} + \dfrac{\sin 2x}{\cos^2 x} + \dfrac{\sin 3x}{\cos^3 x} + \dots + \dfrac{\sin nx}{\cos^n x} = \displaystyle\sum_{k=0}^{n} \dfrac{\sin kx}{\cos^k x}$, $x \notin \left\{ \dfrac{m\pi}{2} \mid m \in \mathbf{Z} \right\}$.

IX. 14. Evaluate the sum

$S_n = 1 + \dfrac{\cos x}{\cos x} + \dfrac{\cos 2x}{\cos^2 x} + \dfrac{\cos 3x}{\cos^3 x} + \dots + \dfrac{\cos nx}{\cos^n x} = \displaystyle\sum_{k=0}^{n} \dfrac{\cos kx}{\cos^k x}$, $x \neq \dfrac{p\pi}{2},\ p \in \mathbf{Z}$.

IX. 15. Evaluate the sum

$S_n = \dfrac{\sin\dfrac{a}{1 \cdot 2}}{\cos a \cos\dfrac{a}{2}} + \dfrac{\sin\dfrac{a}{2 \cdot 3}}{\cos\dfrac{a}{2}\cos\dfrac{a}{3}} + \dots + \dfrac{\sin\dfrac{a}{n(n+1)}}{\cos\dfrac{a}{n}\cos\dfrac{a}{n+1}} = \displaystyle\sum_{k=1}^{n} \dfrac{\sin\dfrac{a}{k(k+1)}}{\cos\dfrac{a}{k}\cos\dfrac{a}{k+1}}$,

where $n \in \mathbf{N}^*$.

IX. 16. Evaluate the sum

$S_n = \dfrac{\sin x}{\cos^2 x} + \dfrac{\sin 3x}{\cos^2 2x \cos^2 x} + \dots + \dfrac{\sin(2n-1)x}{\cos^2 nx \cos^2(n-1)x} =$

$= \displaystyle\sum_{k=1}^{n} \dfrac{\sin(2k-1)x}{\cos^2 kx \cos^2(k-1)x}$.

IX. 17. Evaluate the sum

$$S_n = \tan^2 \frac{x}{2} \tan x + 2 \tan^2 \frac{x}{2^2} \tan \frac{x}{2} + \dots + 2^{n-1} \tan^2 \frac{x}{2^n} \tan \frac{x}{2^{n-1}} =$$

$$= \sum_{k=1}^{n} 2^{k-1} \tan^2 \frac{x}{2^k} \tan \frac{x}{2^{k-1}} .$$

IX. 18. Calculate the sum

$$S_n = \csc 2x + \csc 4x + \dots + \csc 2^n x = \sum_{k=1}^{n} \csc 2^k x , \text{ for admissible value of the}$$
number $x \in \mathbf{R}$.

IX. 19. Calculate the sum

$$S_n = \frac{1}{2^2} \sec^2 \frac{x}{2} + \frac{1}{2^4} \sec^2 \frac{x}{2^2} + \dots + \frac{1}{2^{2n}} \sec^2 \frac{x}{2^n} = \sum_{k=1}^{n} \frac{1}{2^{2k}} \sec^2 \frac{x}{2^k} , \text{ for}$$

admissible value of the number $x \in \mathbf{R}$.

IX. 20. Evaluate the sum

$$S_n = \sum_{i=1}^{n} \left(\frac{1}{1+\tan x_i} + \frac{1}{1-\tan x_i} + \frac{1}{1+\cot x_i} + \frac{1}{1-\cot x_i} \right) \text{ for admissible value}$$

of the real numbers $x_1, x_2, \dots, x_n$ . .

VI. 21. Calculate the sum $S_n = \displaystyle\sum_{k=1}^{n} \arcsin \frac{\sqrt{k+1}-\sqrt{k}}{\sqrt{k+1}\sqrt{k+2}}$.

VI. 22. Calculate the sum

$$S_n = \frac{\arctan \dfrac{1}{\sqrt{3}}}{\arcsin \dfrac{\sqrt{3}}{2}} + \frac{\arctan \dfrac{1}{\sqrt{5}}}{\arcsin \dfrac{\sqrt{5}}{2}} + \dots + \frac{\arctan \dfrac{1}{\sqrt{2n+1}}}{\arcsin \dfrac{\sqrt{2n+1}}{n+1}} = \sum_{k=1}^{n} \frac{\arctan \dfrac{1}{\sqrt{2k+1}}}{\arcsin \dfrac{\sqrt{2k+1}}{k+1}} .$$

VI. 23. Calculate the sum $S_n = \tan \left(\displaystyle\sum_{k=1}^{n} \arctan \frac{k^2+k+2}{k^2+k} \right)$.

VI. 24. Calculate the sum $S_n = \tan \left(\displaystyle\sum_{k=1}^{n} \arctan \frac{2k}{k^4+k^2+2} \right)$.

IX. 25. Calculate the sum

$$S_n = \arctan\left(\frac{1}{2 \cdot 1^2}\right) + \arctan\left(\frac{1}{2 \cdot 2^2}\right) + \dots + \arctan\left(\frac{1}{2 \cdot n^2}\right) = \sum_{k=1}^{n} \arctan\left(\frac{1}{2 \cdot k^2}\right).$$

IX. 26. Calculate the sum

$$S_n = \arctan\left(\frac{1}{1+2x^2}\right) + \arctan\left(\frac{1}{1+6x^2}\right) + \dots + \arctan\left(\frac{1}{1+n(n+1)x^2}\right) =$$

$$= \sum_{k=1}^{n} \arctan\left(\frac{1}{1+k(k+1)x^2}\right).$$

IX. 27. Calculate the sum

$$S_n = \operatorname{arc\,cot} 3 + \operatorname{arc\,cot} 7 + \dots + \operatorname{arc\,cot}\left(1+n+n^2\right) = \sum_{k=1}^{n} \operatorname{arc\,cot}\left(1+k+k^2\right).$$

IX. 28. Calculate the sum

$$S_n = \operatorname{arc\,cot}\left(2 \cdot 1^2\right) + \operatorname{arc\,cot}\left(2 \cdot 2^2\right) + \dots + \operatorname{arc\,cot}\left(2 \cdot n^2\right) = \sum_{k=1}^{n} \operatorname{arc\,cot}\left(2 \cdot k^2\right).$$

IX. 29. In an arithmetic progression

$a_1, a_2, a_3, \dots, a_n$, $n \in \mathbf{N}^*$, where $r > 0$ is the ratio. Calculate

$$S_n = \arctan \frac{r}{1+a_1 a_2} + \arctan \frac{r}{1+a_2 a_3} + \dots + \arctan \frac{r}{1+a_{n-1} a_n}.$$

IX. 30. In an arithmetic progression $a_1, a_2, a_3, \dots, a_n$, $n \in \mathbf{N}^*$, where r

is the ratio, then calculate $S_n = \sum_{k=1}^{n} \arctan \dfrac{\left(a_{n+1} - a_k\right)^2}{a_k^2 - a_k a_{k+1} + a_{k+1}^2}$.

IX. 31. Calculate the sum

$$S_n = \operatorname{h} \cos\frac{x}{2} + \ln\cos\frac{x}{2^2} + \dots + \operatorname{h} \cos\frac{x}{2^n} = \sum_{k=1}^{n} \ln\cos\frac{x}{2^k}, \text{ for admissible value}$$

of the number $x \in \mathbf{R}$.

IX. 32. Calculate the sum

$$S_n = \ln(1 + 2\cos x) + \ln(1 + 2\cos 3x) + \dots + \ln\left(1 + 2\cos 3^{n-1} x\right) =$$

126

$$= \sum_{k=1}^{n} \ln\left(1 + 2\cos 3^{k-1} x\right), \text{ for all admissible value of the number } x \in \mathbf{R}.$$

IX. 33. Calculate the sums:

$$S_1 = \sin x \cos 2y + \sin 2x \cos 3y + \dots + \sin(n-1)x \cos ny = \sum_{k=2}^{n} \sin(k-1)x \cos ky,$$

$$S_2 = \cos x \sin 2y + \cos 2x \sin 3y + \dots + \cos(n-1)x \sin ny = \sum_{k=2}^{n} \cos(k-1)x \sin ky.$$

IX. 34. Calculate the sums:

a) $S_1 = \displaystyle\sum_{k=0}^{n} \cos\frac{(2k+1)\pi}{4n+3} = \cos\frac{\pi}{4n+3} + \cos\frac{3\pi}{4n+3} + \cos\frac{5\pi}{4n+3} + \dots + \cos\frac{(2n+1)\pi}{4n+3}$;

b) $S_2 = \displaystyle\sum_{k=0}^{n} \cos\frac{2k\pi}{4n+3} = 1 + \cos\frac{2\pi}{4n+3} + \cos\frac{4\pi}{4n+3} + \dots + \cos\frac{2n\pi}{4n+3}$, $n\hat{\text{I}}\text{N}$;

c) $S_3 = \displaystyle\sum_{k=1}^{2n} \cos\frac{(2k+1)\pi}{4n+3} = \cos\frac{\pi}{4n+3} + \cos\frac{3\pi}{4n+3} + \cos\frac{5\pi}{4n+3} + \dots + \cos\frac{(4n+1)\pi}{4n+3}$;

d) $S_4 = \displaystyle\sum_{k=1}^{2n+1} (-1)^{k+1} \cos\frac{k\pi}{4n+3}$, $n\hat{\text{I}}\text{N}$.

IX. 35. Calculate the sum

$$S_n = \tan 2x \sum_{k=1}^{n} \tan(2k+1)x \cdot \tan(2k-1)x \ .$$

Trigonometric Products

IX. 36. Calculate the product $\tan 1° \cdot \tan 2° \cdot \tan 2° \cdot \dots \cdot \tan 88° \cdot \tan 89°$.

IV. 37. Find a simply form for the expressions

$$E_1 = \cos x \cos 2x \cos 4x \cdot \dots \cdot \cos 2^n x = \prod_{k=0}^{n} \cos 2^k x \ , \ n \in \mathbf{N} \ \text{ and}$$

$$E_2 = \cos x \cos\frac{x}{2} \cos\frac{x}{2^2} \cdot \dots \cdot \cos\frac{x}{2^n} = \prod_{k=0}^{n} \cos\frac{x}{2^k} \ , \ n \in \mathbf{N} \ .$$

IX. 38. Evaluate the product

$$P_n = (1 + \sin x)(1 + \sin^2 x) \cdots (1 + \sin^{2^n} x) = \prod_{k=0}^{n} \left(1 + \sin^{2^k} x\right), \; n \in \mathbf{N} \; .$$

IX. 39. Evaluate the product

$$P_n = \left(1 - \tan^2 \frac{x}{2}\right)\left(1 - \tan^2 \frac{x}{4}\right) \cdots \left(1 - \tan^2 \frac{x}{2^n}\right) = \prod_{k=1}^{n} \left(1 - \tan^2 \frac{x}{2^k}\right), \; n \in \mathbf{N}^* \; .$$

IX. 40. Evaluate the product

$$P_n = (1 + \sec x)(1 + \sec 2x) \cdots (1 + \sec 2^{n-1} x) = \prod_{k=1}^{n} \left(1 + \sec 2^{k-1} x\right), \; n \in \mathbf{N}^* \; .$$

IX. 41. Evaluate the product

$$P_n = \left(\cos \frac{x}{2} + \cos \frac{y}{2}\right)\left(\cos \frac{x}{4} + \cos \frac{y}{4}\right) \cdots \left(\cos \frac{x}{2^n} + \cos \frac{x}{2^n}\right) =$$

$$= \prod_{k=1}^{n} \left(\cos \frac{x}{2^k} + \cos \frac{x}{2^k}\right), \; n \in \mathbf{N}^* \; .$$

IX. 42. Evaluate the product

$$P_n = (2 \cos x - 1) \cdot (2 \cos 2x - 1) \cdot \ldots \cdot (2 \cos 2^{n-1} x - 1) = \prod_{k=1}^{n} \left(2 \cos 2^k x - 1\right), \; n \in \mathbf{N}^* \; .$$

IX. 43. Evaluate the product

$$P_n = \left(2 \cos \frac{x}{2} - 1\right) \cdot \left(2 \cos \frac{x}{2^2} - 1\right) \cdot \ldots \cdot \left(2 \cos \frac{x}{2^n} - 1\right) = \prod_{k=1}^{n} \left(2 \cos \frac{x}{2^k} - 1\right), \; n \in \mathbf{N}^* \; .$$

IX. 44. Evaluate the product

$$P_n = \left(\frac{\cos 2a}{1 + \cos 2a}\right) \cdot \left(\frac{\cos 2^2 a}{1 + \cos 2^2 a}\right) \cdot \ldots \cdot \left(\frac{\cos 2^n a}{1 + \cos 2^n a}\right) = \prod_{k=1}^{n} \left(\frac{\cos 2^k a}{1 + \cos 2^k a}\right), \; n \in \mathbf{N}^* \; .$$

IX. 45. Evaluate the product

$$P_n = (\cos a + \cos b)(\cos 2a + \cos 2b) \cdot \ldots \cdot (\cos 2^n a + \cos 2^n b) = \prod_{k=0}^{n} \left(\cos 2^k a + \cos 2^k b\right)$$

where $n \in \mathbf{N}^*$.

IX. 46. Evaluate the product

$$P_n = \prod_{k=1}^{n} \frac{1 + \tan^2 2^k x}{\left(1 - \tan^2 2^k x\right)^2} , \quad |x| < \frac{\pi}{2^{n+2}} .$$

IX. 47. Consider the recurrence $A_{k+1} = 2\cos\theta \cdot A_k - A_{k-1}$, $k \in \mathbf{N}, n > 2$ where $A_1 = \cos\theta$ and $A_2 = \cos 2\theta$. Find A_k.

Using Complex Numbers for calculating Trigonometric Sums and Products

IX. 48. Calculate the sums

$$S_1 = 1 + \cos x + \cos 2x + \ldots + \cos nx = \sum_{k=0}^{n} \cos kx \text{ and}$$

$$S_2 = \sin x + \sin 2x + \ldots + \sin nx = \sum_{k=0}^{n} \sin kx.$$

IX. 49. Calculate the sums:

$$S_1 = \frac{1}{2}\cos x - \frac{1}{2^2}\cos 2x + \frac{1}{2^3}\cos 3x - \ldots - \frac{1}{2^{2n}}\cos 2nx = \sum_{k=1}^{2n} \frac{(-1)^{k-1}}{2^k}\cos kx,$$

$$S_2 = \frac{1}{2}\sin x - \frac{1}{2^2}\sin 2x + \frac{1}{2^3}\sin 3x - \ldots - \frac{1}{2^{2n}}\sin 2nx = \sum_{k=1}^{2n} \frac{(-1)^{k-1}}{2^k}\sin kx,$$

$$S_3 = \frac{1}{2}\cos x + \frac{1}{2^2}\cos 2x + \frac{1}{2^3}\cos 3x + \ldots + \frac{1}{2^{2n}}\cos 2nx = \sum_{k=1}^{2n} \frac{1}{2^k}\cos kx,$$

$$S_4 = \frac{1}{2}\sin x + \frac{1}{2^2}\sin 2x + \frac{1}{2^3}\sin 3x + \ldots + \frac{1}{2^{2n}}\sin 2nx = \sum_{k=1}^{2n} \frac{1}{2^k}\sin kx$$

IX. 50. Calculate the sums
$$S_1 = \cos x + 2\cos 2x + \ldots + n\cos 2nx \text{ and}$$
$$S_2 = \sin x + 2\sin 2x + \ldots + n\sin 2nx.$$

IX. 51. Evaluate the sums

$$S_1 = 1 + \frac{\cos x}{\cos x} + \frac{\cos 2x}{\cos^2 x} + \ldots + \frac{\cos nx}{\cos^n x} = \sum_{k=0}^{n} \frac{\cos kx}{\cos^k x} \text{ and}$$

$$S_2 = \frac{\sin x}{\cos x} + \frac{\sin 2x}{\cos^2 x} + \dots + \frac{\sin nx}{\cos^n x} = \sum_{k=0}^{n} \frac{\sin kx}{\cos^k x}, \quad x \neq \frac{\pi}{2} + p\pi, \, p \in \mathbf{Z}.$$

IX. 52. Evaluate the sums

$$S_1 = \cos x \cos x + \cos 2x \cos^2 x + \dots + \cos nx \cos^n x = \sum_{k=1}^{n} \cos kx \cos^k x \text{ and}$$

$$S_2 = \sin x \cos x + \sin 2x \cos^2 x + \dots + \sin nx \cos^n x = \sum_{k=1}^{n} \sin kx \cos^k x.$$

IX. 53. Prove the identities

$$\sum_{k=1}^{n-1} \cos \frac{k\pi}{n} = 0, \, n \in \mathbf{N}, n \geq 2 \text{ and } \sum_{k=1}^{n-1} \sin \frac{k\pi}{n} = \cot \frac{\pi}{2n}, \, n \in \mathbf{N}, n \geq 2 \, ;$$

IX. 54. Prove the identities:

a) $\tan \dfrac{\pi}{2n} \tan \dfrac{2\pi}{2n} \cdot \dots \cdot \tan \dfrac{(n-2)\pi}{2n} \tan \dfrac{(n-1)\pi}{2n} = 1$;

b) $\sin \dfrac{\pi}{4n} \sin \dfrac{3\pi}{4n} \cdot \dots \cdot \sin \dfrac{(2n-3)\pi}{4n} \sin \dfrac{(2n-1)\pi}{4n} = \dfrac{\sqrt{2}}{2^n}$.

IX. 55. Prove the identities:

a) $\displaystyle\prod_{k=0}^{n-1} \sin\left(x + \frac{k\pi}{n}\right) = \frac{\sin nx}{2^{n-1}}, n \in \mathbf{N}, n \geq 2, x \in \mathbf{R};$

b) $\displaystyle\prod_{k=0}^{n-1} \sin \frac{(2k+1)\pi}{n} = \frac{1}{2^{n-1}}, n \in \mathbf{N}, n \geq 2$.

IX. 56. Calculate the sums:

a) $\displaystyle\sum_{k=0}^{n} \binom{n}{k} \sin^{2k} x \cos^{2n-2k} x$;   **b)** $\displaystyle\sum_{k=0}^{n} \binom{n}{k}(-1)^k \cosh^{2k} x \cdot \sinh^{2n-2k} x$.

IX. 57. Prove the identities:

a) $1 - \binom{n}{2} + \binom{n}{4} - \binom{n}{6} + \dots = 2^{\frac{n}{2}} \cos \frac{n\pi}{4}$;

b) $\binom{n}{1} - \binom{n}{3} + \binom{n}{5} - \binom{n}{7} + \dots = 2^{\frac{n}{2}} \sin \frac{n\pi}{4}$;

c) $\left[1 - \binom{n}{2} + \binom{n}{4} - \binom{n}{6} + \dots\right]^2 + \left[\binom{n}{1} - \binom{n}{3} + \binom{n}{5} - \binom{n}{7} + \dots\right]^2 = 2^n$;

d) $\left[1-3\binom{n}{2}+9\binom{n}{4}-27\binom{n}{6}+...\right]^2+3\left[\binom{n}{1}-3\binom{n}{3}+9\binom{n}{5}-27\binom{n}{7}+...\right]^2=2^{2n}.$

IX. 58. Prove the identities:

a) $\binom{n}{0}+\binom{n}{3}+\binom{n}{6}+...=\dfrac{1}{3}\left(2^n+2\cos\dfrac{n\pi}{3}\right);$

b) $1-3\binom{n}{2}+9\binom{n}{4}-\mathbf{2}\ \binom{n}{6}+...=(-1)^n\,2^n\cos\dfrac{2n\pi}{3};$

c) $\binom{n}{1}-3\binom{n}{3}+9\binom{n}{5}-...=\dfrac{(-1)^{n+1}2^{n+1}}{\sqrt{3}}\sin\dfrac{2n\pi}{3}.$

IX. 59. Prove the identity

$$\binom{n}{1}-\frac{1}{3}\binom{n}{3}+\frac{1}{9}\binom{n}{5}-\frac{1}{27}\binom{n}{7}+...=\frac{2^n}{3^{\frac{n-1}{2}}}\sin\frac{n\pi}{6}.$$

IX. 60. Prove the identity $\displaystyle\sum_{k=0}^{3n}\binom{6n}{2k}(-3)^k=2^{6n}.$

IX. 61. Prove the identities:

a) $\binom{n}{0}+\binom{n}{4}+\binom{n}{8}+...=\dfrac{1}{2}\left(2^{n-1}+2^{\frac{n}{2}}\cos\dfrac{n\pi}{4}\right);$

b) $\binom{n}{1}+\binom{n}{5}+\binom{n}{9}+...=\dfrac{1}{2}\left(2^{n-1}+2^{\frac{n}{2}}\sin\dfrac{n\pi}{4}\right);$

c) $\binom{n}{2}+\binom{n}{6}+\binom{n}{10}+...=\dfrac{1}{2}\left(2^{n-1}-2^{\frac{n}{2}}\cos\dfrac{n\pi}{4}\right);$

d) $\binom{n}{3}+\binom{n}{7}+\binom{n}{1}+...=\dfrac{1}{2}\left(2^{n-1}-2^{\frac{n}{2}}\sin\dfrac{n\pi}{4}\right).$

IX. 62. Calculate the sums

$$S_1 = 1 + \binom{n}{1}\cos x + \binom{n}{2}\cos 2x + \ldots + \binom{n}{n}\cos nx \text{ and}$$

$$S_2 = \binom{n}{1}\sin x + \binom{n}{2}\sin 2x + \ldots + \binom{n}{n}\sin nx.$$

IX. 63. Calculate the sums

$$S_1 = \cos x + \binom{n}{1}\cos 2x + \binom{n}{2}\cos 3x + \ldots + \binom{n}{n}\cos(n+1)x \text{ and}$$

$$S_2 = \sin x + \binom{n}{1}\sin 2x + \binom{n}{2}\sin 3x + \ldots + \binom{n}{n}\sin(n+1)x.$$

ANSWERS

Trigonometric Circle

I. 1. a) $\sec 300° = \dfrac{1}{\cos\left(360° - 60°\right)} = \dfrac{1}{\cos 60°} = 2;$

b) $\cos 135° = \cos\left(90° + 45°\right) = -\sin 45° = -\dfrac{\sqrt{2}}{2};$

c) $\sin 330° = \sin\left(360° - 30°\right) = \sin\left(-30°\right) = -\dfrac{1}{2};$

d) $\csc\left(-120°\right) = \dfrac{1}{\sin\left(-120°\right)} = -\dfrac{1}{\sin\left(90 + 30°\right)} = -\dfrac{1}{\cos 30°} = -\dfrac{2}{\sqrt{3}};$

e) $\sec 750° = \dfrac{1}{\cos\left(720° + 30°\right)} = \dfrac{1}{\cos 30°} = \dfrac{2}{\sqrt{3}};$

f) $\tan\left(-300°\right) = \tan\left(-300° + 360°\right) = \tan 60° = \sqrt{3};$

g) $\sin\left(2190°\right) = \sin\left(6 \cdot 360° + 30°\right) = \sin 30° = \dfrac{1}{2};$

h) $\cos\left(-690°\right) = \cos\left(720° - 690°\right) = \cos 30° = \dfrac{\sqrt{3}}{2}.$

I. 2. a) Use $\sin 70° = \cos 20°$, 1;   **b)** Use $\cos 80° = \sin 10°$, 1.

I. 3. a) $\tan\alpha = \dfrac{\tan 45° - \tan 10°}{1 + \tan 45° \tan 10°} = \tan(45° - 10°) = \tan 35°.$ Therefore $\alpha = 35°$;

b) $\tan\alpha = \dfrac{\left(\tan 10° + \tan 11°\right) - \left(1 - \tan 10° \tan 11°\right)}{-\left(\tan 10° + \tan 11°\right) - \left(1 - \tan 10° \tan 11°\right)} =$

$= \dfrac{\dfrac{\tan 10° + \tan 11°}{1 - \tan 10° \tan 11°} - 1}{-\dfrac{\tan 10° + \tan 11°}{1 - \tan 10° \tan 11°} - 1} = \dfrac{1 - \tan 21°}{\tan 21° + 1} = \dfrac{\tan 45° - \tan 21°}{1 + \tan 45° \tan 21°} = \tan 24°.$

Therefore $\alpha = 24°.$ **I. 4.** $131°.$ **I. 5.** $311°.$

I. 7. $\sin\alpha = \dfrac{12}{13},\ \tan\alpha = -\dfrac{12}{5},\ \cot\alpha = -\dfrac{5}{12},\ \sec\alpha = -\dfrac{13}{5},\ \csc\alpha = \dfrac{13}{12}.$

I. 8. $\cos\alpha = -\sqrt{1 - \sin^2\alpha} = -\dfrac{12}{13},\ \tan\alpha = -\dfrac{5}{12},\ \cot\alpha = -\dfrac{12}{5},\ \sec\alpha = -\dfrac{13}{12},$

$\csc\alpha = \dfrac{13}{5}.$ **I. 9.** $\sin^2\alpha + \cos^2\alpha = 1$ and $\sin\alpha = -\dfrac{1}{2}\cos\alpha$. From

$\left(-\dfrac{1}{2}\cos\alpha\right)^2 + \cos^2\alpha = 1$ we obtain $\cos\alpha = \dfrac{2}{\sqrt{5}},\ \sin\alpha = -\dfrac{1}{\sqrt{5}},\ \cot\alpha = -2,$

$\sec\alpha = \dfrac{\sqrt{5}}{2},\ \csc\alpha = -\sqrt{5}$.

I. 10. $\sin a = -\sqrt{1 - \dfrac{2 + \sqrt{2}}{4}} = -\dfrac{\sqrt{2 - \sqrt{2}}}{2};\quad \tan a = -\dfrac{2\left(\sqrt{2} - 1\right)}{2} = 1 - \sqrt{2}$.

I. 11. $\sin\alpha = \dfrac{1}{2},\ \alpha = \dfrac{\pi}{6}.$ **I. 12.** $\cos(x + y) = \dfrac{16}{65}.$ **I. 13.** $\sin(x + y) = -\dfrac{63}{65}.$

I. 14. $\cos a = -\dfrac{4}{5},\ \sin\dfrac{a}{2} = \dfrac{3}{\sqrt{10}},\ \tan\dfrac{a}{2} = -3.$

I. 15. $\dfrac{\cot\alpha + \tan\alpha}{\cot\alpha - \tan\alpha} = \dfrac{1}{\cos 2\alpha} = \dfrac{169}{119}$.

I. 16. $\cos\alpha = -\dfrac{12}{13},\ \tan\alpha = -\dfrac{5}{12}$ and $\cot\alpha = -\dfrac{12}{5}.$ **I. 17.** $\sin\alpha = -\dfrac{1}{2}$.

I. 18. a) $\cos x = -\dfrac{2\sqrt{6}}{5}$; **b)** $\tan x = 2$; **c)** $\tan 2x = -\dfrac{\sqrt{15}}{8}$;

d) $\tan\dfrac{x}{2} = \sqrt{2}$ or $\tan\dfrac{x}{2} = 3 + 2\sqrt{2}$;

e) $1 + \dfrac{1 + \tan x}{1 - \tan x} = \dfrac{2\tan y}{\tan y - 1} \Rightarrow \tan x \tan y = 1 \Rightarrow \tan x = \cot y \Rightarrow$

$\tan x = \tan\left(\dfrac{\pi}{2} - y\right) \Rightarrow x + y = \dfrac{\pi}{2}$.

I. 19. $\cot(a + b) = -1$, $a + b = \arctan(-1) = \dfrac{3\pi}{4}$.

I. 20. $\sin\alpha = -\dfrac{1}{2}$, $\cos(x - y) = -\dfrac{29}{36}$.

I. 21. We use the formula $\cos^2\dfrac{\alpha}{2} = \dfrac{\cos\alpha + 1}{2}$. For each case we

have $\cos^2\dfrac{\alpha}{2} = \dfrac{\dfrac{2}{5} + 1}{2} = \dfrac{7}{10} \Rightarrow \sin^2\dfrac{\alpha}{2} = 1 - \cos^2\dfrac{\alpha}{2} = 1 - \dfrac{7}{10} = \dfrac{3}{10}$

and $\tan^2\dfrac{\alpha}{2} = \dfrac{\sin^2\dfrac{\alpha}{2}}{\cos^2\dfrac{\alpha}{2}} = \dfrac{3}{7}$. Similarly $\tan^2\dfrac{\beta}{2} = \dfrac{1}{3}$ and $\tan^2\dfrac{\gamma}{2} = \dfrac{2}{3}$.

$\tan^2\dfrac{\alpha}{2} + \tan^2\dfrac{\beta}{2} + \tan^2\dfrac{\gamma}{2} = \dfrac{10}{7}$.

I. 22. $\cos\dfrac{x - y}{2}\sin\dfrac{x + y}{2} = -\dfrac{21}{130}$ and $\cos\dfrac{x + y}{2}\cos\dfrac{x - y}{2} = -\dfrac{2}{130}$.

Dividing these two, we get $\tan\dfrac{x + y}{2} = \dfrac{7}{9}$ (1). We also have

$\sin^2\dfrac{x + y}{2} + \cos^2\dfrac{x + y}{2} = 1$ (2). Since $\dfrac{\pi}{2} < x < \pi$ and $-\dfrac{\pi}{2} < y < 0$,

then $0 < x + y < \pi$ or $0 < \dfrac{x + y}{2} < \dfrac{\pi}{2}$. From (1) and (2), we obtain

$\sin\dfrac{x + y}{2} = -\dfrac{7}{\sqrt{130}}$ and $\cos\dfrac{x + y}{2} = -\dfrac{9}{\sqrt{130}}$.

137

I. 23. Transforming $\sin x + \sin y = -\dfrac{27}{65}$ into a product, we obtain

$\cos\dfrac{x-y}{2}\sin\dfrac{x+y}{2}=-\dfrac{27}{130}$ (1). We have the system $\tan\dfrac{x+y}{2}=\dfrac{7}{9}$

and $\sin^2\dfrac{x+y}{2}+\cos^2\dfrac{x+y}{2}=1$. since $0<\dfrac{x+y}{2}<\dfrac{\pi}{2}$, we obtain

$\sin\dfrac{x+y}{2}=\dfrac{7}{\sqrt{130}}$. Replacing this in (1), we obtain $\cos\dfrac{x-y}{2}=-\dfrac{27}{7\sqrt{130}}$.

I. 24. Squaring the both sides, then we have

$\sin^2\dfrac{\alpha}{2}+2\sin\dfrac{\alpha}{2}\cos\dfrac{\alpha}{2}+\cos^2\dfrac{\alpha}{2}=1.96$. Since $\sin 2a = 2\sin a\cos a$, then

$\sin x = 0.96$.

I. 25. $\dfrac{4\tan\dfrac{\alpha}{2}-3(1-\tan^2\dfrac{\alpha}{2})}{8\tan\dfrac{\alpha}{2}-5(1-\tan^2\dfrac{\alpha}{2})}=\dfrac{4\sqrt{3}-3(1-3)}{8\sqrt{3}-5(1-3)}=\dfrac{3+2\sqrt{3}}{5+4\sqrt{3}}=\dfrac{9+2\sqrt{3}}{23}$

$\dfrac{2\sin\alpha-3\cos\alpha}{4\sin\alpha-5\cos\alpha}=\dfrac{2\tan\alpha-3}{4\tan\alpha-5}=\dfrac{2\sqrt{3}-3}{4\sqrt{3}-5}=\dfrac{9-2\sqrt{3}}{24}$.

I. 26. a) $A=-\dfrac{5}{6}$;   **b)** $B=\dfrac{7\sqrt{6}-16}{38}$;   **c)** $C=\dfrac{5}{9}$;   **d)** $D=-\dfrac{9}{13}$;

e) $\sin x=\dfrac{4}{5}$, $\cos x=\dfrac{3}{5}$, $E=2$;   **f)** $F=\dfrac{11}{7}$;   **g)** $G=\sqrt{3}$;   **h)** $H=\dfrac{17}{23}$.

I. 27. $\tan a+\cot a=3$, $\dfrac{\sin a}{\cos a}+\dfrac{\cos a}{\sin a}=3\Leftrightarrow \sin 2a=\dfrac{2}{3}$.

a) $E^2=1+2\sin a\cos a\Leftrightarrow E=-\dfrac{\sqrt{15}}{3}$.

b) $B=\dfrac{\sin^2 a}{\cos^2 a}+\dfrac{\cos^2 a}{\sin^2 a}=\dfrac{\left(\sin^2 a+\cos^2 a\right)^2-2\sin^2 a\cos^2 a}{\cos^2 a\sin^2 a}=7$

c) $C=-6\sqrt{15}$.

I. 28. a) $\tan^2 x+\cot^2 x=(\tan x+\cot x)^2-2=23$;

b) $\tan^3 x+\cot^3 x=(\tan x+\cot x)(\tan^2 x-1+\cot^2 x)=110$;

c) $\tan^4 x + \cot^4 x = \left(\tan^2 x + \cot^2 x\right)^2 - 2 = 527$.

I. 29. a) $\left(\sin x + \cos x\right)^2 = p^2$, then $1 + 2\sin x \cos x = p^2$.

$\sin^3 x + \cos^3 x = \left(\sin x + \cos x\right)\left(\sin^2 x - \sin x \cos x + \cos^2 x\right) =$

$= p\left(1 - \sin x \cos x\right) = \dfrac{3p - p^3}{2}$; **b)** $0.25\left(1 + 6p^2 - 3p^4\right)$; **c)** $\dfrac{2p\left(2 - p^2\right)}{p^2 - 1}$.

I. 30. $\dfrac{p\left(p^2 + 1\right)}{2}$. **I. 31.** $1 + \sin 2\alpha = p^2$, $\dfrac{1}{\sin 2\alpha} = q$.

I. 32. Denote $\sin \alpha = x$ and $\cos \alpha = y$. Solving the system

$\begin{cases} 3x + 4y = -5 \\ x^2 + y^2 = 1 \end{cases}$, we obtain the solution $x = -\dfrac{3}{5}$ and $y = -\dfrac{4}{5}$.

Therefore $\tan \alpha = \dfrac{3}{4}$ and $\cot \alpha = \dfrac{4}{3}$.

I. 33. Let $A = \left(1 + \cot a\right)\sin^3 a + \left(1 + \tan a\right)\cos^3 a =$

$= \left(1 + \dfrac{\cos a}{\sin a}\right)\sin^3 a + \left(1 + \dfrac{\sin a}{\cos a}\right)\cos^3 a =$

$= \left(\sin a + \cos a\right)\sin^2 a + \left(\cos a + \sin a\right)\cos^2 a = \sin a + \cos a$.

Squaring $A = \sin a + \cos a$ in both sides we obtain

$A^2 = \left(\sin a + \cos a\right)^2 \Leftrightarrow A^2 = 1 + \sin a \cos a$ and then $A = \dfrac{\sqrt{6}}{2}$.

I. 34. $\cos t = \sqrt{1 - \left(\dfrac{a^2 - b^2}{a^2 + b^2}\right)^2} \Leftrightarrow \cos t = \dfrac{2ab^2}{a^2 + b^2}$ and

$\tan t = \dfrac{\dfrac{a^2 - b^2}{a^2 + b^2}}{\dfrac{2ab^2}{a^2 + b^2}} = \dfrac{a^2 - b^2}{a^2 + b^2} \cdot \dfrac{a^2 + b^2}{2ab^2} = \dfrac{a^2 - b^2}{2ab^2}$.

I. 35. $\dfrac{aA + bB}{aB + bA} = \dfrac{\dfrac{a}{b} \cdot \dfrac{A}{B} + 1}{\dfrac{a}{b} + \dfrac{A}{B}} = \dfrac{\dfrac{\sin(x - \alpha)}{\sin(x - \beta)} \cdot \dfrac{\cos(x - \beta)}{\cos(x - \alpha)} + 1}{\dfrac{\sin(x - \alpha)}{\sin(x - \beta)} + \dfrac{\cos(x - \beta)}{\cos(x - \alpha)}} =$

$$= \frac{\sin(x-\alpha)\cos(x-\beta)+\sin(x-\beta)\cos(x-\alpha)}{\sin(x-\alpha)\cos(x-\alpha)+\sin(x-\beta)\cos(x-\beta)} = \frac{2\sin(2x-\alpha-\beta)}{\sin(2x-2\alpha)+\sin(2x-2\beta)} =$$

$$= \frac{2\sin(2x-\alpha-\beta)}{2\sin(2x-\alpha-\beta)\cos(\alpha-\beta)} = \frac{1}{\cos(\alpha-\beta)} = \sec(\alpha-\beta).$$

I. 36. $p^2 + q^2$.

I. 37. We have successively $\cos 2\chi = \dfrac{\cos 2\alpha}{\cos 2\beta}$, $1 - 2\sin^2 \chi = \dfrac{\cos 2\alpha}{\cos 2\beta}$,

$$2\sin^2 \chi = 1 - \frac{\cos 2\alpha}{\cos 2\beta}, \quad \frac{1}{\sin 2\chi} = \frac{2\cos 2\beta}{\cos 2\beta - \cos 2\alpha} = \frac{2\cos 2\beta}{2\sin(\alpha+\beta)\sin(\alpha-\beta)} =$$

$$= \frac{\cos(\alpha+\beta)\cos(\alpha-\beta)+\sin(\alpha+\beta)\sin(\alpha-\beta)}{\sin(\alpha+\beta)\sin(\alpha-\beta)} = 1 + \cot(\alpha+\beta)\cot(\alpha-\beta)$$

We used the formulae: $\cos 2x = 1 - 2\sin^2 x$,

$\cos(x-y) = \cos x \cos y + \sin x \sin y$, and $\cos x - \cos y = 2\sin(x+y)\sin(y-x)$

I. 38. $\dfrac{\cos\alpha}{\cos\beta} + \dfrac{\sin\alpha}{\sin\beta} = -1 \Rightarrow \cos\alpha\sin\beta + \cos\beta\sin\alpha = -\cos\beta\sin\beta$

$\Leftrightarrow -2\sin(\alpha+\beta) = \sin(2\beta)$.

$$\frac{\cos^3 \beta}{\cos\alpha} + \frac{\sin^3 \beta}{\sin\alpha} - 1 = \frac{1}{\cos\alpha\sin\alpha}\left(\cos\alpha\sin^3 \beta + \cos^3 \beta\sin\alpha - \cos\alpha\sin\alpha\right) =$$

$$= 2\csc(2\alpha)\left[\cos\alpha\sin\beta(1 - \cos^2 \beta) + \cos\beta\sin\alpha(1 - \sin^2 \beta) - \cos\alpha\sin\alpha\right] =$$

$$= 2\csc(2\alpha)\left(-\cos\alpha\cos^2 \beta\sin\beta - \cos\beta\sin\alpha\sin^2 \beta + \cos\alpha\sin\beta + \cos\beta\sin\alpha - \cos\alpha\sin\alpha\right) =$$

$$= 2\csc(2\alpha)\left[-(\cos\alpha\cos\beta + \sin\alpha\sin\beta)\cos\beta\sin\beta + \sin(\alpha+\beta) - \cos\alpha\sin\alpha\right] =$$

$$= \csc(2\alpha)\left[-\cos(\alpha-\beta)\sin(2\beta) - \sin(2\alpha) + 2\sin(\alpha+\beta)\right] =$$

$$= \csc(2\alpha)\left[2\cos(\alpha-\beta)\sin(\alpha+\beta) - \sin(2\alpha) - \sin(2\beta)\right] =$$

$$= 2\csc(2\alpha)\left[\cos(\alpha-\beta)\sin(\alpha+\beta) - \cos\left(\frac{2\alpha-2\beta}{2}\right)\sin\left(\frac{2\alpha+2\beta}{2}\right)\right] =$$

$$= 2\csc(2\alpha)\left[\cos(\alpha-\beta)\sin(\alpha+\beta) - \cos(\alpha-\beta)\sin(\alpha+\beta)\right] = 0 .$$

I. 39. Adding and subtracting the given relations, we obtain:

$2a\cos\dfrac{\alpha+\beta}{2}\cos\dfrac{\alpha-\beta}{2} + 2b\sin\dfrac{\alpha+\beta}{2}\cos\dfrac{\alpha-\beta}{2} = 2c$. Squaring this relation, it yields to (1) $\cos^2\dfrac{\alpha-\beta}{2} = \dfrac{c^2}{a^2\cos^2\dfrac{\alpha+\beta}{2} + b^2\sin^2\dfrac{\alpha+\beta}{2} + ab\sin(\alpha+\beta)}$

Also $-2a\sin\dfrac{\alpha+\beta}{2}\sin\dfrac{\alpha-\beta}{2} + 2b\sin\dfrac{\alpha-\beta}{2}\cos\dfrac{\alpha+\beta}{2} = 0 \Rightarrow \sin\dfrac{\alpha-\beta}{2} = 0$ or

$\tan\dfrac{\alpha+\beta}{2} = \dfrac{b}{a}$.

Then (2) $\cos^2\dfrac{\alpha+\beta}{2} = \dfrac{1}{1+\tan^2\dfrac{\alpha+\beta}{2}} = \dfrac{a^2}{a^2+b^2}$,

(3) $\sin^2\dfrac{\alpha+\beta}{2} = \dfrac{\tan^2\dfrac{\alpha+\beta}{2}}{1+\tan^2\dfrac{\alpha+\beta}{2}} = \dfrac{b^2}{a^2+b^2}$, and

(4) $\sin(\alpha+\beta) = \dfrac{2\tan\dfrac{\alpha+\beta}{2}}{1+\tan^2\dfrac{\alpha+\beta}{2}} = \dfrac{2ab}{a^2+b^2}$. From (1), (2), (3), and (4) we obtain

$\cos^2\dfrac{\alpha-\beta}{2} = \dfrac{c^2}{a^2+b^2}$.

I. 40. $\sin^3\theta - \cos^3\theta = a \Leftrightarrow (\sin\theta - \cos\theta)(1+\sin\theta\cos\theta) = a$, therefore

(1) $\sin\theta\cos\theta = \dfrac{b-a}{a}$. Squaring the relation $\sin\theta - \cos\theta = a$, we obtain

(2) $\sin\theta\cos\theta = \dfrac{1-a^2}{2}$. Therefore from (1) and (2) $a^3 - 3a + 2b = 0$.

I. 41. $a^3 - 3a + 2b = 0$. **I. 42.** $a^5 - 5a - 4b = 0$. **I. 43.** $a^2 - b - 2 = 0$.

I. 44. $a^2 - b^2 + ac = 0$. **I. 45.** The two given relations form a system of two linear equations with the unknowns $\sin^2\theta$ and $\sin\theta$. Finally $a+b+c=0$.

I. 46. From the second relation $\dfrac{1}{y} = \dfrac{2}{x} + \cot\theta$. Finally

$x^2y^2 + x^2 + 3y^2 - 4xy = 0$. **I. 47.** $\left(x^2 + y^2 + 2ax\right)^2 - 4a^2\left(x^2 + y^2\right) = 0$.

I. 48. $a^4 + 16\,b^2 = 4a^2b^2$. **I. 49.** $\tan c = \dfrac{ab}{b-a}$.

I. 50. Subtract relations (1) and (2) and obtain (4) $x\dfrac{\cos^4 \alpha}{\sin^2 \alpha} + y\dfrac{\cos^4 \beta}{\sin^2 \beta} = 0$.

Substitute (3) in (4) and obtain $x^3 \cos^4 \alpha + y^3 \cos^4 \beta = 0 \Rightarrow$

$\dfrac{\cos^2 \alpha}{\cos^2 \beta} = \sqrt{-\dfrac{y^3}{x^3}}$. Denote $k = \sqrt{-\dfrac{y^3}{x^3}}$, then $\dfrac{1-\sin^2 \alpha}{1-\sin^2 \beta} = k \Rightarrow$

(5) $\sin^2 \alpha - k \sin^2 \beta = 1 - k$.

From (3) we obtain $\dfrac{\sin^2 \alpha}{\sin^2 \beta} = \dfrac{y^2}{x^2} \Leftrightarrow \dfrac{\sin^2 \alpha - k\sin^2 \beta}{\sin^2 \beta} = \dfrac{y^2 - kx^2}{x^2}$.

Then using (5) we obtain $\dfrac{1-k}{\sin^2 \beta} = \dfrac{y^2 - kx^2}{x^2}$. Next use (3) and (1).

I. 51. $y^2 = 4a^2 - 4ax$.

I. 52. Squaring (1) and (2), and adding and subtracting these relations,

we obtain (4) $\cos(\alpha + \beta) = \dfrac{b^2 - a^2}{a^2 + b^2}$, (5) $\cos(\alpha - \beta) = \dfrac{a^2 + b^2 - 2}{2}$.

Multiplying (1) and (2) side by side, we obtain

$\sin 2\alpha + \sin 2\beta + 2\sin(\alpha + \beta) = 2ab \Rightarrow$

$\sin(\alpha + \beta)\cos(\alpha - \beta) + \sin(\alpha + \beta) = ab$.

Now using (4) and (5), we obtain (6) $\sin(\alpha + \beta) = \dfrac{2ab}{a^2 + b^2}$.

Relation (3) is equivalent to $\dfrac{\sin(\alpha + \beta)}{\cos \alpha \cos \beta} = \dfrac{2\sin(\alpha + \beta)}{\cos(\alpha + \beta) + \cos(\alpha - \beta)} = c$.

Finally, using (4), (5), and (6) we obtain

$c\left(a^2 + b^2 + 2a\right)\left(a^2 + b^2 - 2a\right) = 8ab$.

I. 53. $\dfrac{x^2}{a^2} + \dfrac{y^2}{b^2} + \dfrac{z^2}{c^2} = 1$.

I. 54. $\tan^2 x = \dfrac{1 - \dfrac{b}{a}}{1 + \dfrac{b}{a}}$, $\cos^2 y = \dfrac{1}{2} + \dfrac{b}{2a} \Rightarrow \dfrac{b}{a} = \cos 2y$. Substituting

$\dfrac{b}{a} = \cos 2y$ in the first relation, we obtain $\tan x = \pm \tan y$.

I. 55. From $t = \tan x + \cot x$, we obtain $\dfrac{1}{t} = \sin x \cos x$ and

$\sin x + \cos x = \sqrt{1 + \dfrac{2}{t}}$. The proof could be done using mathematical induction.

For $n = 1$ we have

$a_2 = \sqrt{1 + \dfrac{2}{t}} a_1 - \dfrac{1}{t} a_0 = (\sin x + \cos x)(\sin x + \cos x) - 2 \sin x \cos x = 1$, which is

true.

We assume that $a_n = \sqrt{1 + \dfrac{2}{t}} a_{n-1} - \dfrac{1}{t} a_{n-2}$ is true, we have to prove that

$a_{n+1} = \sqrt{1 + \dfrac{2}{t}} a_n - \dfrac{1}{t} a_{n-1}$ is also true.

O the other hand $a_{n+1} = \sin^{n+1} x + \cos^{n+1} x =$

$= \left(\sin^n x + \cos^n x\right)(\sin x + \cos x) - \sin x \cos x\left(\sin^{n-1} x + \cos^{n-1} x\right) =$

$= a_n \sqrt{1 + \dfrac{2}{t}} - \dfrac{1}{t} a_{n-1}$.

I. 56. Squaring $\sin x + \cos x = t$, $\sin x \cos x = \dfrac{t^2 - 1}{2}$, then $a_3 = \dfrac{3t - t^3}{2}$ and

$a_4 = \dfrac{1 + 2t^2 - t^4}{2}$. $a_{n+1} = \sin^{n+1} x + \cos^{n+1} x =$

$= \left(\sin^n x + \cos^n x\right)(\sin x + \cos x) - \sin x \cos x\left(\sin^{n-1} x + \cos^{n-1} x\right) =$

$= a_n t - \dfrac{t^2 - 1}{2} a_{n-1}$.

I. 57. $a_1 = \dfrac{2}{t^2 - 1}$ and $a_2 = \dfrac{2(1 + 2t^2 - t^4)}{(t^2 - 1)^2}$,

$a_{n+1} = \tan^{n+1} x + \cot^{n+1} x =$

$= (\tan^n x + \cot^n x)(\tan x + \cot x) - \tan x \cot x (\tan^{n-1} x + \cot^{n-1} x) =$

$= a_n \dfrac{2}{t^2 - 1} - a_{n-1}.$

I. 58. Use $a + \dfrac{1}{a} \geq 2$, when a is a positive number and $a + \dfrac{1}{a} \leq -2$, when a is a negative number.

$a_{n+1} = \tan^{n+1} x + \cot^{n+1} x =$

$= (\tan^n x + \cot^n x)(\tan x + \cot x) - \tan x \cot x (\tan^{n-1} x + \cot^{n-1} x) =$

$= a_n t - a_{n-1}.$

I. 59. $a_n = \sin(2n+1)x = \sin(2n-1)\cos 2x + \sin 2x \cos(2n-1) =$

$= \sin(2n-1)(1 - 2\sin^2 x) + \sin x [\cos 2nx + \cos 2(n-1)] =$

$= (1 - 2t^2)a_{n-1} + t(b_n + b_{n-1}).$

$b_n = \cos 2nx = \cos 2(n-1)\cos 2x - \sin 2(n-1)\sin 2x =$

$= \cos 2(n-1)(1 - 2\sin^2 x) - \dfrac{1}{2}[\cos 2(n-2)x - \cos 2nx] =$

$= (1 - 2t^2)b_{n-1} - \dfrac{1}{2}(b_{n-2} - b_n).$

UNIT II

Functions, graphs, transformations

II. 1. a) $(x, y) \to (x, 2y + 2)$; **b)** $(x, y) \to \left(x + \dfrac{3\pi}{4}, -2y - 1\right)$;

c) $(x, y) \to \left(\dfrac{1}{3}x - \dfrac{\pi}{6}, y + \dfrac{2}{3}\right)$; **d)** $(x, y) \to \left(\dfrac{1}{3}x - \dfrac{\pi}{3}, y + \dfrac{1}{2}\right)$;

e) $(x, y) \to \left(\dfrac{2}{3}x + \dfrac{\pi}{3}, -3y - 6\right)$; **f)** $(x, y) \to (x - \pi, y - 7)$;

g) $(x, y) \to (2x, y - 5)$; **h)** $(x, y) \to \left(\dfrac{3}{2}x - \dfrac{3\pi}{4}, 3y + 8\right)$;

i) $(x, y) \to \left(x + \dfrac{\pi}{2}, y + 5\right)$; **j)** $(x, y) \to \left(2x - 2\pi, 2y - \dfrac{1}{2}\right)$;

k) $(x, y) \to \left(\dfrac{2}{3}x, y - \dfrac{1}{2}\right)$; **l)** $(x, y) \to \left(x + \dfrac{3\pi}{4}, -2y - 1\right)$;

m) $(x, y) \to \left(\dfrac{1}{3}x - \dfrac{\pi}{6}, y + \dfrac{2}{3}\right)$; **n)** $(x, y) \to (2x - 2\pi, -3y - 1)$.

II. 2. $f(x) = \sqrt{\sin^4 x + 4\left(1 - \sin^2 x\right)} + \sqrt{\cos^4 x + 4\left(1 - \cos^2 x\right)} + 1 =$
$\sqrt{\left(\sin^2 x - 2\right)^2} + \sqrt{\left(\cos^2 x - 2\right)^2} + 1 = 2 - \sin^2 x + 2 - \cos^2 x + 1 = 4$.

II. 3. Since $\{x + 1\} = \{x\}$ for any real number x, we have
$f(x + 1) = 2\{x + 1\} - \cos 2\pi(x + 1) = 2\{x\} - \cos(2\pi x + 2\pi) = 2\{x\} - \cos 2\pi x = f(x)$.

II. 4. From the first equation $\cos^2 x - \sin^2 a > 0$. The second equation has real roots if $\sin^2 a - \cos^2 x > 0$, which is not possible.

II. 5. $\Delta = 4\cos^2 a - 4(1 - \sin a)(1 + \sin a) = 4(\cos^2 a - 1 + \sin^2 a) = 0$, since $1 - \sin a \geq 0$, then $f(x) \geq 0$.

II. 6. We have:

$$f(x) = \frac{m(\sin x \cos^3 x + \cos x \sin^3 x) + \sin 3x \cos x - \cos 3x \sin x}{\sin x \cos x} =$$

$$= \frac{m \sin x \cos x(\cos^2 x + \sin^2 x) + \sin(3x - x)}{\sin x \cos x} = = \frac{m \sin x \cos x + \sin 2x}{\sin x \cos x}$$

$$= \frac{m \sin x \cos x + 2 \sin x \cos x}{\sin x \cos x} = m + 2.$$

II. 7. a) $f(x) = -\dfrac{1}{2}\sin x + \dfrac{1}{4}\cos x$; **b)** $f(x) = 1$.

II. 8. a) $f(x) = 3 - 2(1 + \sin x)^2$, $\{y \in \mathbf{R} \mid -5 \leq x \leq 3\}$;

b) $f(x) = \dfrac{1}{2}(1 - \cos 2x)$, $\{y \in \mathbf{R} \mid 0 \leq x \leq 1\}$;

c) Let $\tan x = t$, where $t > 0$. We observe that $t + \dfrac{3}{t} \geq 2\sqrt{3}$. Indeed $t^2 - 2t\sqrt{3} + 3 = (t - \sqrt{3})^2 \geq 0$. Therefore the range of the given function is $f(x) \in (2\sqrt{3}, \infty)$; **d)** $f(x) = 5 + \left(\dfrac{2}{\sin 2x}\right)^2 \geq 9$;

e) Using the inequality $\dfrac{1}{a} + \dfrac{1}{b} \geq \dfrac{2}{\sqrt{ab}}$, $a, b > 0$, we have

$$f(x) \geq \frac{2}{\sqrt{\cos\left(\dfrac{\pi}{6} - x\right) \cdot \cos\left(\dfrac{\pi}{6} + x\right)}} \geq \frac{2}{\sqrt{\dfrac{1}{4} + \dfrac{\cos 2x}{2}}}.$$ On the other

hand $0 \leq x \leq \dfrac{\pi}{3} \Rightarrow 0 \leq 2x \leq \dfrac{2\pi}{3} \Rightarrow \dfrac{-1}{2} \leq \cos 2x \leq 1$, therefore

$$0 \leq \sqrt{\dfrac{1}{4} + \dfrac{\cos 2x}{2}} \leq \dfrac{\sqrt{3}}{2}.$$ Finally $f(x) \geq \dfrac{4}{\sqrt{3}}$ and $D = \left[\dfrac{4}{\sqrt{3}}, \infty\right)$. Notice:

$f(x) = \dfrac{4}{\sqrt{3}}$ for $x = 0$.

II. 9. a) $D = \mathbf{R} \setminus \left\{ \dfrac{k\pi}{2}, k \in \mathbf{Z} \right\}$. $f(x) = \dfrac{1 + \cos x}{1 + \sin x} \cdot \tan^2 x > 0$, therefore $R = (0, \infty)$;

b) $D = \mathbf{R}$. $f(x) \in [1, 5]$.

II. 10. Let $\theta \in \left(-\dfrac{\pi}{2}, \dfrac{\pi}{2} \right)$ be an angle, such that $\tan\theta = x \in \mathbf{R}$. We have

$$\left| \frac{4x(1 - x^2)}{(1 + x^2)^2} \right| = \left| 2 \frac{2\tan\theta}{1 + \tan^2\theta} \frac{1 - \tan^2\theta}{1 + \tan^2\theta} \right| = |2\sin 2\theta \cos 2\theta| = |\sin 4\theta| \le 1.\ \text{Also}$$

$$\left| \frac{8x(1 - x^2)(x^4 - 6x^2 + 1)}{(1 + x^2)^4} \right| = \left| 4 \frac{2\tan\theta}{1 + \tan^2\theta} \frac{1 - \tan^2\theta}{1 + \tan^2\theta} \frac{\tan^4\theta - 6\tan^2\theta + 1}{(1 + \tan^2\theta)^2} \right| =$$

$$= \left| 4\sin 2\theta \cos 2\theta \left(\sin^4\theta - 6\sin^2\theta\cos^2\theta + \cos^4\theta \right) \right| =$$

$$= \left| 2\sin 4\theta \left[\left(\sin^2\theta + \cos^2\theta \right)^2 - 8\sin^2\theta\cos^2\theta \right] \right| =$$

$$= \left| 2\sin 4\theta \left(1 - 2\sin^2 2\theta \right) \right| = |2\sin 4\theta \cos 4\theta| = |\sin 8\theta| \le 1.$$

Therefore $-\dfrac{1}{4} \le f(x) \le \dfrac{1}{4}$ and $-\dfrac{1}{8} \le g(x) \le \dfrac{1}{8}$.

II. 11. $f(x) = 2\cos\left(3x - \dfrac{\pi}{3} \right) + 1$. Period is $\dfrac{2\pi}{3}$, range is $f(x) \in [-1, 3]$.

II. 12. a) $f(-x) = 2(-x)^4 - 3\cos(-x) = 2x^4 - 3\cos x = f(x)$, it is even;

b) $f(-x) = 3\tan^2(-x) - \cos(-x) + 2(-x)^2 = 3\tan^2 x - \cos x + 2x^2 = f(x)$, it is even;

c) $f(-x) = \cos(-x) - \sin^2(-x) + 2\cot^4(-x) = \cos x - \sin^2 x + 2\cot^2 x = f(x)$, it is even; **d)** $f(-x) = -f(x)$, it is odd;

e) $f(-x) = -\sin x + 2\tan x + 4\cos x \ne f(x)$, it is neither.

II. 13. If P is a period of the function, then $f(x + P) = f(x)$, or $\cos 15(x + P) + \sin 6(x + P) = \cos 15x + \sin 6x$ for any $x \in \mathbf{R}$.

In particular, for $x = 0$ and $x = \dfrac{\pi}{3}$ we obtain the system

$$\begin{cases} \sin 6P + \cos 15P = 1 \\ \sin 6P - \cos 15P = -1 \end{cases} \Leftrightarrow \begin{cases} \sin 6P = 0 \\ \cos 15P = 1 \end{cases}.\ \text{From } \sin 6P = 0,\ P = \frac{k_1 \pi}{6},\ k_1 \in \mathbf{Z}$$

and from $\cos 15P = 1$, $P = \dfrac{2k_2\pi}{15}$, $k_2 \in \mathbf{Z}$. Then $\dfrac{k_1\pi}{6} = \dfrac{2k_2\pi}{15} \Rightarrow k_1 = \dfrac{4k_2}{5}$.

Since $k_1 \in \mathbf{Z}$, it yields $k_2 = 5k$, where $k \in \mathbf{Z}$. Therefore a general period of the function f is $P = \dfrac{2k\pi}{3}$, $k \in \mathbf{Z}$. For $k = 1$, the principal period of the given function is $\dfrac{2\pi}{3}$.

II. 14. a) $\dfrac{\pi}{5}$; b) $\dfrac{\pi}{2}$; c) π; d) 6π; e) $\tan 2x = \tan(2x + 2P) \Rightarrow 2P = \pi$,

therefore the period for $\tan 2x$ is $\dfrac{\pi}{2}$. $\cot 3x = \cot(3x + 3T)$, $3P = \pi$,

therefore the period for $\cot 3x$ is $\dfrac{\pi}{3}$. Finally, the least common multiple

of these periods is π; f) 24; g) $\dfrac{\pi}{2}$; h) $\dfrac{2\pi}{3}$; i) $f(x) = 2(1 + \sin 2x)$,

the period is π;

j) $f(x + 4\pi) = \dfrac{\sin 2(x + 4\pi)}{2\cos\dfrac{x + 4\pi}{2} + \tan\dfrac{x + 4\pi}{2}} = f(x)$.

II. 15. We assume that P is a period of the function. Then
$f(x + P) = f(x)$ for any $x \in \mathbf{R}$. For $x = 0$, $2\sin 4P + \cos P\sqrt{2} = 1$ and for
$x = -P$, $f(0) = f(-P) \Rightarrow -2\sin 4P + \cos P\sqrt{2} = 1$. Therefore $\sin 4P = 0$
and $\cos P\sqrt{2} = 1$, which yields $P = \dfrac{k\pi}{4}$ and $P\sqrt{2} = k'\pi$ or $k = 2k'\sqrt{2}$,
where $k, k' \in \mathbf{Z}$, a false equality.

II. 16. $f(x) \geq 2\sqrt{3}$; $f(x) = 2\sqrt{3}$ when $\tan x = \sqrt{3}$.

II. 17. Let $A = \dfrac{\cot\alpha - \tan\alpha}{1 + \cos 4\alpha} = \dfrac{\cos^2\alpha - \sin^2\alpha}{2\cos^2\dfrac{4\alpha}{2}\sin\alpha\cos\alpha} = \dfrac{\cos 2\alpha}{\cos^2 2\alpha \sin 2\alpha} = \dfrac{2}{\sin 4\alpha}$.

A has a minimum value if $\sin 4\alpha$ has a maximum, this is for $\sin 4\alpha = 1$,
thus $\alpha = \dfrac{\pi}{8}$.

II. 18. $f(x) = \dfrac{3}{2}\sin 2x + \cos 2x + 3$. It is known that

$A\sin\alpha + B\cos\alpha + C \in \left[-\sqrt{A^2+B^2}+C, \ \sqrt{A^2+B^2}+C\right]$. Therefore

$f(x) \in \left[\dfrac{6-\sqrt{13}}{2}, \dfrac{6+\sqrt{13}}{2}\right]$.   **II. 19.** 4.   **II. 20.** $y = 2$.   **II. 21.** $y = f(x)$.

II. 22. $y = 2$.   **II. 23.** $\left(-\dfrac{5\sqrt{3}}{2}, \dfrac{5}{2}\right)$.   **II. 24.** 0.2 inch.   **II. 25.** –1.3 inch

II. 26. a) Amplitude of 0.64, period 60°, horizontal shift to the left 5.4°, vertical translation of 2 units down ;

b) Amplitude of 4, period of 180°, horizontal shift to the left 2°, vertical translation of 3 units up ;

c) Amplitude of 3, reflection in x-axis, period of $\dfrac{\pi}{2}$, horizontal shift to

the left $\dfrac{\pi}{4}$, vertical translation of 5 units up;

d) Amplitude of 3, reflection in y-axis and x-axis, period of $\dfrac{\pi}{2}$,

horizontal shift to the left $\dfrac{\pi}{4}$, vertical translation of 3 units down.

II. 27. a) $y = 2.5\sin(6x + 32.4) + 5.5$.   **II. 28.** $y = -2\sin\pi x - 2$.

II. 29. $y = 2\sin 2x$, $y = 2\cos 2\left(x - \dfrac{\pi}{4}\right)$.

II. 30. a) $y = 4\sin 3\left(x - \dfrac{\pi}{6}\right) - 2$, $y = -4\cos 3x - 2$;

b) $y = 4\sin 4\left(x - \dfrac{\pi}{8}\right) + 1$, $y = -4\cos 4x + 1$;

c) $y = 3\sin\dfrac{4}{5}\left(x - \dfrac{5\pi}{8}\right) + 1$, $y = -3\cos\dfrac{4}{5}x + 1$;

d) $y = 3\sin\dfrac{3}{5}\left(x - \dfrac{5\pi}{6}\right) - 1$, $y = -3\cos\dfrac{3}{5}x - 1$.

II. 31. The phase shift is $\dfrac{3\pi}{2}$ to the right. The period is 4π.

II. 32. a) amplitude 2;

b) period π;

c) phase shift $\dfrac{\pi}{8}$ to the left;

d) vertical translation 1.25 units up;

e) $(x, y) \rightarrow \left(\dfrac{1}{2}x - \dfrac{\pi}{8}, 2y + 1.25 \right)$;

g) $y = -2\sin\left(2x - \dfrac{\pi}{4} \right) + 1.25$.

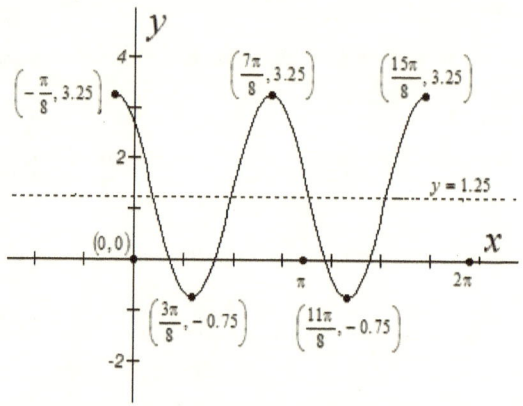

II. 33. $y = 1.75\cos(0.25x) - 3.75$.

II. 34. $y = -\dfrac{7}{2}\sin 2\left(x - \dfrac{\pi}{2} \right) - \dfrac{1}{2}$, $y = \dfrac{7}{2}\cos 2\left(x - \dfrac{\pi}{4} \right) - \dfrac{1}{2}$. **II. 35.** $a = -3$,

$p = \dfrac{5}{2}$. **II. 36.** $a = 3.3$, $b = \dfrac{\pi}{5}$. **II. 37.** $a = -4$, $p = \dfrac{\pi}{4}$, $h = 1$.

II. 38. $y = 4\sin 2\left(x - \dfrac{\pi}{8} \right) - 1$. The mapping transformation is

$(x, y) \rightarrow \left(\dfrac{1}{2}x + \dfrac{\pi}{8}, 4y - 1 \right)$.

Points from the base (parent) function on the interval $[0, 2\pi]$ $y = \sin x$	Transformed function $y = 4\sin 2\left(x - \dfrac{\pi}{8}\right) - 1$
$(0,0)$	$\left(\dfrac{\pi}{8}, -1\right)$
$\left(\dfrac{\pi}{2}, 1\right)$	$\left(\dfrac{3\pi}{8}, 3\right)$
$(\pi, 0)$	$\left(\dfrac{5\pi}{8}, -1\right)$
$\left(\dfrac{3\pi}{2}, -1\right)$	$\left(\dfrac{7\pi}{8}, -5\right)$
$(2\pi, 0)$	$\left(\dfrac{9\pi}{8}, -1\right)$

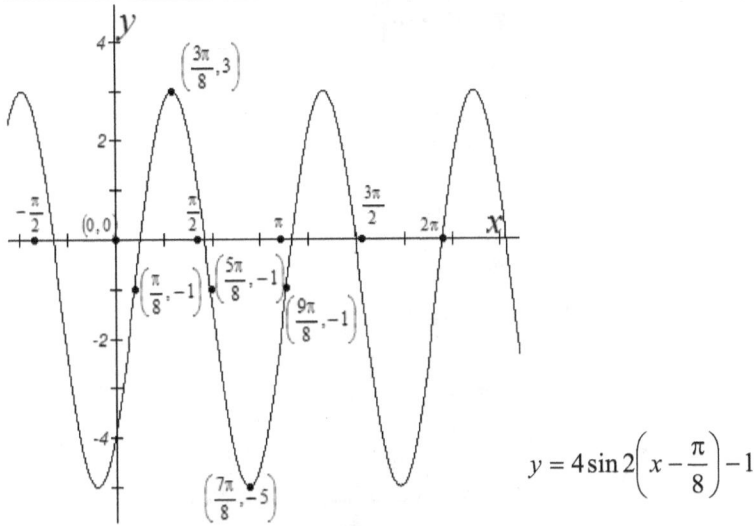

$$y = 4\sin 2\left(x - \frac{\pi}{8}\right) - 1$$

II. 39. a) $y = 5\sin 2x - 1$; b) $y = 3\sin 3(-x + 2\pi) - 1$;

c) $y = -6\sin 12\left(x - \dfrac{2\pi}{5}\right) - 3$; d) $y = -6\sin\pi\left(-x + \dfrac{\pi}{2}\right) - 3$;

e) $y = -4\sin 3\left(x + \dfrac{\pi}{6}\right) - 5$. II. 40. $\left(\dfrac{3\pi}{2}, -2\right)$ and $\left(\dfrac{1}{4}\dfrac{\pi}{4}, -5\right)$.

II. 41. $y = 6\sin\left[10\left(-x - \dfrac{\pi}{4}\right)\right] - 3$, the mapping transformation

is $(x, y) \rightarrow \left(-\dfrac{1}{10}x + \dfrac{\pi}{4}, 6y - 3\right)$. For the point $\left(\dfrac{7\pi}{30}, 0\right)$ equate

$-\dfrac{1}{10}x + \dfrac{\pi}{4} = \dfrac{7\pi}{30}$ and $6y - 3 = 0$, the point is $\left(\dfrac{\pi}{6}, \dfrac{1}{2}\right)$. Similarly, the other

two points are $\left(\dfrac{\pi}{2}, 1\right)$ and $(\pi, 0)$.

II. 42. $y = 4\cos 2\left(x + \dfrac{\pi}{5}\right) + 1$.

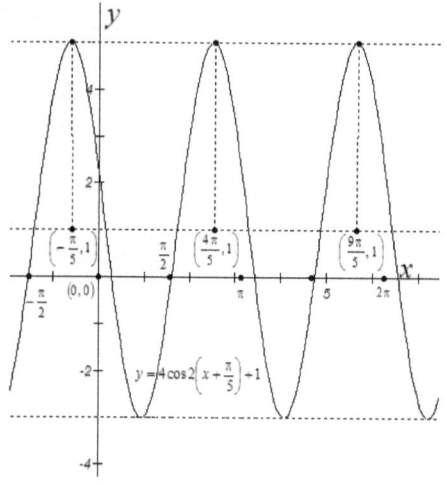

II. 43. $y = 3\cos\dfrac{3}{2}\left(x - \dfrac{5\pi}{6}\right) - 2$ or $y = 3\sin\dfrac{3}{2}\left(x - \dfrac{\pi}{2}\right) - 2$.

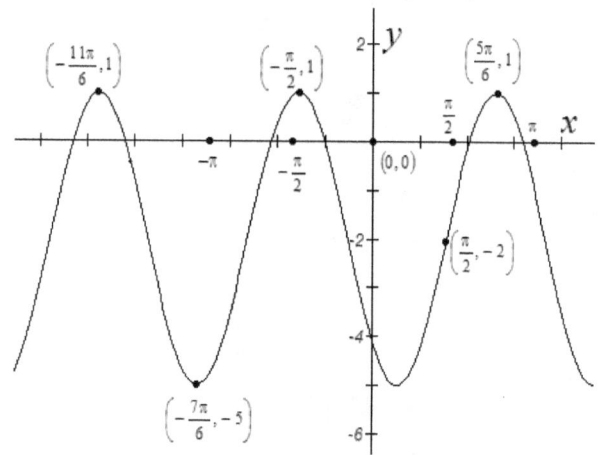

II. 44. $y = 3\sin 3\left(x - \dfrac{\pi}{2}\right) - 1$ or $y = -3\cos 3\left(x - \dfrac{\pi}{3}\right) - 1$.

II. 45. $y = \dfrac{5}{2}\sin\dfrac{8}{5}\left(x - \dfrac{91\pi}{48}\right) - \dfrac{1}{2}$ or $y = -\dfrac{5}{2}\cos\dfrac{8}{5}\left(x - \dfrac{19\pi}{12}\right) - \dfrac{1}{2}$.

II. 46. $y = 3\sin\dfrac{8}{5}\left(x - \dfrac{31\pi}{48}\right) + 2$.

II. 47. $f(x) = -\dfrac{1}{3}\sin 4\left(-x + \dfrac{\pi}{4}\right) + 6$.

II. 48. $f(x) = \dfrac{1}{7}\cos 3\left(-x + \dfrac{\pi}{6}\right) - 10,\ f(0) = -10.$

II. 49.

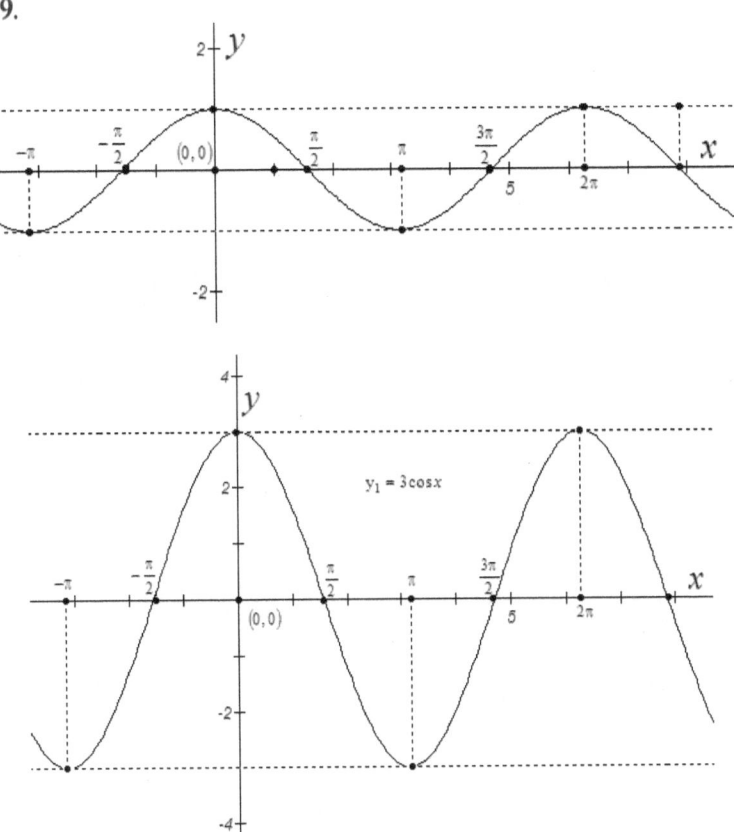

153

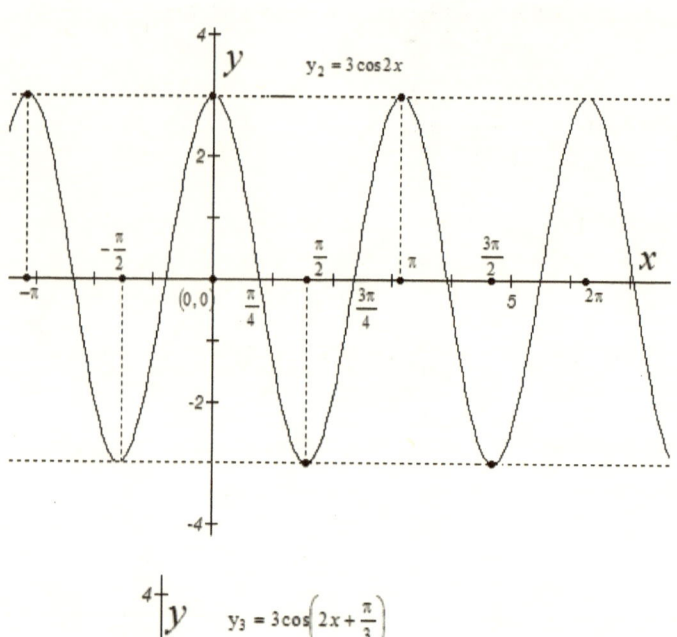

$y_2 = 3\cos 2x$

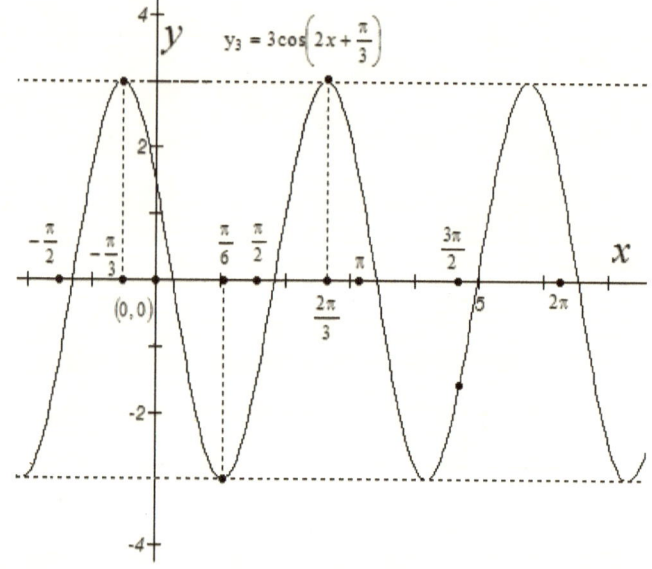

$y_3 = 3\cos\left(2x + \dfrac{\pi}{3}\right)$

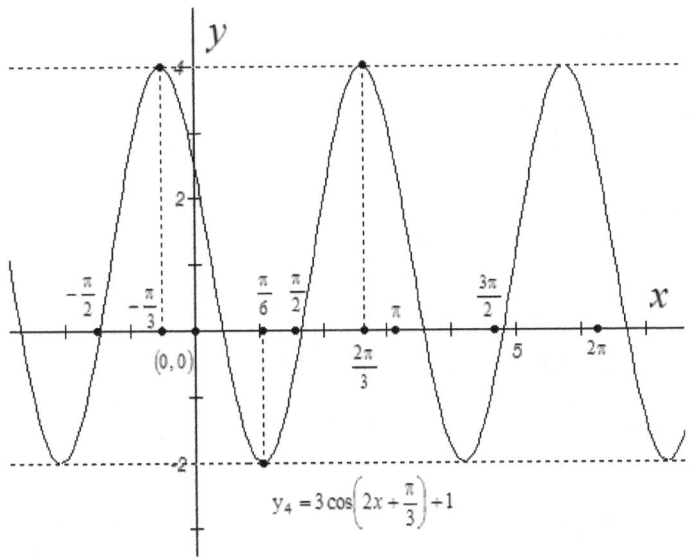

$$y_4 = 3\cos\left(2x + \frac{\pi}{3}\right) + 1$$

II. 50.

a) $\dfrac{f\left(\dfrac{\pi}{6}\right) - f(0)}{\dfrac{\pi}{6} - 0} = \dfrac{\left[3\cos\left(2 \cdot \dfrac{\pi}{6} + \dfrac{\pi}{3}\right) + 1\right] - \left[3\cos\left(2 \cdot 0 + \dfrac{\pi}{3}\right) + 1\right]}{\dfrac{\pi}{6}} = -5.7296$;

b) -1.9098;   c) 2.8648;   d) $\dfrac{18}{\pi}$;   e) $\dfrac{6}{\pi}$;   f) 0.

II. 51. a) $(0, -2)$, $\left(\dfrac{5\pi}{4}, 4\right)$;   **b)** $\left(\dfrac{15\pi}{8}, 1\right)$, $\left(\dfrac{55\pi}{24}, -\dfrac{1}{2}\right)$;

c) $\left(\dfrac{5\pi}{8}, 1\right)$, $\left(\dfrac{5\pi}{12}, \dfrac{2 - 3\sqrt{3}}{2}\right)$.

II. 52. a) $\left(-\dfrac{11\pi}{6}, 1\right)$, $\left(-\dfrac{\pi}{2}, 1\right)$, $\left(-\dfrac{7\pi}{6}, -5\right)$, $\left(\dfrac{\pi}{6}, -5\right)$, $\left(\dfrac{5\pi}{6}, 1\right)$;

b) Any points in the intervals $x \in \left(-\dfrac{11\pi}{6}, -\dfrac{7\pi}{6}\right)$, $x \in \left(-\dfrac{\pi}{2}, \dfrac{\pi}{6}\right)$;

c) Any points in the intervals $x \in \left(-\dfrac{7\pi}{6}, -\dfrac{\pi}{2}\right)$; $x \in \left(\dfrac{\pi}{6}, \dfrac{5\pi}{6}\right)$;

d) at halfway between a maximum and a minimum $\left(-\dfrac{\pi}{6}, -2\right)$, $\left(\dfrac{7\pi}{6}, -2\right)$;

e) at halfway between a minimum and a maximum $\left(-\frac{5\pi}{6},-2\right)$, $\left(\frac{\pi}{2},-2\right)$.
Note: Answers may vary for this question.

II. 53. a) Instantaneous rate of change when $x = 0$ is

$$\frac{f(0.0001)-f(0)}{0.0001-0} = \frac{\sin(0.0001)-\sin(0)}{0.0001} = 0.9999. \cong 1 \text{ and the point is } (0,1)$$

Instantaneous rate of change when $x = \frac{\pi}{2}$ is

$$\frac{f\left(\frac{\pi}{2}+0.0001\right)-f\left(\frac{\pi}{2}\right)}{\frac{\pi}{2}+0.0001-\frac{\pi}{2}} = \frac{\sin\left(\frac{\pi}{2}+0.0001\right)-\sin\left(\frac{\pi}{2}\right)}{0.0001} \cong 0 \text{ and the point is}$$

$\left(\frac{\pi}{2},0\right)$. Similarly we obtain the points $(\pi,-1)$, $\left(\frac{3\pi}{2},0\right)$, and $(2\pi,1)$.
Therefore the instantaneous rate of change cosines function is

$$f'(x) = \cos x .$$

b) The points are $\left(-\frac{\pi}{3},0\right)$, $\left(-\frac{\pi}{12},-6\right)$, $\left(\frac{\pi}{6},0\right)$, $\left(\frac{5\pi}{12},6\right)$, $\left(\frac{2\pi}{3},0\right)$.
Therefore the instantaneous rate of change sines function is

$$f'(x) = -6\sin 2\left(x+\frac{\pi}{3}\right).$$

II. 54. a) $f(x) = \sqrt{2}\cos\left(x-\frac{\pi}{4}\right)$; b) $f(x) = -4\cos\left(2x+\frac{\pi}{6}\right)+1$;

c) $f(x) = 2\cos\left(4x-\frac{\pi}{3}\right)-1$; d) $f(x) = \frac{1}{2}\sin\left(4x-\frac{\pi}{2}\right)+\frac{1}{2}$;

e) $f(x) = -\frac{\sqrt{2}}{2}\cos 2x+1$; f) $f(x) = \frac{1}{4}\sin\frac{3}{2}x-1$;

g) $f(x) = 2\sin\left(x-\frac{\pi}{6}\right)-1$.

II. 55.

$$f(x) = \begin{cases} -1 & \text{if } x < -\dfrac{\pi}{2} \\ \sin x & \text{if } x \in \left[-\dfrac{\pi}{2}, \dfrac{\pi}{2}\right] \\ 1 & \text{if } x > \dfrac{\pi}{2} \end{cases}$$

a)

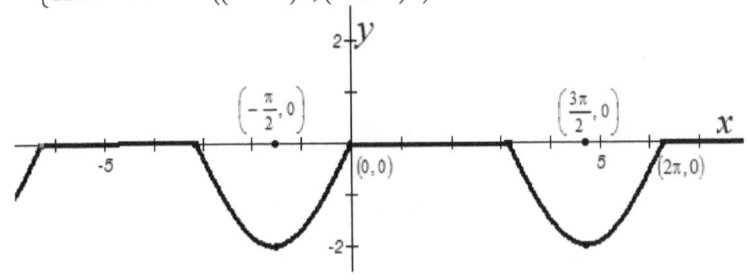

b) $f(x) = \begin{cases} 0 & \text{if } x \in [0, \pi] \\ 2\sin x & \text{if } x \in (\pi, 2\pi) \end{cases}$ or in general

$$f(x) = \begin{cases} 0 & \text{if } x \in [2k\pi, (2k+1)\pi] \\ 2\sin x & \text{if } x \in ((2k+1)\pi, (2k+2)\pi) \end{cases}, \quad k \in \mathbf{Z}.$$

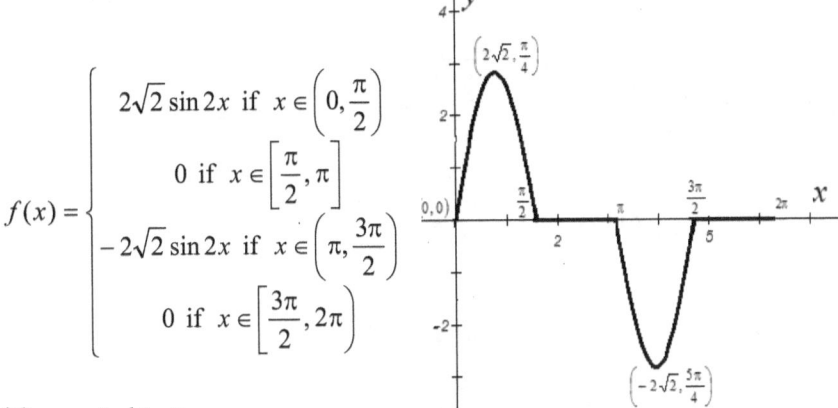

II. 56. $f(x) = \dfrac{\sqrt{2}\sin 2x \cos x}{|\cos x|} + \dfrac{\sqrt{2}\sin x \sin 2x}{|\sin x|}$,

$$f(x) = \begin{cases} 2\sqrt{2}\sin 2x & \text{if } x \in \left(0, \dfrac{\pi}{2}\right) \\ 0 & \text{if } x \in \left[\dfrac{\pi}{2}, \pi\right] \\ -2\sqrt{2}\sin 2x & \text{if } x \in \left(\pi, \dfrac{3\pi}{2}\right) \\ 0 & \text{if } x \in \left[\dfrac{3\pi}{2}, 2\pi\right] \end{cases}$$

The period is 2π.

Identity Transformations of Trigonometric Expressions

III. 1. a) $\sin\left(-1920°\right)=\sin\left(-1920°+360°\cdot 5\right)=\sin\left(-120°\right)=-\sin 120°=-\dfrac{\sqrt{3}}{2}$;

b) $-\dfrac{1}{2}$; **c)** $\tan\left(-1830°\right)=\tan\left(-30°-180°\cdot 10\right)=\tan\left(-30°\right)=-\dfrac{1}{\sqrt{3}}$;

d) $-\dfrac{1}{\sqrt{3}}$; **e)** $\sin 7°30'=\sqrt{\dfrac{1-\cos 15°}{2}}=\sqrt{\dfrac{1-\dfrac{\sqrt{6}-\sqrt{2}}{4}}{2}}=\dfrac{\sqrt{4+\sqrt{2}-\sqrt{6}}}{2\sqrt{2}}$;

f) $\cos 7°30'=\dfrac{\sqrt{4-\sqrt{2}+\sqrt{6}}}{2\sqrt{2}}$;

g) $\tan 7°30'=\dfrac{1-\cos 15°}{\sin 15°}=\dfrac{1-\dfrac{\sqrt{6}-\sqrt{2}}{4}}{\dfrac{\sqrt{6}+\sqrt{2}}{4}}=\left(\sqrt{2}-1\right)\left(\sqrt{3}-\sqrt{2}\right)$;

h) $\cot 7°30'=\left(\sqrt{2}+1\right)\left(\sqrt{3}+\sqrt{2}\right)$.

III. 2. a) $1+\cos\dfrac{\pi}{4}=2\cos^2\dfrac{\pi}{8}$; **b)** $2\sin^2\dfrac{\pi}{8}$;

c) $1-\tan\dfrac{\pi}{3}=\tan\dfrac{\pi}{4}-\tan\dfrac{\pi}{3}=-2\sqrt{2}\sin\dfrac{\pi}{12}$; **d)** $\dfrac{2\sqrt{6}}{3}\sin\dfrac{5\pi}{12}$;

e) $2\sin\dfrac{5\pi}{24}\cos\dfrac{\pi}{24}$; f) $-2\sin\dfrac{5\pi}{24}\sin\dfrac{5\pi}{24}$; g) $\sqrt{2}\sin\dfrac{\pi}{12}$;

h) $2\sqrt{2}\sin\dfrac{7\pi}{12}$; i) $\dfrac{\sqrt{6}}{2}$; j) $\dfrac{\sqrt{6}}{2}$.

III. 3. a) $\sin\left(\dfrac{5\pi}{7}-\dfrac{13\pi}{28}\right)=\dfrac{\sqrt{2}}{2}$; b) $\cos\left(\dfrac{13\pi}{21}-\dfrac{2\pi}{7}\right)=\dfrac{1}{2}$;

c) $\tan\left(\dfrac{\pi}{9}+\dfrac{\pi}{18}\right)=\tan\dfrac{\pi}{6}=\dfrac{1}{\sqrt{3}}$; d) 0.

III. 4. a) 1; b) 1; c) 2; d) $2\tan^2 x$; e) $\cot^2 x$; f) $\tan^4 x$; g) $2\sec^3 x$;

h) $\dfrac{1}{\sin^4 x}-\dfrac{1}{\cos^4 x}=\dfrac{\cos^4 x-\sin^4 x}{\sin^4 x\cos^4 x}=\dfrac{16\left(\cos^2 x-\sin^2 x\right)\left(\cos^2 x+\sin^2 x\right)}{16\sin^4 x\cos^4 x}=$

$\dfrac{16\left(\cos^2 x-\sin^2 x\right)}{\sin^4 2x}=16\cos 2x\csc^4 2x$; i) $2\cot 2x$; j) $2\sin^2 2x$;

k) $\dfrac{\tan^3 x-\tan^5 x}{\cot^3 x-\cot x}=\dfrac{\tan^3 x}{\cot x}\left(\dfrac{1-\tan^2 x}{\cot^2 x-1}\right)=$

$=\tan^4 x\dfrac{\cos^2 x-\sin^2 x}{\cos^2 x}\dfrac{\sin^2 x}{\cos^2 x-\sin^2 x}=\tan^6 x$; l) $\cos a$.

III. 5. a) $\dfrac{\sin 2x}{2}$; b) $\dfrac{\csc^2 x}{2}$; c) $\cos x$; d) $\cos 2x$; e) $\tan^2 x$;

f) $\tan^2 x$; g) $\cos 4x$; h) 1.

III. 6. a) $2\sin 2a\cos a$; b) $-2\sin 4a\sin a$; c) $2\sin b\sin a$;

d) $2\cos a\cos b$; e) $2\sin\left(\dfrac{\pi}{4}+a\right)$; f) $2\cos\dfrac{a}{2}\cos\dfrac{a}{2}$; g) $\dfrac{\sin 2a}{\cos(a+b)\cos(a-b)}$.

III. 7. a) $2\cos x\cos(y+z)$; b) $2\sin\dfrac{z}{2}\cos(x-y)$; c) $-2\sin z\cos(x+y)$.

III. 8. a) $\cos 2x(1+\cos x)$; b) $4\cos x\cos\dfrac{5x}{2}\cos\dfrac{x}{2}$; c) $4\sin 6x\cos 4x\cos x$;

159

d) $4\cos\dfrac{5\pi}{9}\cos\dfrac{2\pi}{9}\cos\dfrac{\pi}{9}$; **e)** $4\cos 6x\cos 4x\cos x$; **f)** $-2\cos x$;

g) 0; **h)** 0.

III. 9. a) $\sin 2x\sin 4y$; **b)** $-\sin 2x\sin 2y$; **c)** $-\cos 2x\cos 2y$;

d) $4\sin^2\left(\dfrac{x+2y}{2}\right)$; **e)** $2\cos x\cos 2y\cos(x-2y)$; **f)** $2\sin 2x\sin y\cos(2x-y)$;

g) $\cos^2\alpha+\cos^2\beta-[\cos(\alpha-\beta)+\cos(\alpha+\beta)]\cos(\alpha-\beta)=$

$=\cos^2\alpha+\cos^2\beta-\cos^2(\alpha-\beta)-\cos(\alpha-\beta)\cos(\alpha+\beta)=$

$=\cos^2\alpha+\cos^2\beta-\cos^2(\alpha-\beta)-\dfrac{1}{2}(\cos 2\alpha+\cos 2\beta)=$

$=\cos^2\alpha+\cos^2\beta-\cos^2(\alpha-\beta)-\dfrac{1}{2}\left(2\cos^2\alpha-1+2\cos^2\beta-1\right)=$

$=1-\cos^2(\alpha-\beta)=\sin^2(\alpha-\beta)$; **h)** $\cos^2(\alpha-\beta)$;

i) $\dfrac{1-\cos 2x-1+\cos 2y}{2\sin(x-y)}=\dfrac{2\sin(x+y)\sin(x-y)}{2\sin(x-y)}=\sin(x+y)$; **j)** $-\tan\alpha\tan\beta$.

III. 10. a) $\dfrac{\sin^2\dfrac{x}{2}+\cos^2\dfrac{x}{2}+2\sin\dfrac{x}{2}\cos\dfrac{x}{2}}{\sin\dfrac{x}{2}\cos\dfrac{x}{2}}=\dfrac{1+\sin x}{\dfrac{1}{2}\sin x}=\dfrac{2\left(1+\cos\left(\dfrac{\pi}{2}-x\right)\right)}{\sin x}=$

$=4\csc x\cos^2\left(\dfrac{\pi}{4}-\dfrac{x}{2}\right)$; **b)** $-\cos x$; **c)** $\dfrac{1}{2}\cot\dfrac{x}{2}$;

d) $\sin\left(\dfrac{\pi}{3}+\dfrac{x}{2}\right)\sin\left(\dfrac{\pi}{3}-\dfrac{x}{2}\right)\sin\dfrac{x}{2}=\dfrac{1}{2}\left(\cos x-\cos\dfrac{2\pi}{3}\right)\sin\dfrac{x}{2}=\dfrac{1}{2}\left(\cos x+\dfrac{1}{2}\right)\sin\dfrac{x}{2}=$

$=\dfrac{1}{2}\left(1-2\sin^2\dfrac{x}{2}+\dfrac{1}{2}\right)\sin\dfrac{x}{2}=\dfrac{1}{4}\left(3-4\sin^2\dfrac{x}{2}\right)\sin\dfrac{x}{2}=\dfrac{1}{4}\sin\dfrac{3x}{2}$;

e) $\dfrac{1+\sin 2x}{\cos 2x\cot\left(x-\dfrac{\pi}{4}\right)}+\cos^2 x=\dfrac{\sin^2 x+\cos^2 x+2\sin x\cos x}{(\cos^2 x-\sin^2 x)\dfrac{\cos x+\sin x}{\sin x-\cos x}}+\cos^2 x=-\sin^2 x$;

f) $-\dfrac{\sqrt{2}}{4}\tan\dfrac{x}{2}$; **g)** $\dfrac{1}{2}\csc^2 x$; **h)** $\cot 2x$; **i)** 2.

III. 11. a) -1; **b)** $\tan^4\alpha$; **c)** -1; **d)** $-\tan^4 x$;

160

e) $\dfrac{\sin^2 x}{\sec^2 x - 1} + \dfrac{\cos^2 x}{\csc^2 x - 1} = \dfrac{\sin^2 x \cos^2 x}{1 - \cos^2 x} + \dfrac{\cos^2 x \sin^2 x}{1 - \sin^2 x} = \sin^2 x + \cos^2 x = 1$;

f) $\sin^2 x$; g) $\tan^2 x$; h) $\dfrac{1}{2}\cot^4 x$.

III. 12. a) $\left(\sqrt{3}\cos\alpha - \sin\alpha\right)\left(\sqrt{3}\cos\alpha + \sin\alpha\right) =$

$= 4\left(\dfrac{\sqrt{3}}{2}\cos\alpha - \dfrac{1}{2}\sin\alpha\right)\left(\dfrac{\sqrt{3}}{2}\cos\alpha + \dfrac{1}{2}\sin\alpha\right) =$

$= 4\left(\cos\dfrac{\pi}{6}\cos\alpha - \sin\dfrac{\pi}{6}\sin\alpha\right)\left(\cos\dfrac{\pi}{6}\cos\alpha + \sin\dfrac{\pi}{6}\sin\alpha\right) =$

$= 4\cos\left(\dfrac{\pi}{6} + \alpha\right)\cos\left(\dfrac{\pi}{6} - \alpha\right)$; b) $\cos 2x$;

c) $\cot^2\alpha - \tan^2\alpha = \dfrac{\cos^4\alpha - \sin^4\alpha}{\sin^2\alpha\cos^2\alpha} = \dfrac{\cos^2\alpha - \sin^2\alpha}{\sin^2\alpha\cos^2\alpha} = \dfrac{4\cos 2\alpha}{\sin^2 2\alpha} =$

$= 4\cot 2\alpha\csc 2\alpha$; d) -2;

e) $\dfrac{\cos\left(\dfrac{3\pi}{2} + 2\alpha\right)}{\left(\cos\dfrac{\alpha}{4} + \sin\dfrac{\alpha}{4}\right)\left(\cos\dfrac{\alpha}{4} - \sin\dfrac{\alpha}{4}\right)\sin\dfrac{\alpha}{2}} = \dfrac{\sin 2\alpha}{\cos\dfrac{\alpha}{2}\sin\dfrac{\alpha}{2}} = \dfrac{4\sin\alpha\cos\alpha}{\sin\alpha} = 4\cos x$;

Solution

f) $\dfrac{1}{2\cot x\sin^2 x} + \dfrac{1+\cos 4x}{\sin^3 2x} - \dfrac{1}{-2\tan x\cos^2 x} = \dfrac{1}{\sin 2x} + \dfrac{2\cos^2 2x}{\sin^3 2x} + \dfrac{1}{\sin 2x} =$

$= 2\csc^3 2x$; g) $\sec^3 x$.

III. 13.

a) $\sin 2x - \cos 2x = \sin 2x - \sin\left(\dfrac{\pi}{2} - 2x\right) = 2\sin\left(2x - \dfrac{\pi}{4}\right)\cos\dfrac{\pi}{4} = 2\sqrt{2}\sin\left(2x - \dfrac{\pi}{4}\right)$;

b) $1 + \sin 2x + \cos 2x = 2\cos^2 x + 2\sin x\cos x = 2\cos x(\cos x + \sin x) =$

$= 2\cos x\left[\cos x + \cos\left(\dfrac{\pi}{2} - x\right)\right] = 4\cos x\cos\dfrac{\pi}{4}\cos\left(x - \dfrac{\pi}{4}\right) = 2\sqrt{2}\cos x\cos\left(x - \dfrac{\pi}{4}\right)$;

c) $2\sqrt{2}\sin x\cos\left(x - \dfrac{\pi}{4}\right)$; d) $2\sqrt{2}\sin x\cos\left(x - \dfrac{\pi}{4}\right)\sec 2x$;

e) $\sin 3x$; f) $\cos 3x$; g) 0; h) $2\sin\left(\dfrac{\pi}{6} - 2x\right)$;

i) $1 - \sin x - \cos x + \tan x = 1 + \tan x - \sin x - \cos x =$

$$= \frac{\sin x + \cos x}{\cos x} - (\sin x + \cos x) = (\sin x + \cos x)\frac{1 - \cos x}{\cos x} =$$

$$= \left[\cos\left(\frac{\pi}{2} - x\right) + \cos x \right] \frac{2\sin^2 \frac{x}{2}}{\cos x} = 2\sqrt{2}\,\sin^2 \frac{x}{2}\cos\left(\frac{\pi}{4} - x\right)\sec x\,;$$

j) $-2\cos x$;   **k)** $\cot x$.

III. 14. a) $\tan x$;   **b)** $\tan 5x$;   **c)** $\cot x$;   **d)** $\cot 4x$;   **e)** $\tan 5x$;

f) $\cot^2\left(x - \frac{\pi}{4}\right)\cot 3x$;   **g)** $\tan 4x$;   **h)** $\tan\dfrac{29x}{2}$;   **j)** $\cot\dfrac{17x}{2}$;   **k)** $\tan\alpha$;

j) $\dfrac{\sin 2x - \sin 6x + \cos 2x - \cos 6x}{\sin 4x - \cos 4x} = \dfrac{2\sin(-2x)\cos 4x + 2\sin 4x \sin 2x}{\sin 4x - \cos 4x} =$

$$= \frac{2\sin 2x(-\cos 4x + \sin 4x)}{\sin 4x - \cos 4x} = 2\sin 2x\,.$$

III. 15. a) $2\sin x$;   **b)** $\sin 2x$;   **c)** $8\cos^4 x$;

d) $1 + \sin 2x + \cos 2x + \tan 2x = \sin 2x + \cos 2x + \left(1 + \dfrac{\sin 2x}{\cos 2x}\right) =$

$$= \left[\cos\left(\frac{\pi}{2} - 2x\right) + \cos 2x\right]\frac{\cos 2x + 1}{\cos 2x} = 4\sqrt{2}\,\cos\left(\frac{\pi}{4} - 2x\right)\frac{\cos^2 x}{\cos 2x}\,;$$

e) $1 + \sin\dfrac{x}{2} - \cos^2\dfrac{x}{4} + \sin^2\dfrac{x}{4} = \sin\dfrac{x}{2} + 2\sin^2\dfrac{x}{4} = 2\sin\dfrac{x}{4}\left(\cos\dfrac{x}{4} + \sin\dfrac{x}{4}\right) =$

$$= 2\sin\frac{x}{4}\left[\sin\left(\frac{\pi}{2} - \frac{x}{4}\right) + \sin\frac{x}{4}\right] = 4\sin\frac{x}{4}\left[\sin\frac{\pi}{4}\cos\left(\frac{\pi}{4} - \frac{x}{4}\right)\right] = 2\sqrt{2}\,\sin\frac{x}{4}\cos\left(\frac{\pi}{4} - \frac{x}{4}\right).$$

f) $\left(2\sin 2x\cos 2x - \cos 4x\dfrac{\sin 2x}{\cos 2x}\right)\left(2\sin 2x\cos 2x + \cos 4x\dfrac{\sin 2x}{\cos 2x}\right) =$

$$= \sin 2x\cos 2x\left(2\cos 2x - \cos 4x\frac{1}{\cos 2x}\right)\left(2\sin 2x + \cos 4x\frac{1}{\sin 2x}\right) =$$

$$= \sin 2x\cos 2x\left(\frac{2\cos^2 2x - \cos 4x}{\cos 2x}\right)\left(\frac{2\sin^2 2x + \cos 4x}{\sin 2x}\right) = 1\,;$$

g) $2\cos^2 x - \sqrt{3}\,\sin 2x - 1 = \cos 2x - \sqrt{3}\,\sin 2x = 2\left(\dfrac{1}{2}\cos 2x - \dfrac{\sqrt{3}}{2}\sin 2x\right) =$

$$= 2\left(\cos\frac{\pi}{6}\cos 2x - \sin\frac{\pi}{6}\sin 2x\right) = 2\cos\left(\frac{\pi}{6} - 2x\right);$$

h) $\left(\cos\alpha - \sqrt{3}\sin\alpha\right)\left(\cos\alpha + \sqrt{3}\sin\alpha\right) = 4\left(\frac{1}{2}\cos\alpha - \frac{\sqrt{3}}{2}\sin\alpha\right)\left(\frac{1}{2}\cos\alpha + \frac{\sqrt{3}}{2}\sin\alpha\right) =$

$$= 4\left(\sin\frac{\pi}{6}\cos\alpha - \cos\frac{\pi}{6}\sin\alpha\right)\left(\sin\frac{\pi}{6}\cos\alpha + \cos\frac{\pi}{6}\sin\alpha\right) = 4\sin\left(\frac{\pi}{6} - \alpha\right)\sin\left(\frac{\pi}{6} + \alpha\right).$$

i) $3 + 4\cos 2x + \left(2\cos^2 2x - 1\right) - 2\left(1 + \cos 2x\right)^2 = 0;$

j) $\dfrac{\sin 4\alpha}{\left(\sin^2\alpha - \cos^2\alpha\right)\left(\sin^2\alpha + \cos^2\alpha\right)} - 2 = \dfrac{2\sin 2\alpha\cos 2\alpha}{\sin^2\alpha - \cos^2\alpha} - 2 = -2\sin 2\alpha - 2 =$

$$= -2\left[\cos\left(\frac{\pi}{2} - 2\alpha\right) + 1\right] = -4\cos^2\left(\frac{\pi}{4} - 2\alpha\right).$$

III. 16. a) $\cos(40° + 2x);$ **b)** $\tan\dfrac{x}{2};$ **c)** $-\dfrac{1}{4}\sin 4\alpha;$

d) $\dfrac{4\cos\alpha}{\cot^2\dfrac{\alpha}{2} - \tan^2\dfrac{\alpha}{2}} = \dfrac{4\cos\alpha\sin^2\dfrac{\alpha}{2}\cos^2\dfrac{\alpha}{2}}{\cos^4\dfrac{\alpha}{2} - \sin^4\dfrac{\alpha}{2}} = \dfrac{\cos\alpha\left(\sin 2\dfrac{\alpha}{2}\right)^2}{\cos^2\dfrac{\alpha}{2} - \sin^2\dfrac{\alpha}{2}} =$

$$= \dfrac{\cos\alpha\sin^2\alpha}{\cos 2\dfrac{\alpha}{2}} = \sin^2\alpha;$$

e) $\dfrac{\sin\alpha\cos\dfrac{\alpha}{2}}{\dfrac{1}{2}\left[1 + \cos\left(\dfrac{\pi}{2} - \dfrac{\alpha}{2}\right)\right]\left(2\cos\dfrac{\alpha}{2} - \sin\alpha\right)} = \dfrac{2\sin\alpha\cos\dfrac{\alpha}{2}}{\left(1 + \sin\dfrac{\alpha}{2}\right)\left(2\cos\dfrac{\alpha}{2} - \sin\alpha\right)} =$

$$= \dfrac{2\sin\alpha\cos\dfrac{\alpha}{2}}{\left(1 + \sin\dfrac{\alpha}{2}\right)\left(2\cos\dfrac{\alpha}{2} - 2\sin\dfrac{\alpha}{2}\cos\dfrac{\alpha}{2}\right)} = \dfrac{2\sin\alpha\cos\dfrac{\alpha}{2}}{\left(1 + \sin\dfrac{\alpha}{2}\right)2\cos\dfrac{\alpha}{2}\left(1 - \sin\dfrac{\alpha}{2}\right)} =$$

$$= \dfrac{\sin\alpha}{1 - \sin^2\dfrac{\alpha}{2}} = \dfrac{2\sin\dfrac{\alpha}{2}\cos\dfrac{\alpha}{2}}{\cos^2\dfrac{\alpha}{2}} = 2\tan\dfrac{\alpha}{2};$$ **f)** $-\tan 5x;$ **g)** $\tan^3 x;$ **h)** $\sin 2x.$

III. 17. a) $3 - 4\sin^2\left(\dfrac{\pi}{2} - \dfrac{x}{2}\right) = 4\left(\dfrac{3}{4} - \cos^2\dfrac{x}{2}\right) = 4\left(\dfrac{\sqrt{3}}{2} - \cos\dfrac{x}{2}\right)\left(\dfrac{\sqrt{3}}{2} + \cos\dfrac{x}{2}\right) =$

$= 4\left(\cos\dfrac{\pi}{6} - \cos\dfrac{x}{2}\right)\left(\cos\dfrac{\pi}{6} + \cos\dfrac{x}{2}\right) = 16\sin\left(\dfrac{x}{4} + \dfrac{\pi}{12}\right)\sin\left(\dfrac{x}{4} - \dfrac{\pi}{12}\right) \cdot$

$\cdot \cos\left(\dfrac{x}{4} + \dfrac{\pi}{12}\right)\cos\left(\dfrac{x}{4} - \dfrac{\pi}{12}\right) = 4\sin\left(\dfrac{x}{2} + \dfrac{\pi}{6}\right)\sin\left(\dfrac{x}{2} - \dfrac{\pi}{6}\right);$

b) $8\sin\left(x - \dfrac{\pi}{6}\right)\cos\left(x + \dfrac{\pi}{6}\right);$ **c)** $8\sin\left(x + \dfrac{\pi}{6}\right)\cos\left(x + \dfrac{\pi}{3}\right);$

d) $4\cos 2x\cos\left(x + \dfrac{\pi}{3}\right)\cos\left(x - \dfrac{\pi}{3}\right)\sec^4 x\,;$

e) $(\tan x + 1)(\tan^2 x - 3) = \dfrac{\sin x + \cos x}{\cos x}\left(\dfrac{\cos x - \sqrt{3}\sin x}{\cos x}\right)\left(\dfrac{\cos x + \sqrt{3}\sin x}{\cos x}\right) =$

$= \dfrac{\sin x + \sin\left(\dfrac{\pi}{2} - x\right)}{\cos x}\left(\dfrac{\dfrac{1}{2}\cos x - \dfrac{\sqrt{3}}{2}\sin x}{\cos x}\right)\left(\dfrac{\dfrac{1}{2}\cos x + \dfrac{\sqrt{3}}{2}\sin x}{\cos x}\right) =$

$= 8\dfrac{\sin\dfrac{\pi}{4}\cos\left(x - \dfrac{\pi}{4}\right)}{\cos x}\left(\dfrac{\cos\dfrac{\pi}{3}\cos x - \sin\dfrac{\pi}{3}\sin x}{\cos x}\right)\left(\dfrac{\cos\dfrac{\pi}{3}\cos x - \sin\dfrac{\pi}{3}\sin x}{\cos x}\right) =$

$= 4\sqrt{2}\cos\left(x - \dfrac{\pi}{4}\right)\cos\left(x + \dfrac{\pi}{3}\right)\cos\left(x - \dfrac{\pi}{3}\right)\sec^3 x\,.$

III. 18. a) $-\sin 4x\,;$

b) $\dfrac{1}{\cos^2 2x} - \tan^2 2x - 1 - \cos x - \sqrt{3}\sin x = -\cos x - \sqrt{3}\sin x =$

$= -2\left(\dfrac{1}{2}\cos x + \dfrac{\sqrt{3}}{2}\sin x\right) = -2\left(\cos\dfrac{\pi}{3}\cos x + \sin\dfrac{\pi}{3}\sin x\right) = -2\cos\left(\dfrac{\pi}{3} - x\right);$

c) $\tan 4x\sec 4x\,;$ **d)** $8\sin^2\dfrac{x}{2}\sin^2 x\,;$

e) $2\sin^2 x + \sqrt{3}\sin 2x - \dfrac{4\tan x\left(1 - \tan^2 x\right)}{\sin 4x\left(1 + \tan^2 x\right)^2} = 2\sin^2 x + \sqrt{3}\sin 2x -$

$$-\frac{4\tan x\left(\cos^2 x-\sin^2 x\right)\cos^2 x}{\sin 4x\left(\cos^2 x+\sin^2 x\right)^2}=2\sin^2 x+\sqrt{3}\sin 2x-\frac{4\tan x\cos 2x\cos^2 x}{2\sin 2x\cos 2x}=$$

$$=2\sin^2 x+\sqrt{3}\sin 2x-1=2\left(\frac{\sqrt{3}}{2}\sin 2x-\frac{1}{2}\cos 2x\right)=2\sin\left(\sin 2x-\frac{\pi}{6}\right);$$

f) $\dfrac{\sqrt{2}}{2}\sin\left(\dfrac{\pi}{4}-x\right)\sin 2x$;   g) $2\sqrt{2}\cos^2\dfrac{x}{2}\sin\left(\dfrac{\pi}{4}-x\right)\sec x$.

III. 19. a) $\sqrt{\sin x}\left(\sqrt{\dfrac{1}{\cos x}+1}-\sqrt{\dfrac{1}{\cos x}-1}\right)=\sqrt{\tan x}\left(\sqrt{1+\cos x}-\sqrt{1-\cos x}\right)=$

$$=\sqrt{2\tan x}\left(\cos\frac{x}{2}-\sin\frac{x}{2}\right)=\sqrt{2\tan x}\left[\cos\frac{x}{2}-\cos\left(\frac{\pi}{2}-\frac{x}{2}\right)\right]=$$

$$=2\sin\frac{\pi}{4}\sin\left(\frac{\pi}{4}-\frac{x}{2}\right)\sqrt{2\tan x}=2\sin\left(\frac{\pi}{4}-\frac{x}{2}\right)\sqrt{\tan x}\;;\quad \textbf{b)}$$

$$\sqrt{\left(\frac{\cos^2 x-\sin^2 x}{\cos^2 x}\right)\left(\frac{\cos^2 x-\sin^2 x}{\sin^2 x}\right)}=\sqrt{\frac{\cos 2x}{\cos^2 x}\cdot\frac{\cos 2x}{\sin^2 x}}=\sqrt{\frac{4\cos^2 2x}{\sin^2 2x}}=2\left|\cot 2x\right|$$

c) $\sqrt{\dfrac{\cos 2x\cos^2 x\sin^2 x}{\cos^4 x-\sin^4 x}}=\sqrt{\dfrac{\cos 2x\cos^2 x\sin^2 x}{\left(\cos^2 x-\sin^2 x\right)\left(\cos^2 x+\sin^2 x\right)}}=-\cos x\sin x\;;$

d) $\sqrt{\left(1+\sin\alpha\sin\beta-\cos\alpha\cos\beta\right)\left(1+\sin\alpha\sin\beta+\cos\alpha\cos\beta\right)}=$

$$=\sqrt{\left(1-\cos(\alpha+\beta)\right)\left(1+\cos(\alpha-\beta)\right)}=\sqrt{4\sin^2\left(\frac{\alpha+\beta}{2}\right)\cos^2\left(\frac{\alpha-\beta}{2}\right)}=$$

$$=\sqrt{\left(\sin\alpha-\sin\beta\right)^2}=\left|\sin\alpha-\sin\beta\right|\;;$$

e) $\sqrt{\dfrac{\cos^2 x+\sin^2 x}{\cos^2 x\sin^2 x}}=\sqrt{\dfrac{4}{4\cos^2 x\sin^2 x}}=2\sqrt{\dfrac{1}{\sin^2 2x}}=2\left|\csc 2x\right|\;;$

f) $\sqrt[3]{\dfrac{-\cos x+\dfrac{1}{\cos x}}{-\sin x+\dfrac{1}{\sin x}}}=\sqrt[3]{\dfrac{-\cos^2 x+1}{\cos x}\cdot\dfrac{\sin x}{-\sin^2 x+1}}=\sqrt[3]{\dfrac{\sin^3 x}{\cos^3 x}}=\tan x\;;$

g) $\sqrt{\dfrac{\cos^2 x-\sin^2 x}{\cos^2 x}\cdot\dfrac{\cos^2 x-\sin^2 x}{\sin^2 x}}=\sqrt{\dfrac{4\cos^2 2x}{4\cos^2 x\sin^2 x}}=2\left|\cot 2x\right|\;;$

h)
$$\frac{\sqrt{\tan x} + \dfrac{1}{\sqrt{\tan x}}}{\sqrt{\tan x} - \dfrac{1}{\sqrt{\tan x}}} = \frac{\tan x + 1}{\tan x - 1} = \frac{\sin x + \cos x}{\sin x - \cos x} =$$

$$= \frac{\sin x + \sin\left(\dfrac{\pi}{2} - x\right)}{\sin x - \sin\left(\dfrac{\pi}{2} - x\right)} = -\tan\left(\frac{\pi}{4} + x\right);$$

i)
$$\sqrt{\frac{2\sin x - 2\sin x\cos x}{2\sin x + 2\sin x\cos x}} = \sqrt{\frac{1 - \cos x}{1 + \cos x}} = \sqrt{\frac{2\sin^2 \dfrac{x}{2}}{2\cos^2 \dfrac{x}{2}}} = \left|\tan\frac{x}{2}\right|;$$

j)
$$\sqrt{\frac{(1 + \sin x)^2}{(1 - \sin x)(1 + \sin x)}} - \sqrt{\frac{(1 - \sin x)^2}{(1 + \sin x)(1 - \sin x)}} = \frac{1 + \sin x}{|\cos x|} - \frac{1 - \sin x}{|\cos x|} = \frac{2\sin x}{|\cos x|};$$

k)
$$\frac{\left(\sqrt{1 + \sin\alpha} + \sqrt{1 - \sin\alpha}\right)\left(\sqrt{1 + \sin\alpha} + \sqrt{1 - \sin\alpha}\right)}{\left(\sqrt{1 + \sin\alpha} - \sqrt{1 - \sin\alpha}\right)\left(\sqrt{1 + \sin\alpha} + \sqrt{1 - \sin\alpha}\right)} = \frac{2 + 2\sqrt{1 - \sin^2\alpha}}{2\sin\alpha} =$$

$$= \frac{1 + |\cos\alpha|}{\sin\alpha}. \text{ If } -\frac{\pi}{2} < \alpha < \frac{\pi}{2} \text{ or } -\frac{\pi}{2} + 2k\pi < \alpha < \frac{\pi}{2} + 2k\pi, \ k \in \mathbf{Z} \text{ (quad-}$$

rant I and IV) the value of the expression is $\dfrac{1 + \cos\alpha}{\sin\alpha} = \dfrac{2\cos^2 \dfrac{\alpha}{2}}{2\sin\dfrac{\alpha}{2}\cos\dfrac{\alpha}{2}} = \cot\dfrac{\alpha}{2}$

and if α is in the second and the third quadrant ($\dfrac{\pi}{2} < \alpha < \dfrac{3\pi}{2}$ or

$\dfrac{\pi}{2} + 2k\pi < \alpha < \dfrac{3\pi}{2} + 2k\pi, \ k \in \mathbf{Z}$) the expression is $\tan\dfrac{\alpha}{2}$;

l)
$$\sqrt{1 + \cos\left(\frac{\pi}{2} - \frac{x}{2}\right)} - \sqrt{1 - \cos\left(\frac{\pi}{2} - \frac{x}{2}\right)} = \sqrt{2\cos^2\left(\frac{\pi}{4} - \frac{x}{4}\right)} - \sqrt{2\sin^2\left(\frac{\pi}{4} - \frac{x}{4}\right)} =$$

Since $0° < x \le 180°$, then $0° < \dfrac{x}{4} \le 45°$ and therefore we have

$$= \sqrt{2}\cos\left(\frac{\pi}{4} - \frac{x}{4}\right) - \sqrt{2}\sin\left(\frac{\pi}{4} - \frac{x}{4}\right) = \sqrt{2}\left[\cos\left(\frac{\pi}{4} - \frac{x}{4}\right) - \cos\left(\frac{\pi}{4} + \frac{x}{4}\right)\right] =$$

$$= 2\sqrt{2}\,\sin\frac{\pi}{4}\sin\frac{x}{4} = 2\sin\frac{x}{4}.$$

III. 31. $x_1 = \dfrac{\cos^2\dfrac{a}{2}}{\cos^2\dfrac{b}{2}}$, $x_2 = \dfrac{\sin^2\dfrac{b}{2}}{\sin^2\dfrac{a}{2}}$, $E = \dfrac{\sin\dfrac{b-a}{2}}{\sin\dfrac{b+a}{2}}$.

UNIT IV

Trigonometric Identities

IV. 7. a) $\left(\sin^2 x + \cos^2 x\right)\left(\sin^4 x - \sin^2 x \cos^2 x + \cos^4 x\right) + 3\sin^2 x \cos^2 x =$

$\left[\left(\sin^2 x + \cos^2 x\right)^2 - 2\sin^2 x \cos^2 x - \sin^2 x \cos^2 x\right] + 3\sin^2 x \cos^2 x = 1;$

b) $\left(\sin^2 x - \cos^2 x\right)\left(\sin^4 x + \sin^2 x \cos^2 x + \cos^4 x\right) = -\cos 2x\left(1 - \sin^2 x \cos^2 x\right) =$

$= -\cos 2x\left(1 - \dfrac{\sin^2 2x}{4}\right) = \dfrac{\sin^2 2x - 4}{4}\cos 2x ;$

c) $\left(\cos^2 x + \sin^2 x\right)\left(\cos^4 x - \cos^2 x \sin^2 x + \sin^4 x\right) =$

$= \left(\cos^2 x + \sin^2 x\right)^2 - 2\cos^2 x \sin^2 x - \cos^2 x \sin^2 x = 1 - \dfrac{3}{4}\sin^2 2x =$

IV. 10. a) $\left(\dfrac{1 + \cos 2x}{2}\right)^2 - \cos^2 y + \dfrac{1}{4}\left(1 - \cos^2 2x\right) =$

$= \dfrac{1}{4}\left(2 + 2\cos 2x\right) - \cos^2 y = \cos^2 x - \cos^2 y =$

$= \left(\cos x - \cos y\right)\left(\cos x + \cos y\right) = -2\sin\dfrac{x-y}{2}\sin\dfrac{x+y}{2}\,2\cos\dfrac{x+y}{2}\cos\dfrac{x-y}{2} =$

$= -\sin(x+y)\sin(x-y);$

f) $\sin x + \sin y + \sin(x+y) = 2\sin\dfrac{x+y}{2}\cos\dfrac{x-y}{2} + 2\sin\dfrac{x+y}{2}\cos\dfrac{x+y}{2} =$

168

$$= 2\sin\frac{x+y}{2}\left(\cos\frac{x-y}{2}+\cos\frac{x+y}{2}\right) = 4\sin\frac{x+y}{2}\cos\frac{x}{2}\cos\frac{y}{2}\,.$$

X. 11. b) We have

$$\frac{\sin 4\alpha}{1+\cos 4\alpha}\frac{\cos 2\alpha}{1+\cos 2\alpha} = \frac{2\sin 2\alpha\cos 2\alpha}{1+2\cos^2 2\alpha -1}\frac{\cos 2\alpha}{1+2\cos^2\alpha -1} = \frac{\sin 2\alpha}{2\cos^2\alpha} = \tan\alpha\,;$$

e) We have

$$\frac{1+\tan x\tan\dfrac{x}{2}}{\cot x+\tan\dfrac{x}{2}} = \frac{\left(\cos x\cos\dfrac{x}{2}+\tan x\tan\dfrac{x}{2}\right)\sin x\cos\dfrac{x}{2}}{\cos x\cos\dfrac{x}{2}\left(\cos x\cos\dfrac{x}{2}+\sin x\sin\dfrac{x}{2}\right)} = \frac{\sin x}{\cos x} = \tan x\,;$$

g) $\dfrac{3+4\cos 2x+2\cos^2 2x-1}{3-4\cos 2x+2\cos^2 2x-1} = \dfrac{2(1+\cos 2x)^2}{2(1-\cos 2x)^2} = \cot^4 x\,;$

h) $\dfrac{1}{\tan 3a-\tan a}-\dfrac{1}{\cot 3a-\cot a} = \dfrac{1}{\dfrac{\sin 3a}{\cos 3a}-\dfrac{\sin a}{\cos a}}-\dfrac{1}{\dfrac{\cos 3a}{\sin 3a}-\dfrac{\cos a}{\sin a}} =$

$$= \frac{\cos a\cos 3a}{\cos a\sin 3a-\cos 3a\sin a}-\frac{\sin 3a\sin a}{\cos 3a\sin a-\cos a\sin 3a} =$$

$$= \frac{\cos a\cos 3a+\sin 3a\sin a}{\cos a\sin 3a-\cos 3a\sin a} = \frac{\cos 2a}{\sin 2a} = \cot 2a\,.$$

IV. 12.

a) $\dfrac{\tan 2x+\cot 3y}{\cot 2x+\tan 3y} = \dfrac{\dfrac{\sin 2x\sin 3x+\cos 2x\cos 3y}{\cos 2x\sin 3x}}{\dfrac{\sin 2x\cos 3y}{\cos 2x\cos 3y+\sin 2x\sin 3x}} = \dfrac{\tan 2x}{\tan 3y}.$

l) Since $\tan(\alpha\pm\beta) = \dfrac{\tan\alpha\pm\tan\beta}{1\mp\tan\alpha\tan\beta}$, then we have

$$\frac{\tan\alpha+\tan\beta}{\tan(\alpha+\beta)}+\frac{\tan\alpha-\tan\beta}{\tan(\alpha-\beta)}+2\tan^2\alpha =$$

$$= 1-\tan\alpha\tan\beta+1+\tan\alpha\tan\beta+2\tan^2\alpha = 2(1+\tan^2\alpha) = \frac{2}{\cos^2\alpha}\,.$$

IV. 13. a) $2\sqrt{2}\sin\dfrac{x}{2}\cos\left(\dfrac{x}{2}-\dfrac{\pi}{4}\right) = 2\sqrt{2}\sin\dfrac{x}{2}\left(\cos\dfrac{x}{2}\cos\dfrac{\pi}{4}+\sin\dfrac{x}{2}\sin\dfrac{\pi}{4}\right) =$

$$= 2\left(\sin\frac{x}{2}\cos\frac{x}{2}+\sin^2\frac{x}{2}\right) = \sin x+1-\cos x\,;\quad or$$

$$1 + \sin x - \cos x = 2\left(\sin^2 \frac{x}{2} + \sin \frac{x}{2}\cos \frac{x}{2}\right) =$$

$$= 2\sin \frac{x}{2}\left(\sin \frac{x}{2} + \cos \frac{x}{2}\right) = 2\sin \frac{x}{2}\left[\sin \frac{x}{2} + \sin\left(\frac{\pi}{2} - \frac{x}{2}\right)\right] =$$

$$= 2\sqrt{2}\,\sin \frac{x}{2}\cos\left(\frac{\pi}{4} - \frac{x}{2}\right);$$

b) $1 + \cos x + \cos 2x = 2\cos^2 x + \cos x = 2\cos x\left(\cos x + \frac{1}{2}\right) =$

$$= 2\cos x\left(\cos x + \cos \frac{\pi}{3}\right) = 4\cos x \cos\left(\frac{\pi}{6} + \frac{x}{2}\right)\cos\left(\frac{\pi}{6} - \frac{x}{2}\right);$$

c) $1 + \sin x + \cos x - \tan x = 2\sin^2 \frac{x}{2} + 2\sin \frac{x}{2}\cos \frac{x}{2} - \dfrac{2\sin \frac{x}{2}\cos \frac{x}{2}}{\cos x} =$

$$= 2\sin^2 \frac{x}{2} + 2\sin \frac{x}{2}\cos \frac{x}{2}\left(1 - \frac{1}{\cos x}\right) = 2\sin^2 \frac{x}{2} + 2\sin \frac{x}{2}\cos \frac{x}{2}\cdot\frac{\cos x - 1}{\cos x} =$$

$$= 2\sin^2 \frac{x}{2}\left(1 - \dfrac{2\sin \frac{x}{2}\cos \frac{x}{2}}{\cos x}\right) = \dfrac{2\sin^2 \frac{x}{2}}{\cos x}(\cos x - \sin x) =$$

$$= \dfrac{2\sin^2 \frac{x}{2}}{\cos x}\left[\sin\left(\frac{\pi}{2} - x\right) - \sin x\right] = \dfrac{4\sin^2 \frac{x}{2}}{\cos x}\sin\left(\frac{\pi}{4} - x\right)\cos \frac{\pi}{4} =$$

$$= \dfrac{2\sqrt{2}\,\sin^2 \frac{x}{2}\sin\left(\frac{\pi}{4} - x\right)}{\cos x} = \dfrac{2\sqrt{2}\,\sin \frac{x}{2}\cos\left(\frac{\pi}{4} + x\right)}{\cos x};$$

d) $\dfrac{\sin x \sin\left(\frac{\pi}{3} - x\right)\sin\left(\frac{\pi}{3} + x\right)}{\cos x \cos\left(\frac{\pi}{3} - x\right)\cos\left(\frac{\pi}{3} + x\right)} = \dfrac{\sin x\left(\cos 2x - \cos \frac{2\pi}{3}\right)}{\cos x\left(\cos 2x + \cos \frac{2\pi}{3}\right)} =$

$$= \dfrac{\sin x\left(1 - 2\sin^2 x + \frac{1}{2}\right)}{\cos x\left(2\cos^2 x - 1 - \frac{1}{2}\right)} = \dfrac{\sin x\left(3 - 4\sin^2 x\right)}{\cos x\left(4\cos^2 x - 3\right)} = \dfrac{\sin 3x}{\cos 3x} = \tan 3x\,.$$

IV. 15. b) $\dfrac{\cos 4x \sin 2x - \cos 2x \sin 4x}{\cos 2x} = \dfrac{\sin(2x - 4x)}{\cos 2x} = -\tan 2x$;

c) $\sin a + \sin 3a + \sin 5a + \sin 7a =$

$= 2\sin 2a \cos a + 2\sin 6a \cos a = 2\cos a(\sin 2a + \sin 6a) =$

$= 2\cos a \, 2\sin 4a \cos 2a = 4\cos a \sin 4a \cos 2a$;

g) $\tan x - \cot x + 2\tan 2x + 4\tan 4x + 8\cot 8x =$

$= \dfrac{\sin^2 x - \cos^2 x}{\sin x \cos x} + 2\tan 2x + 4\tan 4x + 8\cot 8x =$

$= \dfrac{-2\cos 2x}{\sin 2x} + \dfrac{2\sin 2x}{\cos 2x} + 4\tan 4x + 8\cot 8x =$

$= \dfrac{-2(\cos^2 2x - \sin^2 2x)}{\sin 2x \cos 2x} + 4\tan 4x + 8\cot 8x =$

$= \dfrac{-4\cos 4x}{\sin 4x} + \dfrac{\sin 4x}{\cos 4x} + 8\cot 8x = \dfrac{-4(\cos^2 4x - \sin^2 4x)}{\sin 4x \cos 4x} + 8\cot 8x = 0$;

h) $\dfrac{\sin 3x \cos x - \cos 3x \sin x}{\cos 3x \cos x} - \tan 2x = \dfrac{\sin(3x - x)}{\cos 3x \cos x} - \dfrac{\sin 2x}{\cos 2x} =$

$= \sin 2x\left(\dfrac{1}{\cos 3x \cos x} - \dfrac{1}{\cos 2x}\right) = \sin 2x \dfrac{\cos(3x - x) - \cos 3x \cos x}{\cos 3x \cos x} = \tan 3x \tan 2x \tan x.$

X. 17. a) $2\left(\dfrac{1}{2} - \sin 50°\right) = 2(\sin 30° - \sin 50°) = 4\sin(-10°)\cos 40° =$

$= -4\dfrac{\sin 10° \cos 40° \sin 40°}{\sin 40°} = \dfrac{-2\sin 10° \sin 80°}{\sin 40°} =$

$= \dfrac{-2\sin 10° \cos 10°}{\sin 40°} = \dfrac{-\sin 20°}{2\sin 20° \cos 20°} = \dfrac{-1}{2\cos 20°} = \dfrac{1}{2}\sec 160°$;

b) $\tan 20° + 4\sin 20° = \dfrac{\sin 20° + 4\sin 20° \cos 20°}{\cos 20°} = \dfrac{\sin 20° + 2\sin 40°}{\cos 20°} =$

$= \dfrac{(\sin 20° + \sin 40°) + \sin 40°}{\cos 20°} = \dfrac{2\sin 30° \cos 10° + \sin 40°}{\cos 20°} =$

$= \dfrac{\cos 10° + \sin 40°}{\cos 20°} = \dfrac{\sin 80° + \sin 40°}{\cos 20°} = \dfrac{2\sin 60° \cos 20°}{\cos 20°} = \sqrt{3},$

We used $\sin \alpha + \sin \beta = 2\sin \dfrac{\alpha + \beta}{2}\cos \dfrac{\alpha - \beta}{2}$, $\sin \alpha = \cos(90° - \alpha)$;

c)
$$\frac{\left(\dfrac{\sin 1^\circ}{\cos 1^\circ}+\dfrac{\sin 45^\circ}{\cos 45^\circ}\right)\left(\dfrac{\sin 31^\circ}{\cos 31^\circ}+\dfrac{\sin 45^\circ}{\cos 45^\circ}\right)-2}{\left(\dfrac{\sin 1^\circ}{\cos 1^\circ}-\dfrac{\sin 45^\circ}{\cos 45^\circ}\right)\left(\dfrac{\sin 31^\circ}{\cos 31^\circ}-\dfrac{\sin 45^\circ}{\cos 45^\circ}\right)-2}=$$

$$=\frac{\dfrac{\sin 46^\circ}{\cos 1^\circ \cos 45^\circ}\dfrac{\sin 76^\circ}{\cos 31^\circ \cos 45^\circ}-2}{\dfrac{\sin 44^\circ}{\cos 1^\circ \cos 45^\circ}\dfrac{\sin 14^\circ}{\cos 31^\circ \cos 45^\circ}-2}=\frac{\sin 46^\circ \sin 76^\circ-\cos 1^\circ \cos 31^\circ}{\sin 44^\circ \sin 14^\circ-\cos 1^\circ \cos 31^\circ}=$$

$$=\frac{\cos 122^\circ+\cos 32^\circ}{\cos 58^\circ+\cos 32^\circ}=\tan 13^\circ;$$

d) $\sqrt{1+\sin 20^\circ}=\sqrt{\left(\sin 10^\circ+\cos 10^\circ\right)^2}=\sin 10^\circ+\cos 10^\circ.$

IV. 18. a) $\csc x+\cot x=\dfrac{1+\cos x}{\sin x}=\dfrac{2\cos^2\dfrac{x}{2}}{2\sin\dfrac{x}{2}\cos\dfrac{x}{2}}=\cot\dfrac{x}{2};$

b) $(\tan x+\cot x)+(\tan 3x+\cot 3x)=\dfrac{2}{\sin 2x}+\dfrac{2}{\sin 6x}=\dfrac{2(\sin 6x+\sin 2x)}{\sin 6x\sin 2x}=\dfrac{8\cos^2 2x}{\sin 6x}.$

c) $\sec 2x-\tan 2x=\dfrac{1}{\cos 2x}-\dfrac{\sin 2x}{\cos 2x}=\dfrac{(\cos x-\sin x)^2}{\cos^2 x-\sin^2 x}=\dfrac{\cos x-\sin x}{\cos x+\sin x}=$

$$=\frac{\cos x-\cos\left(\dfrac{\pi}{2}-x\right)}{\cos x+\cos\left(\dfrac{\pi}{2}-x\right)}=\tan\left(\frac{\pi}{4}-x\right)=\tan\left(\frac{5\pi}{4}-x\right);$$

d) $\sin 2x\left(\dfrac{\cos x\cos 2x-\cos x+\sin x\sin 2x}{\cos x\cos 2x}\right)+\dfrac{1+\cos\left(\dfrac{\pi}{2}-x\right)}{1-\cos\left(\dfrac{\pi}{2}-x\right)}=$

$$\sin 2x\frac{\cos(x-2x)-\cos x}{\cos x\cos 2x}+\frac{2\cos^2\left(\dfrac{\pi}{4}-\dfrac{x}{2}\right)}{2\sin^2\left(\dfrac{\pi}{4}-\dfrac{x}{2}\right)}=\cot^2\left(\frac{\pi}{4}-\frac{x}{2}\right)=\tan^2\left(\frac{\pi}{4}+\frac{x}{2}\right);$$

e) $2\sin x\left(\cos 2x-\cos\dfrac{2\pi}{3}\right)=2\sin x\left(\cos 2x+\dfrac{1}{2}\right)=2\sin x\left(1-2\sin^2 x+\dfrac{1}{2}\right)=$

$$= \sin x\left(3 - 4\sin^2 x\right) = \sin 3x \; ;$$

g) $\left[(1 + \tan x) + \sec x\right]\left[(1 + \tan x) - \sec x\right] = (1 + \tan x)^2 - \sec^2 x =$

$$= 1 + 2\tan x + \tan^2 x - \sec^2 x = 2\tan x \; .$$

IV. 19. a) $\dfrac{2\sin a - 2\sin a \cos a}{2\sin a + 2\sin a \cos a} = \dfrac{2\sin a(1 - \cos a)}{2\sin a(1 + \cos a)} = \dfrac{1 - \cos a}{1 + \cos a} =$

$$= \dfrac{2\sin^2 \dfrac{a}{2}}{2\cos^2 \dfrac{a}{2}} = \tan^2 \dfrac{a}{2}; \quad \textbf{b)} \; \dfrac{1 + \sin 2a + \cos 2a}{1 + \sin 2a - \cos 2a} = \dfrac{(\sin a + \cos a)^2 + \cos^2 a - \sin^2 a}{(\sin a + \cos a)^2 - \cos^2 a + \sin^2 a} =$$

$$= \dfrac{(\sin a + \cos a)(\sin a + \cos a + \cos a - \sin a)}{(\sin a + \cos a)(\sin a + \cos a - \cos a + \sin a)} = \dfrac{2\cos a}{2\sin a} = \cot a \; ;$$

c) $\dfrac{2\sin 3a \cos 4a + 2\sin 3a}{2\sin 5a \cos 4a + 2\sin 5a} = \dfrac{2\sin 3a(\cos 4a + 1)}{2\sin 5a(\cos 4a + 1)} = \dfrac{\sin 3a}{\sin 5a} \; ;$

d) $\dfrac{(\sin 2x + \sin 4x) - \sin 3x}{(\cos 2x + \cos 4x) - \cos 3x} = \dfrac{2\sin 3x \cos 2x - \sin 3x}{2\cos 3x \cos 2x - \cos 3x} = \tan 3x \; .$

IV. 20. a) $1 - \dfrac{1 - \tan 2x}{1 + \tan 2x}\cos 4x = 1 - \dfrac{\cos 2x - \sin 2x}{\cos 2x + \sin 2x}\left(\cos^2 2x - \sin^2 2x\right) =$

$$= 1 - \left(\cos 2x - \sin 2x\right)^2 = \sin 4x \; ;$$

b) $\dfrac{\cos x \cos 2x + \sin x \sin 2x}{\cos x \cos 2x} \dfrac{\sin x \cos x}{\cos^2 x + \sin^2 x} = \dfrac{\cos(x - 2x)}{\cos x \cos 2x} \dfrac{\sin x \cos x}{1} = \dfrac{1}{2}\tan 2x \; ;$

d) $2\left[\csc 4x + \cot 4x\right] + \tan x = \dfrac{2(1 + \cos 4x)}{\sin 4x} + \tan x = \dfrac{4\cos^2 2x}{\sin 4x} + \tan x =$

$$= \dfrac{2\cos 2x}{\sin 2x} + \dfrac{\sin x}{\cos x} = \dfrac{2\cos 2x + 2\sin^2 x}{2\sin x \cos x} = \dfrac{2\cos^2 x - 1}{2\sin x \cos x} = \cot 2x \; ;$$

e) $\tan x \tan\left(x + \dfrac{\pi}{3}\right)\tan\left(x + \dfrac{2\pi}{3}\right) = \tan x \dfrac{\sin\left(x + \dfrac{\pi}{3}\right)\sin\left(x + \dfrac{2\pi}{3}\right)}{\cos\left(x + \dfrac{\pi}{3}\right)\cos\left(x + \dfrac{2\pi}{3}\right)} =$

$$= \tan x \dfrac{\cos\dfrac{\pi}{3} - \cos(2x + \pi)}{\cos\dfrac{\pi}{3} + \cos(2x + \pi)} = \tan x \dfrac{1 + 2\cos 2x}{1 - 2\cos 2x} =$$

$$= \frac{\sin x}{\cos x} \cdot \frac{3 - 4\sin^2 x}{-4\cos^2 x + 3} = \frac{\sin 3x}{-\cos 3x} = -\tan 3x \,.$$

IV. 21. e) $\cos^2\left(\frac{\pi}{8} + x\right) - \sin^2\left(\frac{\pi}{8} - x\right) = \frac{1}{2}\left[1 + \cos\left(\frac{\pi}{4} + 2x\right) - 1 + \cos\left(\frac{\pi}{4} - 2x\right)\right] =$

$$= \frac{\cos 2x}{\sqrt{2}}; \quad \textbf{f)} \ \cos^2\left(\frac{5\pi}{8} + \frac{x}{2}\right) - \cos^2\left(\frac{3\pi}{8} + \frac{x}{2}\right) =$$

$$= \frac{1}{2}\left[1 + \cos\left(\frac{5\pi}{4} + x\right) - 1 - \cos\left(\frac{3\pi}{4} - x\right)\right] = \frac{\sin x}{\sqrt{2}}\,;$$

g) $\sin^2 2x + \frac{1}{2}\left(\cos 4x - \cos\frac{\pi}{3}\right) = \sin^2 2x + \frac{1}{2}\left(1 - 2\sin^2 2x - \frac{1}{2}\right) = \frac{1}{4}\,.$

IV. 22. a) $\dfrac{2\sin\dfrac{2x - 3x}{2}\cos\dfrac{2x + 3x}{2} + 2\sin\dfrac{5x}{2}\cos\dfrac{5x}{2}}{2\cos\dfrac{x + 4x}{2}\cos\dfrac{x - 4x}{2}} = \dfrac{\sin\dfrac{2x - 3x}{2} + \sin\dfrac{5x}{2}}{\cos\dfrac{x - 4x}{2}} =$

$$= \frac{2\sin\dfrac{1}{2}\left(-\dfrac{x}{2} + \dfrac{5x}{2}\right)\cos\dfrac{1}{2}\left(-\dfrac{x}{2} - \dfrac{5x}{2}\right)}{\cos\dfrac{3x}{2}} = 2\sin x\,;$$

b) $\dfrac{\sin 4x \sin 2x + \cos 4x \cos 2x}{\sin 2x} + \dfrac{1}{2}\left(\dfrac{\sin^2 x - \cos^2 x}{\sin^2 x}\right)\tan x =$

$$= \frac{\cos(4x - 2x)}{\sin 2x} - \frac{1}{2}\left(\frac{\cos 2x}{\sin x}\right)\frac{1}{\cos x} = 0\,;$$

c) $\dfrac{\sin 7x - 2(\sin x \cos 2x + \sin x \cos 4x + \sin x \cos 6x)}{\sin x} =$

$$= \frac{\sin 7x - \sin(x + 2x) - \sin(x - 2x) - \sin(x + 4x) - \sin(x - 4x) - \sin(x + 6x) - \sin(x - 6x)}{\sin x} = 1.$$

IV. 26. e) $\dfrac{1 + \sin 2x - 2\sin^2 x}{1 + \cot^2 x} = \dfrac{\sin^2 x + \cos^2 x + \sin 2x - 2\sin^2 x}{1 + \dfrac{\cos^2 x}{\sin^2 x}} =$

$$= \sin^2 x(\sin 2x + \cos 2x) = \sin^2 x\left[\sin 2x + \sin\left(\frac{\pi}{2} - 2x\right)\right] =$$

$$= 2\sin^2 x\left[\sin\frac{\pi}{4}\cos\left(\frac{\pi}{4}-2x\right)\right] = \sqrt{2}\sin^2 x\cos\left(\frac{\pi}{4}-2x\right).$$

IV. 27. a) $\dfrac{\sin 2x}{\cot\left(x+\dfrac{\pi}{4}\right)(1+\sin 2x)} = \dfrac{\sin 2x\tan\left(x+\dfrac{\pi}{4}\right)}{(1+\sin 2x)} =$

$$= \frac{\sin 2x(1+\tan x)}{(1-\tan x)\left(\sin^2 x+\cos^2 x+2\sin x\cos x\right)} = \frac{\sin 2x(\sin x+\cos x)}{(\cos x-\sin x)(\sin x+\cos x)^2} =$$

$$= \frac{\sin 2x}{\cos^2 x-\sin^2 x} = \tan 2x\,.$$

IV. 28. a) $2(1+\cos 2x)^2 - 4\cos 2x - \cos 4x =$

$$= 2+4\cos 2x+2\cos^2 2x-4\cos 2x-\left(2\cos^2 2x-1\right)= 3\,;$$

b) $\sin x\left(16\sin^4 x-20\sin^2 x+5\right)= \sin x\left(16\sin^2 x\left(1-\cos^2 x\right)\right)-20\sin^2 x+5\right)=$

$\sin x\left(-4\sin^2 2x-4\sin^2 x+5\right)= \sin x\left[-2(1-\cos 4x)-2(1-\cos 2x)+5\right]=$

$= \sin x(2\cos 4x+2\cos 2x+1)= 2\sin x\cos 4x+2\sin x\cos 2x+\sin x =$

$= 2\sin x\cos 4x+2\sin x\cos 2x+\sin x = \sin 5x-\sin 3x+\sin 3x-\sin x+\sin x = \sin 5x.$

IV. 29. a) $\left(\sin^2\dfrac{\pi}{8}\right)^2+\left(\cos^2\dfrac{3\pi}{8}\right)^2+\left(\sin^2\dfrac{5\pi}{8}\right)^2+\left(\cos^2\dfrac{7\pi}{8}\right)^2 =$

$$= \left(\frac{1-\cos\dfrac{\pi}{4}}{2}\right)^2+\left(\frac{1+\cos\dfrac{3\pi}{4}}{2}\right)^2+\left(\frac{1-\cos\dfrac{5\pi}{4}}{2}\right)^2+\left(\frac{1+\cos\dfrac{7\pi}{4}}{2}\right)^2 =$$

$$= \frac{1}{4}\left[\left(1-\frac{\sqrt{2}}{2}\right)^2+\left(1-\frac{\sqrt{2}}{2}\right)^2+\left(1+\frac{\sqrt{2}}{2}\right)^2+\left(1+\frac{\sqrt{2}}{2}\right)^2\right]=\frac{3}{2}\,.$$

IV. 30. b) We have $\dfrac{\cos 7°+\sin 7°+\sqrt{3}\left(\cos 7°-\sin 7°\right)}{2\sin\left(60°+8°\right)} =$

$$= \frac{\sqrt{2}\cos\left(45°-7°\right)+\sqrt{6}\sin\left(45°-7°\right)}{2\sin 68°} = \frac{\sqrt{2}\cos 38°+\sqrt{6}\sin 38°}{2\sin 68°} =$$

$$= \frac{\sqrt{2}\sin\left(30°+38°\right)}{\sin 68°} = \sqrt{2}\,;$$

IV. 31. c) $\cos^2 20° \sin^2 50° \cos^2 80° = \dfrac{1}{4}\cos^2 20°(\sin 130° - \sin 30°)^2 =$

$$= \frac{1}{4}\left(\cos 40° \cos 20° - \frac{\cos 20°}{2}\right)^2 = \frac{1}{4}\left(\frac{\cos 60° + \cos 20°}{2} - \frac{\cos 20°}{2}\right)^2 = \frac{1}{64}.$$

IV. 32.

c) $\sin 40° + 4\cos 20°(\sin 50° - \sin 30°) = \cos 50° + 2(\sin 70° + \sin 30°) - 2\cos 20° =$
$= \cos 50° + 2\cos 20° + 1 - 2\cos 20° = \cos 50° + 1 = 2\cos^2 25°;$

d) $\cos 70° + 4(\cos 60° + \cos 20°)\cos 80° = \cos 70° + 2\cos 80° + 4\cos 20° \cos 80° =$
$= \cos 70° + 2\cos 80° + 2\cos 100° + 2\cos 60° = \cos 70° + 2\cos 80° + 2\cos 100° + 1 =$
$= \cos 70° + 2\sin 10° - 2\sin 10° + 1 = 2\cos^2 35°;$

e) $\dfrac{\sin 9°}{\cos 9°} - \dfrac{\sin 27°}{\cos 27°} + \dfrac{\cos 9°}{\sin 9°} - \dfrac{\cos 27°}{\sin 27°} =$

$$= \frac{\sin 9° \cos 27° - \cos 9° \sin 27°}{\cos 9° \cos 27°} + \frac{\cos 9° \sin 27° - \sin 9° \cos 27°}{\sin 9° \sin 27°} =$$

$$= \frac{\sin(9° - 27°)}{\cos 9° \cos 27°} + \frac{\sin(27° - 9°)}{\sin 9° \sin 27°} = \sin 18°\left(\frac{-1}{\cos 9° \cos 27°} + \frac{1}{\sin 9° \sin 27°}\right) =$$

$$= \sin 18°\left(\frac{\cos(27° + 9°)}{\cos 9° \sin 9° \sin 27° \cos 27°}\right) = 4\sin 18°\left(\frac{\cos 36°}{\sin 18° \sin 54°}\right) = 4;$$

f) $\tan 30° + \tan 50° + \tan 40° + \tan 60° = \dfrac{\sin(30° + 50°)}{\cos 30° \cos 50°} + \dfrac{\sin(40° + 60°)}{\cos 40° \cos 60°} =$

$$= \cos 10°\left(\frac{1}{\cos 30° \cos 50°} + \frac{1}{\cos 40° \cos 60°}\right) = \cos 10°\frac{\sin 50° \sin 30° + \cos 30° \cos 50°}{\cos 30° \cos 50° \cos 40° \cos 60°} =$$

$$= \cos 10°\frac{\cos 20°}{\cos 30° \cos 50° \sin 50° \sin 30°} = 4\cos 10°\frac{\sin 70°}{\sin 60° \sin 100°} = \frac{8}{\sqrt{3}}\sin 70°.$$

IV. 33. Denote $x_1 = \tan 5°$. Since $\tan 3x = \dfrac{3\tan x - \tan^3 x}{1 - 3\tan^2 x}$, we have

$$\tan 15° = \frac{3\tan 5° - \tan^3 5°}{1 - 3\tan^2 5°} = \frac{3x_1 - x_1^3}{1 - 3x_1^2} \text{ or }$$

$$x_1^3 - 3\tan 15° x_1^2 - 3x_1 + \tan 15° = 0.$$

Therefore $x_1 = \tan 5°$ is a root of the equation

$$x^3 - 3\tan 15° x^2 - 3x + \tan 15° = 0 \ (1).$$

The equation (1) can also be written

$$x^3 - 3\left(\frac{3x_1 - x_1^3}{1 - 3x_1^2}\right)x^2 - 3x + \frac{3x_1 - x_1^3}{1 - 3x_1^2} = 0 \ (2).$$ Dividing (2) by $x - x_1$

we obtain the equation $\left(1 - 3x_1^2\right)x^2 - 8x_1x + x_1^2 - 3 = 0$ with the roots

$$x_2 = \frac{x_1 + \sqrt{3}}{1 - \sqrt{3}x_1} \text{ and } x_3 = \frac{x_1 - \sqrt{3}}{1 + \sqrt{3}x_1}.$$

On the other hand $\tan 25° = \tan\left(30° - 5°\right) = \dfrac{\tan 30° - \tan 5°}{1 + \tan 30° \tan 5°} = \dfrac{1}{x_2}$,

therefore $x_2 = \cot 25°$. Similarly $x_3 = -\cot 35°$. According to Vieta's

formulae in the equation (1) $\tan 15° = -x_1 x_2 x_3$ and the result follows.

IV. 34. a) $\dfrac{\cos x - \sin x}{\cos x + \sin x}\left(\sin^2 x + \cos^2 x + 2\sin x \cos x\right)\dfrac{1}{\cos 2x} + 2\cos 4x =$

$$= \frac{\cos x - \sin x}{\cos x + \sin x}\left(\sin x + \cos x\right)^2 \frac{1}{\cos^2 x - \sin^2 x} + 2\cos 4x =$$

$$= 1 + 2\cos 4x = \frac{\sin 2x\left(1 + 2\cos 4x\right)}{\sin 2x} = \frac{\sin 2x + 2\sin 2x \cos 4x}{\sin 2x} = \frac{\sin 6x}{\sin 2x};$$

c) $\sin 3x \sin^3 x - \cos 3x \cos^3 x = \sin 3x \dfrac{3\sin x - \sin 3x}{4} - \cos 3x \dfrac{3\cos x + \cos 3x}{4} =$

$$= \frac{1}{4}\left(3\sin x \sin 3x - \sin^2 3x + 3\cos x \cos 3x + \cos^2 3x\right) =$$

$$= \frac{1}{4}\left(3\cos(3x - x) + \cos 6x\right) = \frac{1}{4}\left(3\cos 2x + 4\cos^2 2x - 3\cos 2x\right) = \cos^3 2x.$$

d) $2(1 + \cos 4x)^2 + \cos 6x + 3\cos 2x - 4(1 + \cos 4x) - 3\cos 2x + 1 =$

$$= 2 + 4\cos 4x + 2\cos^2 4x + \cos 6x - 4(1 + \cos 4x) + 1 =$$

$$= 1 + \cos 8x + \cos 6x - 1 = 2\cos 7x \cos 2x;$$

IV. 35. a) $\sin\dfrac{3\pi}{10} - \sin\dfrac{\pi}{10} = \dfrac{1}{2\cos\dfrac{\pi}{10}}\left(2\cos\dfrac{\pi}{10}\sin\dfrac{3\pi}{10} - 2\cos\dfrac{\pi}{10}\sin\dfrac{\pi}{10}\right) =$

$$= \frac{1}{2\cos\dfrac{\pi}{10}}\left(\sin\frac{2\pi}{5} + \sin\frac{\pi}{5} - \sin\frac{\pi}{5}\right) = \frac{\cos\left(\dfrac{\pi}{2} - \dfrac{2\pi}{5}\right)}{2\cos\dfrac{\pi}{10}} = \frac{1}{2};$$

b) $\cos\dfrac{1}{5}\dfrac{\pi}{5} - \cos\dfrac{2\pi}{5} = \cos\left(2\pi - \dfrac{\pi}{5}\right) - \sin\left(\dfrac{\pi}{2} - \dfrac{2\pi}{5}\right) = \cos\dfrac{\pi}{5} - \sin\dfrac{\pi}{10} =$

$$= \dfrac{2\sin\dfrac{\pi}{5}\cos\dfrac{\pi}{5} - 2\sin\dfrac{\pi}{5}\sin\dfrac{\pi}{10}}{2\sin\dfrac{\pi}{5}} = \dfrac{\sin\dfrac{\pi}{10} - \left(\cos\dfrac{\pi}{10} - \cos\dfrac{3\pi}{10}\right)}{2\sin\dfrac{\pi}{5}} =$$

$$= \dfrac{\sin\dfrac{\pi}{10} - \left(\cos\dfrac{\pi}{10} - \cos\dfrac{3\pi}{10}\right)}{2\sin\dfrac{\pi}{5}} = \dfrac{\sin\dfrac{\pi}{10} - \left[\sin\left(\dfrac{\pi}{2} - \dfrac{\pi}{10}\right) - \sin\left(\dfrac{\pi}{2} - \dfrac{3\pi}{10}\right)\right]}{2\sin\dfrac{\pi}{5}} =$$

$$= \dfrac{\sin\dfrac{\pi}{10} - \left(\sin\dfrac{\pi}{10} - \sin\dfrac{\pi}{5}\right)}{2\sin\dfrac{\pi}{5}} = \dfrac{1}{2};$$

c) $4\left(\cos\dfrac{2\pi}{3} + \cos\dfrac{2\pi}{9}\right)\cos\dfrac{\pi}{9} = 4\left(-\dfrac{1}{2}\cos\dfrac{\pi}{9} + \cos\dfrac{2\pi}{9}\cos\dfrac{\pi}{9}\right) =$

$$= 2\left(-\cos\dfrac{\pi}{9} + 2\cos\dfrac{2\pi}{9}\cos\dfrac{\pi}{9}\right) = 2\left(-\cos\dfrac{\pi}{9} + \cos\dfrac{\pi}{3} + \cos\dfrac{\pi}{9}\right) = 1;$$

d) $\dfrac{2\cos\dfrac{\pi}{14}\left(\sin\dfrac{\pi}{14} + \cos\dfrac{2\pi}{14} - \sin\dfrac{3\pi}{14}\right)}{2\cos\dfrac{\pi}{14}} =$

$$\dfrac{\sin\dfrac{2\pi}{14} + \cos\dfrac{3\pi}{14} + \cos\dfrac{\pi}{14} - \sin\dfrac{4\pi}{14} - \sin\dfrac{2\pi}{14}}{2\cos\dfrac{\pi}{14}} = \dfrac{1}{2}, \text{ where } \cos\dfrac{3\pi}{14} = \sin\dfrac{4\pi}{14};$$

e) $\dfrac{2\cos\dfrac{\pi}{14}\left(\sin\dfrac{\pi}{14}\cos\dfrac{2\pi}{14}\sin\dfrac{3\pi}{14}\right)}{2\cos\dfrac{\pi}{14}} = \dfrac{2\sin\dfrac{2\pi}{14}\cos\dfrac{2\pi}{14}\sin\dfrac{3\pi}{14}}{4\cos\dfrac{\pi}{14}} = \dfrac{\sin\dfrac{4\pi}{14}\sin\dfrac{3\pi}{14}}{4\cos\dfrac{\pi}{14}} =$

$$= \dfrac{\cos\dfrac{3\pi}{14}\sin\dfrac{3\pi}{14}}{4\cos\dfrac{\pi}{14}} = \dfrac{\sin\dfrac{6\pi}{14}}{8\cos\dfrac{\pi}{14}} = \dfrac{1}{8}; \quad \textbf{g)} \ \cos\dfrac{\pi}{5} + \cos\dfrac{2\pi}{5} + \cos\dfrac{4\pi}{5} + \cos\dfrac{6\pi}{5} =$$

$$= \dfrac{1}{2\sin\dfrac{\pi}{5}}\left(2\sin\dfrac{\pi}{5}\cos\dfrac{\pi}{5} + 2\sin\dfrac{\pi}{5}\cos\dfrac{2\pi}{5} + 2\sin\dfrac{\pi}{5}\cos\dfrac{4\pi}{5} + 2\sin\dfrac{\pi}{5}\cos\dfrac{6\pi}{5}\right) =$$

178

$$= \frac{1}{2\sin\frac{\pi}{5}}\left(\sin\frac{2\pi}{5}+\sin\frac{3\pi}{5}-\sin\frac{\pi}{5}+\sin\pi-\sin\frac{3\pi}{5}+\sin\frac{7\pi}{5}-\sin\pi\right)=$$

$$= \frac{1}{2\sin\frac{\pi}{5}}\left(\sin\frac{2\pi}{5}-\sin\frac{\pi}{5}+\sin\frac{7\pi}{5}\right)=\frac{1}{2\sin\frac{\pi}{5}}\left(\sin\frac{2\pi}{5}-\sin\frac{\pi}{5}-\sin\frac{2\pi}{5}\right)=-\frac{1}{2};$$

h) Multiply on both sides by $2\cos\dfrac{\pi}{22}$.

IV. 36. a) $\dfrac{2\sin\dfrac{\pi}{33}\cos\dfrac{\pi}{33}\cos\dfrac{2\pi}{33}\cos\dfrac{4\pi}{33}\cos\dfrac{8\pi}{33}\cos\dfrac{16\pi}{33}}{2\sin\dfrac{\pi}{33}} =$

$$= \frac{2\sin\dfrac{2\pi}{33}\cos\dfrac{2\pi}{33}\cos\dfrac{4\pi}{33}\cos\dfrac{8\pi}{33}\cos\dfrac{16\pi}{33}}{4\sin\dfrac{\pi}{33}} = \frac{2\sin\dfrac{4\pi}{33}\cos\dfrac{4\pi}{33}\cos\dfrac{8\pi}{33}\cos\dfrac{16\pi}{33}}{8\sin\dfrac{\pi}{33}} =$$

$$= \frac{2\sin\dfrac{8\pi}{33}\cos\dfrac{8\pi}{33}\cos\dfrac{16\pi}{33}}{16\sin\dfrac{\pi}{33}} = \frac{2\sin\dfrac{16\pi}{33}\cos\dfrac{16\pi}{33}}{32\sin\dfrac{\pi}{33}} = \frac{\sin\dfrac{32\pi}{33}}{32\sin\dfrac{\pi}{33}} = \frac{\sin\left(\pi-\dfrac{32\pi}{33}\right)}{32\sin\dfrac{\pi}{33}} = \frac{1}{32}.$$

IV. 37. a) $\dfrac{1-\cos 2\alpha}{2}-\dfrac{1+\cos 2\beta}{2}+2\cos\alpha\cos\beta\cos(\alpha-\beta)=$

$$= -\frac{\cos 2\alpha+\cos 2\beta}{2}+2\cos\alpha\cos\beta\cos(\alpha-\beta)=$$
$$= -\cos(\alpha+\beta)\cos(\alpha-\beta)+2\cos\alpha\cos\beta\cos(\alpha-\beta)=$$
$$= -\cos(\alpha-\beta)[\cos(\alpha+\beta)-2\cos\alpha\cos\beta]=$$
$$= -\cos(\alpha-\beta)[\cos\alpha\cos\beta-\sin\alpha\sin\beta-2\cos\alpha\cos\beta]=\cos^2(\alpha-\beta);$$

e) $(\sin\alpha+\sin\beta)^2+(\cos\alpha+\cos\beta)^2=(\sin^2\alpha+\cos^2\alpha)+(\sin^2\beta+\cos^2)+$

$+2(\cos a\cos b+\sin a\sin b)=2+2\cos(\alpha-\beta)=4\cos^2\dfrac{\alpha-\beta}{2}$, we used the

relation $1+\cos\alpha=2\cos^2\dfrac{\alpha}{2}$.

IV. 38. f) $2\sin\dfrac{a+b}{2}\cos\dfrac{a-b}{2} - 2\sin\dfrac{a+b}{2}\cos\dfrac{a+b+c}{2} =$

$= 2\sin\dfrac{a+b}{2}\left(\cos\dfrac{a-b}{2} - \cos\dfrac{a+b+c}{2}\right) =$

$= 2\sin\dfrac{a+b}{2}\left(2\sin\dfrac{2a+2c}{4}\sin\dfrac{2b+2c}{4}\right) =$

$= 4\sin\dfrac{a+b}{2}\sin\dfrac{a+c}{2}\sin\dfrac{b+c}{2}\,.$

IV. 39. b) $\sqrt{2+2\cos x} = \sqrt{2(1+\cos x)} = \sqrt{2\cdot 2\cos^2\dfrac{x}{2}} = \cos\dfrac{x}{2}\,;$

c) $\left(\dfrac{\sin x + \cos x}{\sqrt{\tan x} + \sqrt{\cot x}}\right)^2 = \dfrac{(\sin x + \cos x)^2}{\tan x + \cot x + 2} = \dfrac{(\sin x + \cos x)^2 \tan x}{(\tan x + 1)^2} =$

$= \dfrac{(\sin x + \cos x)^2 \cos^2 x \tan x}{(\sin x + \cos x)^2} = \sin x \cos x\,;$

g) We have $\sqrt[4]{1+\sin 2x} = \sqrt[4]{(\sin x + \cos x)^2} = \sqrt{\sin x + \cos x}\,,$

$\sqrt[4]{1-\sin 2x} = \sqrt[4]{(\sin x - \cos x)^2} = \sqrt{\sin x - \cos x}$ because $\sin x > \cos x\,.$

Therefore $\dfrac{\sqrt[4]{1+\sin 2x} + \sqrt[4]{1-\sin 2x}}{\sqrt[4]{1+\sin 2x} - \sqrt[4]{1-\sin 2x}} = \dfrac{\sqrt{\sin x + \cos x} + \sqrt{\sin x - \cos x}}{\sqrt{\sin x + \cos x} - \sqrt{\sin x - \cos x}} =$

$= \dfrac{2\sin x + 2\sqrt{\sin^2 x - \cos^2 x}}{2\cos x} = \tan x + \sqrt{\dfrac{\sin^2 x - \cos^2 x}{\cos^2 x}} = \tan x + \sqrt{\tan^2 x - 1}\,;$

h) $2\sin 3x + 3\left(\sin x + \sqrt{3}\cos x\right) = -2\sin 3\left(x+\dfrac{\pi}{3}\right) + 6\sin\left(x+\dfrac{\pi}{3}\right) = 8\sin^3\left(x+\dfrac{\pi}{3}\right).$

Similarly $2\cos 3x + 3\left(\sqrt{3}\sin x - \cos x\right) = -8\cos^3\left(x+\dfrac{\pi}{3}\right).$

UNIT V

Trigonometric Equations

V. 1. a) $x = 90°$; **b)** $x_1 = 30°$, $x_2 = 150°$; **c)** $x = 180°$;

d) $x_1 = 135°$, $x_2 = 225°$;

e) $x_1 = 135°$, $x_2 = 221.81°$; **f)** $x_1 = 78.46°$, $x_2 = 70.39°$, $x_3 = 99.61°$, $x_4 = 160.39°$, $x_5 = 189.61°$, $x_6 = 250.39°$, $x_7 = 279.61°$, $x_8 = 340.39°$;

g) $x_1 = 94.8°$, $x_2 = 175.2°$, $x_3 = 274.8°$, $x_4 = 355.2°$;

h) $x_1 = 40°$, $x_2 = 80°$, $x_3 = 160°$, $x_4 = 200°$, $x_5 = 280°$, $x_6 = 320°$;

i) $x_1 = 5°$, $x_2 = 25°$, $x_3 = 65°$, $x_4 = 85°$, $x_5 = 125°$, $x_6 = 145°$, $x_7 = 185°$, $x_8 = 205°$, $x_9 = 245°$, $x_{10} = 265°$, $x_{11} = 305°$, $x_{12} = 325°$;

j) $x_1 = 14.04°$, $x_2 = 194.04°$;

k) $x_1 = 71.23°$, $x_2 = 108.77°$, $x_3 = 131.23°$, $x_4 = 168.7°$, $x_5 = 191.23°$, $x_6 = 228.77°$, $x_7 = 251.23°$, $x_8 = 288.77°$, $x_9 = 311.23°$, $x_{10} = 348.77°$;

l) $x_1 = 104.04°$, $x_2 = 284.04$;

m) $x_1 = 37.27°$, $x_2 = 97.27°$, $x_3 = 157.27°$, $x_4 = 217.27°$, $x_5 = 277.27°$, $x_6 = 337.27°$;

n) $x_1 = 49.87°$, $x_2 = 85.13°$, $x_3 = 139.87°$, $x_4 = 175.13°$, $x_5 = 229.87°$, $x_6 = 265.13°$, $x_7 = 319.87°$, $x_8 = 355.13°$;

o) $x_1 = 46.20°$, $x_2 = 73.80°$, $x_3 = 166.20°$, $x_4 = 193.80°$, $x_5 = 286.20°$, $x_6 = 313.80°$.

V. 2. a) Multiplying both side of this equation by $\dfrac{\sqrt{2}}{2}$, we obtain

$$\frac{\sqrt{2}}{2}\sin x + \frac{\sqrt{2}}{2}\cos x = \frac{\sqrt{2}}{2} \Rightarrow$$

$$\sin\frac{\pi}{4}\sin x + \cos\frac{\pi}{4}\cos x = \frac{\sqrt{2}}{2} \Rightarrow \cos\left(x - \frac{\pi}{4}\right) = \frac{\sqrt{2}}{2} \Rightarrow$$

$$x - \frac{\pi}{4} = \pm\arccos\frac{\sqrt{2}}{2} + 2k\pi \Rightarrow x - \frac{\pi}{4} = \pm\frac{\pi}{4} + 2k\pi .$$ And we have

1) $x - \dfrac{\pi}{4} = \dfrac{\pi}{4} + 2k\pi \Rightarrow x_1 = \dfrac{\pi}{2} + 2k\pi,\ \ k \in \mathbf{Z}$ and 2) $x_2 = 2k\pi,\ \ k \in \mathbf{Z}$;

b) $6\sin\dfrac{x}{2}\cos\dfrac{x}{2} - 2\left(\cos^2\dfrac{x}{2} - \sin^2\dfrac{x}{2}\right) = 3\left(\cos^2\dfrac{x}{2} + \sin^2\dfrac{x}{2}\right) \Rightarrow$

$$6\tan\frac{x}{2} - 2\left(1 - \tan^2\frac{x}{2}\right) = 3\left(1 + \tan^2\frac{x}{2}\right) \Rightarrow \tan^2\frac{x}{2} - 6\tan\frac{x}{2} + 5 = 0 .$$

Therefore $\tan\dfrac{x}{2} = 5$ and $\tan\dfrac{x}{2} = 1$.

The solutions are $x_1 = 2\operatorname{arccot}5 + 2k\pi$ and $x_2 = \dfrac{\pi}{2} + 2k\pi$;

c) $\cot^2 x - 6\cot x + 5 = 0$. Denote $\cot x = t$. Then $t^2 - 6t + 5 = 0$. The

roots are $t = 5$ and $t = 1$. The solutions are $x_1 = \arctan\dfrac{1}{5} + k\pi$ and

$x_2 = \dfrac{\pi}{4} + k\pi$.

d) $8\sin\dfrac{x}{2}\cos\dfrac{x}{2} + \cos^2\dfrac{x}{2} - \sin^2\dfrac{x}{2} = 4\cos^2\dfrac{x}{2} + 4\sin^2\dfrac{x}{2} \Rightarrow$

$$3\cot^2\frac{x}{2} - 8\cot\frac{x}{2} + 5 = 0 .$$ Denote $\cot\dfrac{x}{2} = t$. Then $3t^2 - 8t + 5 = 0$. The

roots are $t = \dfrac{5}{3}$ and $t = 1$. The solutions are $x_1 = 2\arctan\dfrac{3}{5} + 2k\pi$ and

$x_2 = \dfrac{\pi}{2} + 2k\pi$.

e) $24\sin\dfrac{5x}{2}\cos\dfrac{5x}{2} + \left(\cos^2\dfrac{5x}{2} - \sin^2\dfrac{5x}{2}\right) = 9\left(\cos^2\dfrac{5x}{2} + \sin^2\dfrac{5x}{2}\right)$, or

$$5\tan^2\frac{5x}{2} - 12\tan\frac{5x}{2} + 4 = 0 .$$ Denote $\tan\dfrac{5x}{2} = t$. Then $5t^2 - 12t + 4 = 0$

and the roots are $t = 2$ and $t = \dfrac{2}{5}$. The solutions are $x_1 = \dfrac{2}{5}\arctan 2 + \dfrac{2k\pi}{5}$

and $x_2 = \dfrac{2}{5}\arctan\dfrac{2}{5} + \dfrac{2k\pi}{5}$.

f) $6\sin\dfrac{7x}{2}\cos\dfrac{7x}{2} - 2\cos^2\dfrac{7x}{2} + 2\sin^2\dfrac{7x}{2} = 3\cos^2\dfrac{7x}{2} + 3\sin^2\dfrac{7x}{2}$,

$5\cot^2\dfrac{7x}{2} - 6\cot\dfrac{7x}{2} + 1 = 0$. Denote $\cot\dfrac{7x}{2} = t$. Then the roots are $t = 1$ and

$t = \dfrac{1}{5}$. The solutions are $x_1 = \dfrac{\pi}{14} + \dfrac{2k\pi}{7}$ and $x_2 = \dfrac{2}{7}\arctan 5 + \dfrac{2k'\pi}{7}$, $k, k' \in \mathbf{Z}$.

g) $2\sin x\cos x = 2\sqrt{3}\cos^2 x$. Then $\cos x = 0$ and $\tan x = \sqrt{3}$. The solutions

are $x_1 = \dfrac{\pi}{2} + k\pi$ and $x_2 = \dfrac{\pi}{3} + k\pi$;

h) $\dfrac{1}{\sqrt{2}}\sin 4\alpha - \dfrac{1}{\sqrt{2}}\cos 4\alpha = \dfrac{\sqrt{3}}{2}$ or $\cos\dfrac{\pi}{4}\sin 4\alpha - \sin\dfrac{\pi}{4}\cos 4\alpha = \dfrac{\sqrt{3}}{2}$ or

$\sin\left(4\alpha - \dfrac{\pi}{4}\right) = \dfrac{\sqrt{3}}{2}$ or $4\alpha - \dfrac{\pi}{4} = (-1)^k\dfrac{\pi}{3} + k\pi$, $k \in \mathbf{Z}$. For $k = 2n$ the

solutions are $\alpha = \dfrac{7\pi}{48} + \dfrac{n\pi}{2}$, $n \in \mathbf{Z}$ and for $k = 2n+1$ the solutions are

$\alpha = \dfrac{11\pi}{48} + \dfrac{n\pi}{2}$, $n \in \mathbf{Z}$.

V. 3. a) $\dfrac{1}{2}\cos x - \dfrac{\sqrt{3}}{2}\sin x = \cos 3x$ or $\cos\left(\dfrac{\pi}{3} - x\right) = \cos 3x$. The solutions are

$x_1 = \dfrac{\pi}{12} + \dfrac{k\pi}{2}$, $k \in \mathbf{Z}$, $x_2 = -\dfrac{\pi}{6} + k'\pi$, $k' \in \mathbf{Z}$, $k, k' \in \mathbf{Z}$;

b) We have $\dfrac{1}{2}\cos x - \dfrac{\sqrt{3}}{2}\sin x = \dfrac{\cos 3x}{2} \Rightarrow \cos x\cos\dfrac{\pi}{3} - \sin x\sin\dfrac{\pi}{3} = \dfrac{\cos 3x}{2}$

$2\cos\left(x + \dfrac{\pi}{3}\right) - \cos 3x = 0 \Rightarrow \cos\left(x + \dfrac{\pi}{3}\right) + \cos\left(x + \dfrac{\pi}{3}\right) - \cos 3x = 0 \Rightarrow$

$\cos\left(x + \dfrac{\pi}{3}\right) - 2\sin\left(2x + \dfrac{\pi}{6}\right)\cos\left(\dfrac{\pi}{3} + x\right) = 0 \Rightarrow$

$\cos\left(x + \dfrac{\pi}{3}\right)\left[1 - 2\sin\left(2x + \dfrac{\pi}{6}\right)\right] = 0$. The solutions are $x_1 = \dfrac{\pi}{6} + k\pi, k \in \mathbf{Z}$,

$x_2 = \dfrac{\pi}{3} + n\pi, n \in \mathbf{Z}$, $x_2 = l\pi, l \in \mathbf{Z}$;

c) $\dfrac{\sqrt{3}}{2}\sin 4\alpha + \dfrac{1}{2}\cos 4\alpha = -\sin 3\alpha$ or $\sin\left(4\alpha + \dfrac{\pi}{6}\right) = \sin(-3\alpha)$ or

$4\alpha + \dfrac{\pi}{6} = (-1)^k(-3\alpha) + k\pi,\ k \in \mathbf{Z}$. For $k = 2n$ the solutions

are $\alpha = -\dfrac{\pi}{42} + \dfrac{2n\pi}{7},\ n \in \mathbf{Z}$ and for $k = 2n+1$ the solutions are

$\alpha = \dfrac{5\pi}{6} + 2n\pi,\ n \in \mathbf{Z}$;

d) $\cos 9x - \cos 5x = \sqrt{3}\sin 2x,\ -2\sin 7x\sin 2x = \sqrt{3}\sin 2x$. We obtain

$\sin 2x = 0$ and $\sin 7x = -\dfrac{\sqrt{3}}{2}$. The solutions are $x_1 = \dfrac{k\pi}{2},\ k \in \mathbf{Z}$ and

$x_2 = (-1)^{k'+1}\dfrac{\pi}{21} + \dfrac{k'\pi}{7},\ k' \in \mathbf{Z}$;

e) We have $\cos\dfrac{\pi}{4}\sin 2x + \sin\dfrac{\pi}{4}\cos 2x = \sin 3x$ or

$\sin\left(\dfrac{\pi}{4} + 2x\right) = \sin 3x$ Implying that $\dfrac{\pi}{4} + 2x = (-1)^k 3x + k\pi,\ k \in \mathbf{Z}$.

The final solutions are $x = \dfrac{\pi}{4} + 2n\pi,\ n \in \mathbf{Z}$ and $x = \dfrac{3\pi}{20} + \dfrac{2n\pi}{5},\ n \in \mathbf{Z}$.

f) $\dfrac{1}{2}\cos 3x + \dfrac{\sqrt{3}}{2}\sin 3x = \dfrac{1}{2}\sin 2x + \dfrac{\sqrt{3}}{2}\cos 2x$ or

$\sin\dfrac{\pi}{6}\cos 3x + \cos\dfrac{\pi}{6}\sin 3x = \cos\dfrac{\pi}{3}\sin 2x + \sin\dfrac{\pi}{3}\cos 2x$ or

$\sin\left(\dfrac{\pi}{6} + 3x\right) = \sin\left(2x + \dfrac{\pi}{3}\right)$. Then $\dfrac{\pi}{6} + 3x = (-1)^n\left(2x + \dfrac{\pi}{3}\right) + n\pi,\ k \in \mathbf{Z}$.

For $n = 2k$ the solutions are $x = \dfrac{\pi}{6} + 2k\pi,\ k \in \mathbf{Z}$ and for $n = 2k+1$ the

solutions are $x = \dfrac{\pi}{10} + \dfrac{2k'\pi}{5},\ k' \in \mathbf{Z}$.

g) $2\cos 3\alpha = 2\sin\alpha\cos 3\alpha + 2\sin 3\alpha\sin\alpha$ or $\cos 3\alpha = \sin 4\alpha$ or

$\cos 3\alpha = \cos\left(\dfrac{\pi}{2} - 4\alpha\right)$ or $3\alpha = \pm\left(\dfrac{\pi}{2} - 4\alpha\right) + 2k\pi,\ k \in \mathbf{Z}$. Finally the

solutions are $\alpha = \dfrac{\pi}{14} + \dfrac{2k\pi}{7},\ k \in \mathbf{Z}$ and $\alpha = \dfrac{\pi}{2} + 2k'\pi,\ k' \in \mathbf{Z},\ k,k' \in \mathbf{Z}$;

h) $-2\cos 5x\sin 2x = \sqrt{3}\sin 2x$. We obtain $\sin 2x = 0$ and $\cos 5x = -\dfrac{\sqrt{3}}{2}$.

The solutions are $x_1 = \dfrac{k\pi}{2}$, $k \in \mathbf{Z}$ and $x_2 = \pm\dfrac{\pi}{6} + \dfrac{2k`\pi}{5}$, $k` \in \mathbf{Z}$;

i) $\sqrt{3}\sin\left(\dfrac{3x}{2} - \dfrac{x}{2}\right) + \sin 2x = 0$ or $\sqrt{3}\sin x + 2\sin x\cos x = 0$. From $\sin x = 0$,

$x = k\pi$, $k \in \mathbf{Z}$ and from $\cos x = -\dfrac{\sqrt{3}}{2}$, $x = \pm\dfrac{5\pi}{6} + 2k\pi$, $k \in \mathbf{Z}$.

V. 4. a) $2\sin 4x\cos x = \sin 4x$, then from $\sin 4x = 0$, $x = \dfrac{k\pi}{4}$, $k \in \mathbf{Z}$ and

from $\cos x = \dfrac{1}{2}$, $x = \pm\dfrac{\pi}{3} + 2k\pi$, $k \in \mathbf{Z}$;

b) $2\sin 2\alpha\sin\alpha = \sin 2\alpha$, then from $\sin 2\alpha = 0$, $\alpha = \dfrac{k\pi}{2}$, $k \in \mathbf{Z}$ and from

$\sin\alpha = \dfrac{1}{2}$, $\alpha = (-1)^k\dfrac{\pi}{6} + k\pi$, $k \in \mathbf{Z}$;

c) $x_1 = \dfrac{\pi}{4} + k\pi$, $k \in \mathbf{Z}$, $x_2 = \dfrac{3\pi}{2} + 2k\pi$, $k \in \mathbf{Z}$, $x_3 = \dfrac{2k\pi}{3}$, $k \in \mathbf{Z}$;

d) $3\sin x - 4\sin^3 x + 2 = 2(1 - 2\sin^2 x)$. Denote $\sin x = t$, then

$3t - 4t^3 + 2 = 2 - 4t^2$ or $4t^3 - 4t^2 - 3t = 0$. We obtain $t_1 = 0$ or $\sin x = 0$

with the solutions $x_1 = k`\pi$, $k` \in \mathbf{Z}$ and $4t^2 - 4t - 3 = 0$. $t = -\dfrac{1}{2}$

is a convenient solution. Therefore from $\sin x = -\dfrac{1}{2}$ we obtain

$x_2 = (-1)^{k+1}\dfrac{\pi}{6} + k\pi$, $k \in \mathbf{Z}$;

e) $\sin 9x - \sin 3x = \sin 3x$, $2\cos 6x\sin 3x = \sin 3x$. Then $\sin 3x = 0$ and

$\cos 6x = \dfrac{1}{2}$. The solutions are $x_1 = \dfrac{k\pi}{3}$, $k \in \mathbf{Z}$ and $x_2 = \pm\dfrac{\pi}{18} + \dfrac{k`\pi}{3}$, $k` \in \mathbf{Z}$;

f) $\cos 9x - \cos x = \cos 7x - \cos 3x$, $\sin 5x\sin 4x = \sin 5x\sin 2x$. Then

$\sin 5x = 0$ and $\sin 2x = \sin 4x$. The solutions are $x_1 = \dfrac{k\pi}{5}$, $k \in \mathbf{Z}$ and

$x_2 = \dfrac{k`\pi}{3} + \dfrac{\pi}{6}$, $k` \in \mathbf{Z}$;

g) $\cos 3x \neq 0$. $2\sin^2 3x = \tan 3x$, $2\sin^2 3x\cos 3x = \sin 3x$. Then

$\sin 3x = 0$ and $\sin 6x = 1$. The solutions are $x_1 = \dfrac{k\pi}{3}$, $k \in \mathbf{Z}$ and

$x_2 = \dfrac{\pi}{12} + \dfrac{k'\pi}{3}$, $k' \in \mathbf{Z}$;

h) $1 + \cos 3x = -(\cos x + \cos 2x)$ or $\cos^2 \dfrac{3x}{2} = -\cos \dfrac{3x}{2} \cos \dfrac{x}{2}$.

Then $\cos \dfrac{3x}{2} = 0$ with the solutions $x_1 = \dfrac{\pi}{3} + \dfrac{2k\pi}{3}$, $k' \in \mathbf{Z}$ or

$\cos \dfrac{3x}{2} = -\cos \dfrac{x}{2} \Leftrightarrow \cos x \cos \dfrac{x}{2} = 0$. With the solutions $x_2 = \dfrac{\pi}{2} + k\pi$, $k \in \mathbf{Z}$.

Notice: The solutions for $\cos \dfrac{x}{2} = 0$, $x_3 = \pi + 2k\pi$, $k \in \mathbf{Z}$ are included in

the solutions $x_1 = \dfrac{\pi}{3} + \dfrac{2k\pi}{3}$, $k' \in \mathbf{Z}$;

i) $(\sin x + \cos x)^2 - (\sin x + \cos x) = 0$, $(\sin x + \cos x)(\sin x + \cos x - 1) = 0$.

Then $\sin x + \cos x = 0$ with the solutions $x_1 = -\dfrac{\pi}{4} + k\pi$, $k \in \mathbf{Z}$ and

$\sin x + \cos x = 1$ or $\cos\left(x - \dfrac{\pi}{4}\right) = \dfrac{1}{\sqrt{2}}$ with the solutions $x_2 = 2k'\pi$, $k' \in \mathbf{Z}$

and $x_3 = \dfrac{\pi}{2} + 2k'\pi$, $k' \in \mathbf{Z}$;

j) $(\cos x - \sin x)^2 = \cos x - \sin x$. Then $\cos x - \sin x = 0$ and $\cos x - \sin x = 1$.

The solutions are $x_1 = \dfrac{\pi}{4} + k\pi$, $x_2 = -\dfrac{\pi}{2} + 2k\pi$, and $x_3 = 2k\pi$;

k) $(\sin x + \cos x)(\sin x + \cos x + 1 + \cos x - \sin x) = 0$,

$(\sin x + \cos x)(1 + 2\cos x) = 0$. Then $\sin x + \cos x = 0$ and $1 + 2\cos x = 0$.

Therefore the solutions are $x_1 = -\dfrac{\pi}{4} + k\pi$ and $x_2 = \pm\dfrac{2\pi}{3} + 2k'\pi$, $k, k' \in \mathbf{Z}$;

l) $2\sin 2z \cos z + \sin 2z = 2\cos 2z \cos z + \cos 2z \Rightarrow$

$(2\cos z + 1)(\sin 2z - \cos 2z) = 0$. The solutions are $x_1 = \dfrac{\pi}{8} + \dfrac{k\pi}{2}$, $k \in \mathbf{Z}$,

$x_2 = \pm\dfrac{2\pi}{3} + 2n\pi$, $n \in \mathbf{Z}$.

V. 5. a) $6\cos^2 x + 4\sin x \cos x = 5 \Rightarrow \dfrac{6}{1+\tan^2 x} + 2\dfrac{2\tan x}{1+\tan^2 x} = 5 \Rightarrow$

$5\tan^2 x - 4\tan x - 1 = 0$. Therefore the solutions are $x_1 = \dfrac{\pi}{4} + k\pi$ and

$x_2 = -\arctan\dfrac{1}{5} + k\pi$;

b) $\dfrac{\sin 3x}{\cos 3x} - \dfrac{\sin x}{\cos x} = 4\sin x$, $\dfrac{\sin x \cos x}{\cos x \cos 3x} = 2\sin x$. Then $\sin x = 0$ and

$\cos 3x = \dfrac{1}{2}$. The solutions are $x_1 = k\pi$ and $x_2 = \pm\dfrac{\pi}{9} + \dfrac{2k'\pi}{3}$, $k, k' \in \mathbf{Z}$;

c) $\dfrac{\sin x}{\cos x} + \dfrac{\cos x}{\sin x} = \dfrac{2}{\cos 4x}$, $\cos 4x = \cos\left(\dfrac{\pi}{2} - 2x\right)$. The solutions are

$x_1 = \dfrac{\pi}{12} + \dfrac{k\pi}{3}$ and $x_2 = -\dfrac{\pi}{4} + k'\pi$, $k, k' \in \mathbf{Z}$;

Notice: Since the solutions from x_2 are included into the solutions

$x_1 = \dfrac{\pi}{12} + \dfrac{k\pi}{3}$, the final solutions are $x_1 = \dfrac{\pi}{12} + \dfrac{k\pi}{3}$, $k \in \mathbf{Z}$;

d) $\dfrac{\cos\dfrac{3x}{2}}{\sin\dfrac{3x}{2}} - \dfrac{\cos 3x}{\sin 3x} = \dfrac{1}{\cos x}$, $\dfrac{\sin\dfrac{3x}{2}}{\sin\dfrac{3x}{2}\sin 3x} = \dfrac{1}{\cos x}$. Then $\cos x = \sin 3x$ or

$\cos x = \cos\left(\dfrac{\pi}{2} - 3x\right)$. The solutions are $x_1 = \dfrac{\pi}{8} + \dfrac{k\pi}{2}$ and $x_2 = \dfrac{\pi}{4} + k'\pi$,

$k, k' \in \mathbf{Z}$;

e) $4\sin x = \dfrac{\sqrt{3}\cos x + \sin x}{\cos x}$, $2\sin x \cos x = \cos\dfrac{\pi}{6}\cos x + \sin\dfrac{\pi}{6}\sin x$

or $\sin 2x = \sin\left(\dfrac{\pi}{3} + x\right)$. Then the solutions are $x_1 = \dfrac{\pi}{3} + 2k\pi$ and

$x_2 = \dfrac{2\pi}{9} + \dfrac{2k'\pi}{3}$;

f) $3\sin x - 4\sin^3 x - 4\sin x\left(1 - 2\sin^2 x\right) = 0$, $4\sin^3 x - \sin x = 0$.

Denote $\sin x = t$. Then $4t^3 - t = 0$ or $t(2t+1)(2t-1) = 0$. The roots

187

are $t = 0$, $t = -\dfrac{1}{2}$, and $t = \dfrac{1}{2}$. The solutions are $x_1 = k\pi,\ k \in \mathbf{Z}$,

$x_2 = \pm\dfrac{\pi}{6} + k\pi,\ k \in \mathbf{Z}$;

g) $\cos 6x = \dfrac{\sin^2 3x}{\cos 3x} + 1$, $\sin^2 3x\left(\dfrac{1}{\cos 3x} + 2\right) = 0$. Then $\sin 3x = 0$ and

$\cos 3x = -\dfrac{1}{2}$. The solutions are $x_1 = \dfrac{k\pi}{3}$ and $x_2 = \pm\dfrac{2\pi}{9} + \dfrac{2k'\pi}{3}$, $k, k' \in \mathbf{Z}$.

h) The equation becomes $2\left(\tan\dfrac{x}{2} - 1\right) = \dfrac{1 - \tan^2\dfrac{x}{2}}{1 + \tan^2\dfrac{x}{2}}$ or $\dfrac{\tan\dfrac{x}{2} - 1}{1 + \tan^2\dfrac{x}{2}} = 2$ and

$\tan\dfrac{x}{2} = 1$. Then $x = \dfrac{\pi}{2} + 2k\pi,\ k \in \mathbf{Z}$. $\dfrac{\tan\dfrac{x}{2} - 1}{1 + \tan^2\dfrac{x}{2}} = 2$ is not a valid solution;

i) $\cos 2x + \cos x - \cos 2x + \cos 4x - \cos x + \cos 5x = 0$, $\cos 4x = \cos(5x - \pi)$.

Then $4x = \pm(5x - \pi) + 2k\pi$. The solutions are $x_1 = \pi + 2k\pi$ and

$x_2 = \dfrac{\pi}{9} + \dfrac{2k'\pi}{9}$.

Notice: x_1 is included in $x_2 = \dfrac{\pi}{9} + \dfrac{2k'\pi}{9}$, $k, k' \in \mathbf{Z}$;

j) $\cos 2x \cos 5x + \sin 2x \sin 5x = \sqrt{2}\,\sin 2x \cos 3x$, $\cos 3x = \sqrt{2}\,\sin 2x \cos 3x$.

Then $\cos 3x = 0$ and $\sin 2x = \dfrac{\sqrt{2}}{2}$. Therefore the solutions are

$x_1 = \pm\dfrac{\pi}{6} + k\pi$ and $x_2 = (-1)^{k'}\dfrac{\pi}{8} + \dfrac{k'\pi}{2}$, $k, k' \in \mathbf{Z}$;

k) We have $2\cos 4x - \sin 4x + \sin 2x = \sin 4x + \sin 2x \Rightarrow$

$\cos 4x - \sin 4x = 0 \Rightarrow \tan 4x = 1$. Therefore $x = \dfrac{\pi}{16} + \dfrac{k\pi}{4},\ k \in \mathbf{Z}$.

l) $\sqrt{3}\,\dfrac{\cos 2x \cos 3x + \sin 2x \sin 3x}{\cos 2x \cos 3x} = \dfrac{\sin 2x}{\cos 2x \cos 3x}$, $\sqrt{3}\cos x = 2\sin x \cos x$.

Then $\cos x = 0$ and $\sin x = \dfrac{\sqrt{3}}{2}$. The solutions are $x_1 = \dfrac{\pi}{2} + k\pi$ (not a

188

solution) and $x_2 = (-1)^{k'} \dfrac{\pi}{3} + k'\pi$, $k, k' \in \mathbf{Z}$.

V. 6. a) $\sin 3(2x) = 2\sin 2x$, $3\sin 2x - 4\sin^3 2x = 2\sin 2x$,

$4\sin^3 2x - \sin 2x = 0$. Denote $\sin 2x = t$. Then $t(4t^2 - 1) = 0$. We obtain

$t = 0$ and $t = \pm\dfrac{1}{2}$. The solutions are $x_1 = \dfrac{k\pi}{2}$, $x_2 = \pm\dfrac{\pi}{12} + \dfrac{k\pi}{2}$;

b) $\cos 4\alpha - \cos 8\alpha = \cos 4\alpha + \cos 2\alpha$, $\cos(8\alpha - \pi) = \cos 2\alpha$. Then

$8\alpha - \pi = \pm 2\alpha + 2k\pi$. The solutions are $\alpha_1 = \dfrac{\pi}{6} + \dfrac{k\pi}{3}$, $k \in \mathbf{Z}$ and

$\alpha_2 = \dfrac{\pi}{10} + \dfrac{k'\pi}{5}$, $k` \in \mathbf{Z}$;

c) $\sin x \sin 3x = -\sin 4x \sin 8x$, $\cos 2x = \cos 12x$. Then $2x = \pm 12x + 2k\pi$.

The solutions are $x_1 = -\dfrac{k\pi}{5}$, $k \in \mathbf{Z}$ and $x_2 = \dfrac{k'\pi}{7}$, $k` \in \mathbf{Z}$;

d) $\cos 9x = \cos 1\ x$. Then $9x = \pm 1\ x + 2k\pi$. The solutions are $x_1 = -k\pi$

and $x_2 = \dfrac{k'\pi}{10}$. Since the set of solutions x_1 is included in x_2 , then the

general solution for the given equation is $x_2 = \dfrac{k\pi}{10}$, $k \in \mathbf{Z}$;

e) $x = \dfrac{k\pi}{8}$; **f)** $\sin 3x \cos x - \sin x \cos 3x = 2\sin 2x \cos 3x$,

$\sin 2x = 2\sin 2x \cos 3x$. Then $\sin 2x = 0$ and $\cos 3x = \dfrac{1}{2}$. The solutions are

$x_1 = \dfrac{k\pi}{2}$ and $x_2 = \pm\dfrac{\pi}{9} + \dfrac{2k'\pi}{3}$, $k, k' \in \mathbf{Z}$;

g) $2\cos 2x(\sin 4x + \sin 2x) = \sin 2x$ or $2\cos 2x \sin 4x + 2\cos 2x \sin 2x = \sin 2x$

or $\sin 6x + \sin 2x + \sin 4x = \sin 2x$; $\sin 6x = \sin(-4x)$; $6x = (-1)^n(-4x) + n\pi$.

For $n = 2k$ the solution is $x = \dfrac{k\pi}{5}$ and for $n = 2k+1$ the solution is

$x = \dfrac{(2k+1)\pi}{2}$, $k \in \mathbf{Z}$;

h) $2\sin 2x \cdot 2\sin 6x \cos 4x + \cos 12x = 0$, $2\sin 2x(\sin 10x + \sin 2x) + \cos 12x = 0$

or $\cos 4x(\cos 4x - 1) = 0$. Then $\cos 4x = 0$ and $\cos 4x = \dfrac{1}{2}$. The solutions

are $x_1 = \dfrac{\pi}{8} + \dfrac{k\pi}{4}$ and $x_2 = \pm\dfrac{\pi}{12} + \dfrac{k'\pi}{2}$, $k, k' \in \mathbf{Z}$;

i) $2\cos 2x(\cos 4x + \cos 2x) = \cos 6x$ or $\cos 6x + \cos 2x + 2\cos^2 2x = \cos 6x$, it

follows that $\cos 2x = 0$ and $\cos 2x = -\dfrac{1}{2}$. The solutions are $x_1 = \dfrac{\pi}{4} + \dfrac{k\pi}{2}$

and $x_2 = \pm\dfrac{\pi}{3} + k'\pi$, $k, k' \in \mathbf{Z}$;

j) Multiply the equation in both sides by $2\sin x$ and obtain

$\sin 2x \cos 2x \cos 4x \cos 8x = \dfrac{1}{8}\sin x$, $\sin 4x \cos 4x \cos 8x = \dfrac{1}{4}\sin x$ or $\sin 16x = \sin x$

with the solution $16x = (-1)^k x + k\pi$. Then $x_1 = \dfrac{2p\pi}{15}$, $p \in \mathbf{Z}$

$(p \neq 15q, \ q \in \mathbf{Z})$ for k an even number and $x_2 = \dfrac{(2n+1)\pi}{17}$, $n \in \mathbf{Z}$

$(n \neq 17l + 8, \ l \in \mathbf{Z})$ for k an odd number;

k) $2\sin 2x \cos 5x + \sin 3x \cos 5x - \cos 3x \sin 5x = 0$ or
$2\sin 2x \cos 5x - \sin 2x = 0$.

The solutions are $x_1 = \dfrac{k\pi}{2}$ and $x_2 = \pm\dfrac{\pi}{15} + \dfrac{2k'\pi}{5}$, $k, k' \in \mathbf{Z}$.

V. 7. a) $\dfrac{2\tan x}{1 + \tan^2 x} + \dfrac{2}{\tan x} = 3$, $4\tan^2 x + 2 = 3\tan x + 3\tan^3 x$.

Then $\tan x = 1$ and $3\tan^2 x - \tan x + 2 = 0$ (which is invalid). The

solutions are $x = \dfrac{\pi}{4} + k\pi$, $k \in \mathbf{Z}$; **b)** $\dfrac{4\tan x}{1 + \tan^2 x} + 5\tan x = 7$,

$5\tan^3 x - 2\tan^2 x + 4\tan x - 7 = 0$. Then $\tan x = 1$ and

$5\tan^2 x + 3\tan x + 7 = 0$ (invalid). The solutions are $x = \dfrac{\pi}{4} + k\pi$, $k \in \mathbf{Z}$;

c) $\dfrac{\sin x}{\cos x} - \dfrac{\cos x}{\sin x} = \dfrac{1}{2}$, $\cot 2x = -\dfrac{1}{4}$. Therefore the solutions are

190

$$x = \frac{1}{2}\operatorname{arccot}\left(-\frac{1}{4}\right) + k\pi, \ k \in \mathbf{Z};$$

d) $\dfrac{\sin 2x \sin 3x - \cos 2x \cos 3x}{\cos 2x \sin 3x} + \dfrac{\cos 5x}{\sin 5x} = 0, \quad -\dfrac{\cos 5x}{\cos 2x \sin 3x} + \dfrac{\cos 5x}{\sin 5x} = 0.$

Then $\cos 5x = 0$ and $\sin 2x \cos 3x = 0$. Therefore the solutions are

$$x_1 = \frac{\pi}{10} + \frac{k\pi}{5}, \ x_2 = \frac{k'\pi}{2} \ \text{(only for } k' \text{ odd number), and } x_3 = \frac{\pi}{6} + \frac{k`\pi}{3};$$

e) $\dfrac{\sin 5x \cos 3x - \sin 3x \cos 5x}{\cos 3x \cos 5x} = \dfrac{2\sin 2x}{\cos 2x}, \ (2\cos 3x \cos 5x - \cos 2x)\sin 2x = 0.$

Then $\cos 8x = 0$ and $\sin 2x = 0$. Therefore the solutions are $x_1 = \dfrac{\pi}{16} + \dfrac{k\pi}{8}$ and $x_2 = k'\pi$, $k, k' \in \mathbf{Z};$

f) $\dfrac{\sin x}{\cos\dfrac{x}{2}\cos\dfrac{3x}{2}} = \dfrac{2}{3}\sin x, \ \sin x\left(2\cos\dfrac{x}{2}\cos\dfrac{3x}{2} - 3\right) = 0$ or

$\sin x(\cos 2x + \cos x - 3) = 0$ or . Then $\sin x = 0$ and $2\cos^2 x + \cos x - 4 = 0$.

The solutions are $x = 2k\pi, \ k \in \mathbf{Z};$

g) $\dfrac{\sin 2x - \sin x}{\sin x \sin 2x} = \dfrac{1}{\sin 4x}, \ \sin 2x \sin 4x - \sin x \sin 4x = \sin x \sin 2x,$

$\cos 2x + \cos 5x = \cos x + \cos 6x$ or $\cos\dfrac{7x}{2}\cos\dfrac{3x}{2} = \cos\dfrac{7x}{2}\cos\dfrac{5x}{2}.$

Then $\cos\dfrac{7x}{2} = 0$ and $\cos\dfrac{3x}{2} = \cos\dfrac{5x}{2}$. The only solutions are

$x = \dfrac{\pi}{7} + \dfrac{2k\pi}{7}, \ k \in \mathbf{Z}, \ k \neq 7p+3;$ \quad h) $\dfrac{\cos^2 x - \sin^2 x}{\sin x \cos x} = \sin x + \cos x.$

Then $\sin x + \cos x = 0$. Then one of the solutions is $x_1 = \dfrac{3\pi}{4} + k\pi.$

Also $\dfrac{\cos x - \sin x}{\cos x \sin x} = 1, \ \cos x - \sin x = \sin x \cos x.$ Squaring both sides,

we obtain $1 - \sin 2x = \dfrac{\sin^2 2x}{4}$ or $\sin^2 2x + 4\sin 2x - 4 = 0.$ The roots are

$\sin 2x = -2 + 2\sqrt{2}$ and $\sin 2x = -2 - 2\sqrt{2}$ (invalid). Therefore the solutions are $x_2 = (-1)^k \frac{1}{2}\arcsin\left(-2+2\sqrt{2}\right)+\frac{k\pi}{2}$.

i) $x_1 \neq \frac{k\pi}{2}$, $x_1 = 2\arctan 2 + 2k\pi$, and $x_3 = -2\arctan\frac{2}{3} + 2k\pi$;

j) $\tan x + \cot x = \cos 4x + 3 \Rightarrow \frac{1}{\cos x \sin x} = \left(1 - 2\sin^2 2x\right) + 3$,

$\frac{2}{\sin 2x} + 2\sin^2 2x - 4 = 0$ or $2\sin^3 2x - 4\sin 2x + 2 = 0$. Then $\sin 2x = 1$

and $\sin 2x = \frac{-1+\sqrt{5}}{2}$. Therefore the solutions are $x = \frac{\pi}{4} + \frac{k\pi}{2}$, $k \in \mathbf{Z}$ and

$x = (-1)^k \frac{1}{2}\arcsin\frac{-1+\sqrt{5}}{2} + \frac{k\pi}{2}$, $k \in \mathbf{Z}$;

k) $x \neq \frac{k\pi}{6}$, $k \in \mathbf{Z}$, $\frac{1}{\sin 6x} = \frac{\sin\frac{x}{2}}{\cos\frac{x}{2}} + \frac{\cos x}{\sin x}$, $\sin 6x = \sin x$. Therefore the

solutions are $x_1 = \frac{\pi}{7} + \frac{2k'\pi}{7}$ and $x_2 = \frac{k\pi}{5}$, $k \neq 5p$, $k, p \in \mathbf{Z}$.

V. 8. a) $(\cos x - \sin x)(\cos x + \sin x - \sqrt{2}) = 0$. Then $\cos x - \sin x = 0$ with

the solutions $x = \frac{\pi}{4} + k\pi$ and $\sin x + \cos x = \sqrt{2}$ or $\cos\left(x - \frac{\pi}{4}\right) = 1$ with the

solutions $x = \frac{\pi}{4} + 2k\pi$. The final set of solutions is $x = \frac{\pi}{4} + k\pi$, $k \in \mathbf{Z}$;

b) We have $\sin x + \cos x = \left(\sqrt{3} - 1\right)\left(\cos^2 x - \sin^2 x\right) \Rightarrow$

$(\sin x + \cos x)\left[1 - \left(\sqrt{3} - 1\right)\left(\cos x - \sin x\right)\right] = 0$. From $\sin x + \cos x = 0 \Rightarrow \tan x = -1$,

therefore $x_1 = -\frac{\pi}{4} + k\pi$, $k \in \mathbf{Z}$ and from $\cos x - \sin x = \frac{1}{\sqrt{3} - 1} \Rightarrow$

$\cos x - \sin x = \frac{\sqrt{3}+1}{2} \Rightarrow \sin\left(\frac{\pi}{2} - x\right) - \sin x = \frac{\sqrt{3}-1}{2} \Rightarrow 2\cos\frac{\pi}{4}\sin\left(\frac{\pi}{4} - x\right) = \frac{\sqrt{3}+1}{2} \Rightarrow$

$\cos\left(\frac{\pi}{4} + x\right) = \frac{\sqrt{6}+\sqrt{2}}{4}$.

Since $\cos\dfrac{\pi}{12} = \cos\left(\dfrac{\pi}{3} - \dfrac{\pi}{4}\right) = \cos\dfrac{\pi}{3}\cos\dfrac{\pi}{4} + \sin\dfrac{\pi}{3}\sin\dfrac{\pi}{4} = \dfrac{\sqrt{6}+\sqrt{2}}{4}$ then

$\cos\left(x+\dfrac{\pi}{4}\right) = \cos\dfrac{\pi}{12} \Rightarrow x+\dfrac{\pi}{4} = \pm\dfrac{\pi}{12} + 2k\pi,\ k \in \mathbf{Z}$. The solutions are

$x_2 = -\dfrac{\pi}{3} + 2k\pi \quad k \in \mathbf{Z}$ and $x_3 = -\dfrac{\pi}{6} + 2k\pi,\ k \in \mathbf{Z}$;

c) $(\sin x + \cos x)^2 + 4(\sin x + \cos x) = 0 \Rightarrow$

$(\sin x + \cos x)[(\sin x + \cos x) + 4] = 0 \Rightarrow \sin x + \cos x = 0 \Rightarrow \tan x = -1$.

Therefore $x = -\dfrac{\pi}{4} + k\pi \quad k \in Z$;

d) $4\sin\dfrac{x}{2}\left(2\sin\dfrac{x}{2} + 2\cos\dfrac{x}{2} - \cos\dfrac{x}{2}\cos x\right) = 0$. Then $\sin\dfrac{x}{2} = 0$ and

$2\sin\dfrac{x}{2} + 2\cos\dfrac{x}{2} - \cos\dfrac{x}{2}\cos x = 0 \Rightarrow \left(\tan\dfrac{x}{2} + 1\right)\left(2 - \dfrac{1-\tan\dfrac{x}{2}}{1+\tan^2\dfrac{x}{2}}\right) = 0$. Then

$\tan\dfrac{x}{2} = -1$ and $2\tan^2\dfrac{x}{2} + \tan\dfrac{x}{2} + 1 = 0$ (impossible). Therefore the

solutions are $x_1 = 2k\pi$ and $x_2 = -\dfrac{\pi}{2} + 2k'\pi,\ k, k' \in \mathbf{Z}$;

e) Denote $\sin x + \cos x = t$. Then $3t^2 - 7t + 2 = 0$. The roots

are $t = 2$ (impossible) and $t = \dfrac{1}{3}$. Therefore the solutions are

$x = \pm\arccos\dfrac{1}{3\sqrt{2}} + \dfrac{\pi}{4} + 2k\pi,\ k \in \mathbf{Z}$;

f) $x = \pm\arccos\dfrac{1}{2\sqrt{2}} + \dfrac{\pi}{4} + 2k\pi,\ k \in \mathbf{Z}$;

g) Denote $\sin x - \cos x = t$. Then $2t^2 - 5t + 2 = 0$. The roots

are $t = 2$ (impossible) and $t = \dfrac{1}{2}$. Therefore the solutions are

$x = (-1)^k \arcsin\dfrac{1}{2\sqrt{2}} + \dfrac{\pi}{4} + k\pi,\ k \in \mathbf{Z}$;

h) $\cos x \cdot \dfrac{\cos x - \sin x}{\cos x}(\sin x + \cos x) = \sin x$, $\cos 2x = \cos\left(\dfrac{\pi}{2} - x\right)$. The solutions

are $x_1 = -\dfrac{\pi}{2} + 2k\pi$ (impossible) and $x_2 = \dfrac{\pi}{6} + \dfrac{2k'\pi}{3}$, $k' \neq 3l - 1$, $l \in \mathbf{Z}$;

i) $\left(\sin x + \sqrt{3}\cos x\right)\left(\dfrac{1}{\cos x} + \dfrac{1}{\sin x} - 1\right) = 0$. Then $\tan x = -\sqrt{3}$ with the

solutions $x_1 = -\dfrac{\pi}{3} + k\pi$, $k \in \mathbf{Z}$, and from $\sin x + \cos x = \sin x \cos x$ where

$x \neq \dfrac{k\pi}{2}$, $k \in \mathbf{Z}$ squaring in both sides, we obtain $\sin^2 2x - 4\sin 2x - 4 = 0$

.with the solutions $x_2 = \dfrac{1}{2}(-1)^k \arcsin\left(2 - 2\sqrt{2}\right) + \dfrac{k\pi}{2}$, $k \in \mathbf{Z}$;

j) $x_1 = \dfrac{\pi}{4} + k\pi$, $\cos\left(\dfrac{\pi}{4} - x\right) = -\dfrac{ab}{\sqrt{2}\left(a^2 + b^2\right)} \Rightarrow \sin\left(\dfrac{\pi}{4} + x\right) = \dfrac{ab}{\sqrt{2}\left(a^2 + b^2\right)} \Rightarrow$

$\dfrac{\pi}{4} + x = (-1)^k \arcsin\dfrac{ab}{\sqrt{2}\left(a^2 + b^2\right)} + k\pi$, $k \in \mathbf{Z}$.

V. 9. a) $x \neq \dfrac{\pi}{2} + k\pi$. We have $\left(\sin\dfrac{x}{2} + \cos\dfrac{x}{2}\right)^2 = \dfrac{1 - \cos^2 x}{\cos x} \cdot \dfrac{\cos\dfrac{x}{2} + \sin\dfrac{x}{2}}{\cos\dfrac{x}{2} - \sin\dfrac{x}{2}} \Rightarrow$

$\cos\dfrac{x}{2} + \sin\dfrac{x}{2} = 0$ (impossible) or $\cos\dfrac{x}{2} + \sin\dfrac{x}{2} = \dfrac{\sin^2 x}{\cos x} \cdot \dfrac{1}{\cos\dfrac{x}{2} - \sin\dfrac{x}{2}} \Rightarrow$

$\cos x = \dfrac{\sin^2 x}{\cos x}$ or $\tan x = \pm 1$. Therefore the solutions are $x = \dfrac{\pi}{4} + \dfrac{k\pi}{2}$;

b) If $\tan x \neq 1$, we have $2\left(\sin^2 x + \cos^2 x + \sin 2x\right) - \dfrac{\cos x + \sin x}{\cos x - \sin x} = 0 \Rightarrow$

$2\left(\cos x + \sin x\right)^2 - \dfrac{\cos x + \sin x}{\cos x - \sin x} = 0 \Rightarrow$

$\left(\cos x + \sin x\right)\left[2\left(\cos x + \sin x\right) - \dfrac{1}{\cos x - \sin x}\right] = 0$. We have two cases:

I) $\sin x + \cos x = 0 \Rightarrow \tan x = -1$ so $x_1 = -\dfrac{\pi}{4} + k\pi$, $k \in Z$ and

II) $2\left(\cos x + \sin x\right) - \dfrac{1}{\cos x - \sin x} = 0 \Rightarrow 2\cos 2x - 1 = 0 \Rightarrow \cos 2x = \dfrac{1}{2} \Rightarrow$

$$x_2 = \pm\frac{\pi}{6} + k\pi, \quad k \in \mathbf{Z};$$

c) $\dfrac{\cos x - \sin x}{\sin x}(\cos x + \sin x)^2 = \dfrac{\cos x + \sin x}{\sin x}$,

$\cos 2x(\cos x + \sin x) = \cos x + \sin x$. Then $\cos x + \sin = 0$ and $\cos 2x = 1$.

Therefore the solutions are $x_1 = -\dfrac{\pi}{4} + k\pi$ and $x_2 = k'\pi$ (impossible);

d) $\dfrac{\sin 2x}{\cos 2x} - \dfrac{\cos x}{\sin x} = -4\cos x \cos 3x$ or $\dfrac{-\cos 3x}{\cos 2x \sin x} = -4\cos x \cos 3x$. Then

$\cos 3x = 0$, $x = \dfrac{\pi}{6} + \dfrac{k\pi}{3}$, $k \in \mathbf{Z}$ and $\dfrac{1}{\cos 2x \sin x} = 4\cos x$ or $\sin 4x = 1$,

$x = \dfrac{\pi}{8} + \dfrac{k\pi}{2}$, $k \in \mathbf{Z}$;

e) $4\cos^2 2x = \dfrac{\sin 2x \sin 4x + \cos 2x \cos 4x}{\sin 2x \cos 4x}$, $4\cos^2 2x = \dfrac{\cos 2x}{\sin 2x \cos 4x}$. Then

$\cos 2x = 0$ and $4\sin 2x \cos 2x \cos 4x = 1$. The solutions are $x_1 = \dfrac{\pi}{4} + \dfrac{k\pi}{2}$

and $x_2 = \dfrac{\pi}{16} + \dfrac{k'\pi}{4}$, $k, k' \in \mathbf{Z}$;

f) If $\sin x \neq 0$ we have $\cot x + \sin x \left(\dfrac{\cos x \cos\frac{x}{2} + \sin x \sin\frac{x}{2}}{\cos x \cos\frac{x}{2}} \right) = 4 \Rightarrow$

$\cot x + \sin x \dfrac{\cos\left(x - \frac{x}{2}\right)}{\cos x \cos\frac{x}{2}} = 4 \Rightarrow \dfrac{\cos x}{\sin x} + \dfrac{\sin x}{\cos x} = 4 \Rightarrow$

$\dfrac{\cos^2 x + \sin^2 x}{\sin x \cos x} = 4 \Rightarrow \sin 2x = \dfrac{1}{2}$. Therefore $x = (-1)^k \dfrac{\pi}{12} + \dfrac{k\pi}{2}$, $k \in \mathbf{Z}$;

g) $(\sin x - 1)(\cos^2 x - 2\cos x + 1) = 0$. The solutions are $x_1 = \dfrac{\pi}{2} + 2k\pi$ and

$x_2 = 2k'\pi$.

Notice: Since $\cos x \neq 0$, x_1 is impossible;

h) $x \neq \dfrac{k\pi}{3}$, $k \in \mathbf{Z}$, $x \neq \dfrac{k\pi}{2}$, $k \in \mathbf{Z}$

$$2\tan x + \tan\frac{x}{2} - \cot\frac{x}{2} + 4\cot 2x = \cot 3x - \cot\frac{x}{2},$$

$2\tan x - 2\cot x + 4\cot 2x = \cot 3x - \cot\frac{x}{2}$. Then $\cot 3x - \cot\frac{x}{2} = 0$. The

solutions are $x = \dfrac{2k\pi}{5}$, $k \in \mathbf{Z}$, $k \neq 5p$, $p \in \mathbf{Z}$;

i) $\dfrac{\sin x}{\cos x} - 2\sin x \cos x - \cos 2x(1 - 2\sec x) = 0 \Rightarrow$

$\tan x(1 - 2\cos^2 x) - \cos 2x(1 - 2\sec x) = 0$ then

$-\cos 2x(\tan x + 1 - 2\sec x) = 0 \Rightarrow \dfrac{\cos 2x}{\cos x}(\sin x + \cos x - 2) = 0$. The solutions

are $x = \dfrac{\pi}{4} + \dfrac{k\pi}{2}$, $k \in \mathbf{Z}$;

V. 10. a) $x = \dfrac{\pi}{4} + \dfrac{k\pi}{2}$, $k \in \mathbf{Z}$; **b)** We have $\sin 2x = \sin 6x \Rightarrow \cos 4x \sin 2x = 0$.

The solutions are $x_1 = \pm\dfrac{\pi}{8} + \dfrac{k\pi}{2}$, $k \in \mathbf{Z}$, $x_2 = \dfrac{n\pi}{2}$, $n \in \mathbf{Z}$;

c) $\cos 2x(3\cos 2x - 7) = 0$. Denote $\cos 2x = t$. Then $t(3t - 7) = 0$.

The roots are $t = 0$ and $t = \dfrac{7}{3}$ (invalid). Therefore $\cos 2x = 0$. The

solutions are $x = \dfrac{\pi}{4} + \dfrac{k\pi}{2}$, $k \in \mathbf{Z}$; **d)** $2(1 - \sin^2 x) + 7\sin x - 5 = 0$,

$2\sin^2 x - 7\sin x + 3 = 0$, $(2\sin x - 1)(\sin x - 3) = 0$. The set of solutions is

$x = (-1)^k \dfrac{\pi}{6} + k\pi$, $k \in \mathbf{Z}$;

e) $25\cos^2 x - 90\cos x + 56 = 0$, $(5\cos x - 14)(5\cos x - 4) = 0$, $x = \pm\arccos\dfrac{4}{5} + 2k\pi$;

f) $1 - 2\sin^2 x + \sin x + \dfrac{3}{4} = 1 - \sin^2 x$, $4\sin^2 x - 4\sin x - 3 = 0$. Denote

$\sin x = t$. Then $4t^2 - 4t - 3 = 0$. The roots are $t = \dfrac{3}{2}$ and $t = -\dfrac{1}{2}$.

Therefore the solutions $x = (-1)^{k+1}\dfrac{\pi}{6} + k\pi$, $k \in \mathbf{Z}$; **g)** $\alpha = \pm\dfrac{\pi}{3} + k\pi$, $k \in \mathbf{Z}$;

h) $x = \dfrac{\pi}{4} + \dfrac{k\pi}{2}$, $k \in \mathbf{Z}$;

i) $-2\tan x - 2\tan^2 x = \tan^2 x$ or $\tan x (3\tan x + 2) = 0$. Therefore the

solutions are $x_1 = k\pi$ and $x_2 = -\arctan\dfrac{2}{3} + k'\pi$, $k, k' \in \mathbf{Z}$;

j) $2\sin^2 x + \dfrac{\sin^2 x}{1 - \sin^2 x} = 2$, $2\sin^4 x - 5\sin^2 x + 2 = 0$. Denote $\sin^2 x = t \geq 0$.

Then $2t^2 - 5t + 2 = 0$. The roots are $t = 2$ (impossible) and $t = \dfrac{1}{2}$. The

solutions are $x = \dfrac{\pi}{4} + \dfrac{k\pi}{2}$, $k \in \mathbf{Z}$;   **k)** $x_1 = \dfrac{\pi}{2} + k\pi$ and $x_2 = -\arctan 2 + k'\pi$.

V. 11. a) We have $2\sin 2x \cos x = 4\cos^3 x \Rightarrow$

$4\sin x \cos^2 x = 4\cos^3 x$. Then $\cos x = 0 \Rightarrow x_1 = \dfrac{\pi}{2} + k\pi$, $k \in \mathbf{Z}$

and $\sin x = \cos x$ or $\tan x = 1$ $x_2 = \dfrac{\pi}{4} + k\pi$, $k \in \mathbf{Z}$;

b) $\cos x = \sin 3x$ or $\cos x = \cos\left(\dfrac{\pi}{2} - 3x\right)$. The solutions are $x_1 = \dfrac{\pi}{8} + \dfrac{k\pi}{2}$

and $x_2 = \dfrac{\pi}{4} + k'\pi$, $k, k' \in \mathbf{Z}$;

c) $\cos x - \sin x = 2\sin 2x \sin x$, $\cos 3x = \sin x$ or $\cos 3x = \cos\left(\dfrac{\pi}{2} - x\right)$.

Therefore the solutions are $x_1 = \dfrac{\pi}{8} + \dfrac{k\pi}{2}$ and $x_2 = -\dfrac{\pi}{4} + k'\pi$, $k, k' \in \mathbf{Z}$;

d) $\cos^2 2x - 9\cos 2x + 8 = 0$. Denote $\cos 2x = t$. Then $t^2 - 9t + 8 = 0$. The
convenient root is $t = 1$. Therefore the solutions are $x = k\pi$, $k \in \mathbf{Z}$;

e) $\cos 3(3x) - 2\cos 2(3x) - 2 = 0$; $4\cos^3 3x - 3\cos 3x - 2(2\cos^2 3x - 1) - 2 = 0$.

Denote $\cos 3x = t$. Then $4t^3 - 4t^2 - 3t = 0$ or $t(2t + 1)(2t + 3) = 0$.

Therefore $x = \dfrac{\pi}{6} + \dfrac{k\pi}{3}$ and $x = \pm\dfrac{2\pi}{9} + \dfrac{2k\pi}{3}$, $k \in \mathbf{Z}$.

197

The equation $\cos 3x = -\dfrac{3}{2}$ is impossible;

f) $1 - \cos^2 4x = 3\cos^2 4x$, $\cos^2 4x = \dfrac{1}{4}$. Then $\cos 4x = \pm\dfrac{1}{2}$. Therefore the

solutions are $x = \pm\dfrac{\pi}{12} + \dfrac{2k\pi}{4}$;

g) Dividing both side of the equation by $\cos^2 x$

$$\left(\cos x \neq 0 \Leftrightarrow x = \frac{\pi}{2} + k\pi, \;\; k \in Z\right)$$

we have $\dfrac{\sin^2 x}{\cos^2 x} - 2\dfrac{\sin x}{\cos x} = 3$. Denoting $\tan x = t$, then $t^2 - 2t - 3 = 0$ and

$t_1 = -1$ and $t_2 = 3$. Thus the solutions of the initial equation reduces to

that of two elementary trigonometric equations $\tan x = -1$ and $\tan x = 3$.

So $x_1 = -\dfrac{\pi}{4} + k\pi$, $k \in Z$ and $x_2 = \arctan 3 + k\pi$, $k \in Z$.

h) $\tan^2 x + 4\tan x - 5 = 0$. Therefore $\tan x = -5$ and $\tan x = 1$. The

solutions are $x_1 = -\arctan 5 + k\pi$ and $x_2 = \dfrac{\pi}{4} + k'\pi$;

i) The equation becomes $4\sin^2 x + \sin x \cos x - 3\cos^2 x = 0$.

Dividing both side of the equation by $\cos^2 x$

$$\left(\cos x \neq 0 \Leftrightarrow x = \frac{\pi}{2} + k\pi, \;\; k \in Z\right)$$

we obtain $4\dfrac{\sin^2 x}{\cos^2 x} + \dfrac{\sin x}{\cos x} - 3 = 0$. Denoting $\tan x = t$ we get $4t^2 + t - 3 = 0$,

$t_1 = -1$ and $t_2 = \dfrac{3}{4}$. Thus from $\tan x = -1$, $x_1 = -\dfrac{\pi}{4} + k\pi$, $k \in Z$ and

$\tan x = \dfrac{3}{4}$, $x_2 = \arctan\dfrac{3}{4} + k\pi$, $k \in Z$;

j) $3\sin 2x - 4\sin^3 2x + \sin 2x = 4\sin^3 2x$, $\sin 2x \cos 4x = 0$. Then $\sin 2x = 0$

and $\cos 4x = 0$. Therefore the solutions are $x_1 = \dfrac{k\pi}{2}$ and $x_2 = \pm\dfrac{\pi}{8} + \dfrac{k\pi}{2}$.

Notice: x_2 can also be written in the form $x_2 = \dfrac{\pi}{8} + \dfrac{k'\pi}{4}$, $k' \in \mathbf{Z}$;

k) $\alpha = \pm\dfrac{\pi}{6} + \dfrac{k\pi}{2}$, $k \in \mathbf{Z}$; l) $x = \pm\dfrac{\pi}{12} + \dfrac{k\pi}{3}$, $k \in \mathbf{Z}$.

V. 12. a) $\sin x\left(\sin x - \sqrt{2}\right) = 0$. Then $\sin x = 0$ and $\sin x = \sqrt{2}$ (impossible). Therefore the solutions are $x = k\pi$;

b) $\cos 4x + \cos 2x = 0$ or $\cos 4x = \cos(\pi - 2x)$. Then $4x = \pm(\pi - 2x) + 2k\pi$

with the solutions $x = \dfrac{k\pi}{3} + \dfrac{\pi}{6}$, $k \in \mathbf{Z}$;

Notice: The set of solutions $x = -\dfrac{\pi}{2} + k\pi$, $k \in \mathbf{Z}$ is included in the set of

solutions $x = \dfrac{k\pi}{3} + \dfrac{\pi}{6}$, $k \in \mathbf{Z}$;

c) $2\cos^2 4x \sin 4x = \cos^2 4x$. Then $\cos 4x = 0$ and $\sin 4x = \dfrac{1}{2}$. The

solutions are $x_1 = \dfrac{k`\pi}{4} + \dfrac{\pi}{8}$, $k` \in \mathbf{Z}$ and $x_2 = (-1)^k \dfrac{\pi}{24} + \dfrac{k\pi}{4}$, $k \in \mathbf{Z}$;

d) $\dfrac{(\sin x + \cos x)}{\sin x \cos x}(\sin x + \cos x) + 2 = 0 \Rightarrow \dfrac{1 + \sin 2x}{\sin x \cos x} + 2 = 0$. Then

$\sin 2x = -\dfrac{1}{2}$.

The solutions are $x = (-1)^k \dfrac{7\pi}{12} + \dfrac{k\pi}{2}$; **e)** $x = \dfrac{\pi}{2} + 2k\pi$, $k \in \mathbf{Z}$;

f) $1 + \cos 5x = 1 - \sin 3x$, $\cos 5x = \cos\left(\dfrac{\pi}{2} + 3x\right)$. The solutions are

$x_1 = \dfrac{\pi}{4} + k\pi$, $k \in \mathbf{Z}$ and $x_2 = \dfrac{3\pi}{16} + \dfrac{k`\pi}{4}$, $k` \in \mathbf{Z}$;

g) $x_1 = \dfrac{\pi}{8} + \dfrac{k\pi}{4}$, $k \in \mathbf{Z}$ and $x_2 = -\dfrac{\pi}{4} + k`\pi$, $k` \in \mathbf{Z}$, $k, k' \in \mathbf{Z}$;

h) $2\cos 4\alpha \cos 2\alpha = \cos 2\alpha$.

From $\cos 4\alpha = \dfrac{1}{2}$,

(1) $x_1 = \pm\dfrac{\pi}{12} + \dfrac{k\pi}{2}$, $k \in \mathbf{Z}$

The terminal points (solutions) for x_1 are the points

$$M_1\left(\frac{\pi}{12}\right), M_3\left(\frac{5\pi}{12}\right), M_4\left(\frac{7\pi}{12}\right),$$

$$M_6\left(\frac{11\pi}{12}\right), M_7\left(\frac{13\pi}{12}\right),$$

$$M_9\left(\frac{17\pi}{12}\right), M_{10}\left(\frac{19\pi}{12}\right), M_{12}\left(\frac{23\pi}{12}\right).$$

$\cos 2\alpha = 0$, (2) $x_2 = \dfrac{\pi}{4} + k`\pi,\ k` \in \mathbf{Z}$.

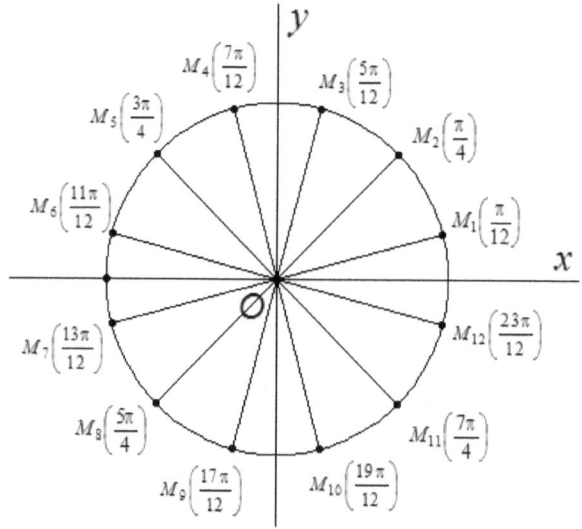

The terminal points (solutions) for x_2 are the points $M_2\left(\dfrac{\pi}{4}\right)$, $M_5\left(\dfrac{3\pi}{4}\right)$,

$M_8\left(\dfrac{5\pi}{4}\right)$, $M_{11}\left(\dfrac{7\pi}{4}\right)$.

Therefore the general solution for the given equation is

$\alpha = \dfrac{\pi}{12} + \dfrac{k\pi}{6}$, $k \in \mathbf{Z}$ which includes the sets of solutions (1) and (2).

i) $\sin x + \cos x + \sqrt{2} \sin 7x = 0 \Rightarrow \sin 7x = \sin\left(-\dfrac{\pi}{4} - x\right) \Rightarrow$

$7x = (-1)^k\left(-\dfrac{\pi}{4} - x\right) + k\pi$. Therefore the solutions are $x_1 = -\dfrac{\pi}{32} + \dfrac{n\pi}{4}$ and

$$x_2 = \frac{5\pi}{24} + \frac{n'\pi}{3};$$

j) $\sin 2x \cos 2x \sin x + \sin 2x \sin x - (1 + \cos 2x) = 0$, $(\sin 2x \sin x - 1)(1 + \cos 2x) = 0$.

Then $\cos 2x = -1$.

The solutions are $x = \frac{\pi}{2} + k\pi$, $k \in \mathbf{Z}$;

k) $x = \frac{k\pi}{4}$, $k \in \mathbf{Z}$, $k \neq 4m$, $m \in \mathbf{Z}$;

l) $2 \sin x \cos x - 2 \cos^2 x + 4\left(\sin x - \cos x + \frac{\sin x - \cos x}{\cos x} \right) = 0$,

$(\sin x - \cos x)\left(\cos x + 2 + \frac{2}{\cos x} \right) = 0$. Then $\sin x - \cos x = 0$ or $\tan x = 1$.

Therefore the solutions are $x = \frac{\pi}{4} + k\pi$, $k \in \mathbf{Z}$;

m) If $\sin 2x \neq 0$, we have

$$\sin 2x(1 - \cos 2x) - \sin^2 2x - (1 - \cos 2x) - \frac{\cos 2x}{\sin 2x} = 0 \Rightarrow$$

$$\sin 2x - \sin 2x \cos 2x - \sin^2 2x - 1 + \cos 2x - \frac{\cos 2x}{\sin 2x} = 0 \Rightarrow$$

$$\sin^2 2x - \sin^2 2x \cos 2x - \sin^3 2x - \sin 2x + \sin 2x \cos 2x - \cos 2x = 0 \Rightarrow$$

$(\sin 2x + \cos 2x)(\sin^2 2x - \sin 2x + 1) = 0$ but $\sin^2 2x - \sin 2x + 1 \neq 0$.

Then $\sin 2x + \cos 2x = 0 \Rightarrow \tan 2x = -1$ so $x = -\frac{\pi}{8} + \frac{k\pi}{2}$ $k \in \mathbf{Z}$.

V. 13. a) $11 \tan x \dfrac{3 - \tan^2 x}{1 - 3 \tan^2 x} = \tan x$. Then $\tan x = 0$ or $11 \dfrac{3 - \tan^2 x}{1 - 3 \tan^2 x} = 1$ or

$\tan x = \pm 2$. Therefore the solutions are $x_1 = k\pi$ and $x_2 = \pm \arctan 2 + k'\pi$,

$k, k' \in \mathbf{Z}$; **b)** $x = k\pi$, $k \in \mathbf{Z}$;

c) $\dfrac{\sin x}{\cos x} + \dfrac{\sin 2x}{\cos 2x} = \dfrac{\sin 3x}{\cos 3x}$, $\dfrac{\sin x \cos 2x + \cos x \sin 2x}{\cos x \cos 2x} = \dfrac{\sin 3x}{\cos 3x}$. Then

$\sin 3x = 0$ and $\cos x \cos 2x = \cos 3x$ (see the restrictions).

The solution is $x = \dfrac{k\pi}{3}$, $k \in \mathbf{Z}$.;

201

d) $\dfrac{6\cos 2x}{\sin 2x}+4\sin 2x=0$, $2\cos^2 2x-3\cos 2x-2=0$. Denote $\cos 2x=t$.

Then $2t^2-3t-2=0$. The roots are $t=-\dfrac{1}{2}$ and $t=2$. Thus $\cos 2x=-\dfrac{1}{2}$.

The solutions are $x=\pm\dfrac{2\pi}{3}+k\pi,\ k\in\mathbf{Z}$;

e) $\tan x\dfrac{2\tan x}{1-\tan^2 x}=\tan x+\dfrac{2\tan x}{1-\tan^2 x}\Rightarrow \tan x=0$ with the solutions $x_1=k\pi$ or

$\dfrac{2\tan x}{1-\tan^2 x}=1+\dfrac{2}{1-\tan^2 x}\Rightarrow \tan^2 x+2\tan x-3=0$. The solutions are

$x_2=\dfrac{\pi}{4}+k\pi$ (not valid) and $x_3=-\arctan 3+k\pi$;

j) $2\tan^3 x-3\tan^2 x+2\tan x-1=0$. Then $\tan x=1$ and

$2\tan^2 x-\tan x+1=0$ (impossible). Therefore the solutions are

$x=\dfrac{\pi}{4}+k\pi,\ k\in\mathbf{Z}$;

g) $\dfrac{\sin^2 x-\cos^2 x}{\cos x\sin x}=\dfrac{\cos x-\sin x}{\cos x\sin x}$ or $(\sin x-\cos x)(\sin x+\cos x)=-(\sin x-\cos x)$.

Then $\sin x-\cos x=0$ and $\sin x+\cos x=-1$. Therefore the solutions are

$x_1=\dfrac{\pi}{4}+k\pi$, $x_2=\dfrac{\pi}{2}+2k\pi$ (impossible), and $x_3=2k\pi$ (impossible);

h) $2\cot 2x-2\tan 2x-4\tan 4x+8=0$, $4\cot 4x-4\tan 4x+8=0$,

$8\cot 8x+8=0$ or $\cot 8x=-1$. Therefore the solutions are

$x=-\dfrac{\pi}{32}+\dfrac{k\pi}{8},\ k\in\mathbf{Z}$.

i) $3\tan x(1+\tan 3x)+\tan 2x(1+\tan 3x)=0$, $(1+\tan 3x)(3\tan x+\tan 2x)=0$.

Then $1+\tan 3x=0$ and $3\tan x+\tan 2x=0$. Therefore the solutions are

$x_1=-\dfrac{\pi}{12}+\dfrac{k\pi}{3}$, $x_2=k\pi$, and $x_3=\pm\arctan\sqrt{\dfrac{5}{3}}+k\pi$;

j) $\tan(ax+b) = \tan\left(\dfrac{\pi}{2} - ax + b\right)$. Therefore the set of solutions is

$$x = \dfrac{\pi}{4a} + \dfrac{k\pi}{2a},\ k \in \mathbf{Z};$$

k) $\tan(2x+1) = \tan(x+3)$. Therefore the solutions are $x = 2 + k\pi,\ k \in \mathbf{Z}$.

V. 14. a) $\cos^2 x \sin x + \cos^3 x - 3 \sin x = 3 \sin^2 x \cos x - 3 \sin x \cos^2 x$,

$\cos^3 x - 3 \sin x = 3 \cos x - 3 \cos^3 x - 4 \sin x \cos^2 x$,

$4 \cos^2 x (\cos x + \sin x) = 3(\cos x + \sin x)$. Then the solutions are

$$x_1 = -\dfrac{\pi}{4} + k\pi \ \text{ and } \ x_2 = \pm\dfrac{\pi}{6} + k\pi; \quad \textbf{b)} \ \cot^2 2x - \dfrac{4 \cdot 2 \sin 2x \sin x}{2 \sin x \cos 2x} + 3 = 0.$$

$\cot^2 2x - \dfrac{4}{\cot 2x} + 3 = 0$. Denote $\cot 2x = t$. Then $t^2 - \dfrac{4}{t} + 3 = 0$;

$(t-1)(t^2 + t + 4) = 0$. Therefore $x = \dfrac{\pi}{8} + \dfrac{k\pi}{2},\ k \in \mathbf{Z}$;

c) $\dfrac{\cos x - \sin x}{\sin x}(\sin x + \cos x)^2 = \dfrac{\sin x + \cos x}{\sin x}$, $\sin x \neq 0$. Then

$\sin x + \cos x = 0$ and $\tan x = -1$. Therefore we obtain $x = -\dfrac{\pi}{4} + k\pi,\ k \in \mathbf{Z}$.

The equation $\cos 2x = 1$ is impossible;

d) $\dfrac{(\cos x - \sin x)(\cos x + \sin x)}{\sin x \cos x} = \dfrac{\cos x - \sin x}{\sin x \cos x}$. Then $\cos x - \sin x = 0$ and

$\cos x + \sin x = 1$. The only solutions are $x = \dfrac{\pi}{4} + k\pi,\ k \in \mathbf{Z}$;

e) $\dfrac{8 \cos x \sin^2 \dfrac{x}{2} \cdot \cos^2 \dfrac{x}{2}}{\left(\cos^4 \dfrac{x}{2} - \sin^4 \dfrac{x}{2}\right)} = 1 - \sin 2x$ or $\dfrac{2 \cos x \sin^2 x}{\cos x} = 1 - \sin 2x$, then

$1 - \cos 2x = 1 - \sin 2x$. Therefore $x = \dfrac{\pi}{8} + \dfrac{k\pi}{2},\ k \in \mathbf{Z}$;

f) $\dfrac{5 - \dfrac{5}{\cos x}}{1 + \dfrac{1}{\cos x}} + 4(1 - \cos x) = 0$, $\sin x \neq 0$

$$\frac{5(\cos x - 1)}{\cos x + 1} + 3(1 - \cos x) = 0; \quad \cos x = 1 \text{ which is impossible. Then}$$

$$\frac{5}{\cos x + 1} = 3 \text{ with the solutions } x = \pm \arccos \frac{2}{3} + 2k\pi, \ k \in \mathbf{Z};$$

g) $\dfrac{3(\cos 2x \sin 2x + \cos 2x)}{\cos 2x - \cos 2x \sin 2x} - 2(1 + \sin 2x) = 0; \quad \dfrac{3(1 + \sin 2x)}{1 - \sin 2x} - 2(1 + \sin 2x) = 0.$

Then $1 + \sin 2x = 0$ with the solution $x = -\dfrac{\pi}{4} + k\pi, \ k \in \mathbf{Z}$ and

$$\sin 2x = -\frac{1}{2} \text{ with the solution } x = (-1)^{k+1}\frac{\pi}{12} + \frac{k\pi}{2}, \ k \in \mathbf{Z};$$

h) $\tan 7x\left(1 - \tan^2 3x\right) = 2\tan 3x$ or $\tan 7x \dfrac{\cos 6x}{\cos^2 3x} = 2\dfrac{\sin 3x}{\cos 3x};$

$\tan 7x \cos 6x = 2 \sin 3x \cos 3x$ or $\tan 7x = \tan 6x$ or $7x = 6x + k\pi$. Therefore

$x = k\pi, \ k \in \mathbf{Z};$

i) $\dfrac{2(1 + \cos 2x) - 1}{\sin x} = \dfrac{\cos x}{\sin x}(1 + 2\cos 2x), \ \sin x \neq 0.$ Then

$1 + 2\cos 2x = \cos x(1 + 2\cos 2x)$. Since $\cos x \neq 1$, therefore we obtain,

$$\cos 2x = -\frac{1}{2}, \quad x = \pm \frac{\pi}{3} + k\pi, \ k \in \mathbf{Z};$$

j) $1 + \cot^2 2x - \dfrac{2\cos 2x}{\sin 2x} - 4 = 0$ or $\cot^2 2x - 2\cot 2x - 3 = 0$, $\cot 2x = t$,

then $\cot 2x = -1$ and $\cot 2x = 3$. Therefore $x = -\dfrac{\pi}{8} + \dfrac{k\pi}{2}, \ k \in \mathbf{Z}$ and

$$x = \frac{1}{2}\arctan\frac{1}{3} + \frac{k\pi}{2}, \ k \in \mathbf{Z};$$

k) $\cos 2x \neq 0$, $\cos 3x \neq 0$.

$$4\frac{\sin 3x \cos 2x - \sin 2x \cos 3x}{\cos 3x \cos 2x} = \tan 3x \frac{1}{\cos^2 2x} \text{ or } \frac{4\sin x}{\cos 3x} = \frac{\sin 3x}{\cos 3x} \cdot \frac{1}{\cos 2x};$$

$4\sin x \cos 2x = \sin 3x$ or $2(\sin 3x - \sin x) = \sin 3x$ or $\sin 3x = 2\sin x$

$3\sin x - 4\sin^3 x = 2\sin x$. Then $\sin x = 0$ or $x = k\pi, \ k \in \mathbf{Z}$ and

$3 - 4\sin^2 x = 2$, $4\sin^2 x = 1$. Therefore $\sin x = \pm\dfrac{\sqrt{2}}{2}$, which is impossible

due to initial conditions.

V. 15. a) $\dfrac{4\sin^2 x\left(\cos^2 x-1\right)}{4\cos^2 x\left(\sin^2 x-1\right)}+1-2\tan^2 x=0$, $\tan^4 x-2\tan^2 x+1=0$

or $\left(\tan^2 x-1\right)^2=0$. Then $\tan x=\pm1$. Therefore the solutions are

$x=\dfrac{\pi}{4}+\dfrac{k\pi}{2}$, $k\in\mathbf{Z}$;

b) $\dfrac{\cot^2 x\left(\sin^2 x-1\right)}{\tan^2 x\left(\cos^2 x-1\right)}+2\cot^3 x+1=0$ or $\cot^6 x+2\cot^3 x+1=0$. Then

$\cot x=-1$. The solutions are $x=-\dfrac{\pi}{4}+k\pi$, $k\in\mathbf{Z}$;

c) The domain is $x\neq\dfrac{k\pi}{4}$, $k\in\mathbf{Z}$.

$-2\cot 2x+2\tan 2x+\cot^2 4x+3=0\Rightarrow\cot^2 4x-4\cot 4x+3=0$. The roots

are $\cot 4x=1$ and $\cot 4x=3$. Therefore the solutions are $x_1=\dfrac{\pi}{16}+\dfrac{k\pi}{4}$

and $x_2=\dfrac{1}{4}\arctan\dfrac{1}{3}+\dfrac{k'\pi}{4}$;

d) $\tan x=\dfrac{2\sin x\cos x-\cos^2 x}{\cos^2 x-2\sin x\cos x+\sin^2 x+\cos^2 x}$, $\tan x=\dfrac{2\tan x-1}{2-2\tan x+\tan^2 x}$

or $\left(\tan x-1\right)\left(\tan^2 x-\tan x-1\right)=0$. Therefore the solutions are $x_1=\dfrac{\pi}{4}+k\pi$,

$x_2=\arctan\dfrac{1+\sqrt5}{2}+k\pi$, and $x_3=\arctan\dfrac{1-\sqrt5}{2}+k'\pi$;

e) $x=\pm\dfrac{\pi}{3}+k\pi$, $k\in\mathbf{Z}$, $x=\pm\dfrac{\pi}{4}+k'\pi$, $k'\in\mathbf{Z}$;

f) We have $\dfrac{4\dfrac{\cos x}{\sin x}}{\dfrac{\cos^2 x}{\sin^2 x}+1}+2=1-\sin^2 2x\Rightarrow4\dfrac{\cos x}{\sin x}\cdot\dfrac{\sin^2 x}{\cos^2 x+\sin^2 x}+\sin^2 2x+1=0\Rightarrow$

$\sin^2 2x+2\sin 2x+1=0\Rightarrow\left(\sin 2x+1\right)^2=0$ then $\sin 2x=-1\Rightarrow$

$x=-\dfrac{\pi}{4}+k\pi$, $k\in\mathbf{Z}$;

g) $\dfrac{3\cos^2 x-\sin^2 x}{\cos^2 x-3\sin^2 x}=\sin 6x\cot x$, $\dfrac{3-4\sin^2 x}{4\cos^2 x-3}\cdot\dfrac{\sin x}{\cos x}=\sin 6x$ or

$\tan 3x = \sin 6x$. The solutions are $x_1 = \dfrac{k\pi}{3}$ and $x_2 = \dfrac{\pi}{12} + \dfrac{k'\pi}{6}$, $k, k' \in \mathbf{Z}$;

h) $\tan^2 x - \dfrac{1 + \cos 2x}{1 - \cos 2x} = 2\cot 2x$, $\tan^2 x - \cot^2 x = 2\cot 2x$;

$\dfrac{\cos^4 x - \sin^4 x}{\sin^2 x \cos^2 x} = 2\cot 2x$, $\dfrac{\cos^2 x - \sin^2 x}{\sin^2 x \cos^2 x} = \dfrac{2\cos 2x}{\sin 2x}$, then only $\cos 2x = 0$

with the solution $x = \dfrac{\pi}{4} + \dfrac{k\pi}{2}$, $k \in \mathbf{Z}$.

i) $\dfrac{2\tan x}{1 - \tan^2 x}\left(3 - \tan^2 x\right)\cos x \cos 2x = 2\sin 5x$,

$\tan 2x\left(3\cos^2 x - \sin^2 x\right)\cos 2x \cdot \dfrac{1}{\cos x} = 2\sin 5x \Rightarrow \sin x\left(3 - 4\sin^2 x\right) = \sin 5x$.

Then $\sin 3x = \sin 5x$. Therefore the solutions are $x_1 = k\pi$ (not a solution)

and $x_2 = \dfrac{\pi}{8} + \dfrac{k\pi}{4}$;

j) $\sin x \neq 0$, $\cos x \neq 0 \Rightarrow x \neq \dfrac{k\pi}{2}$, $2 - \sec^2 x \neq 0 \Rightarrow x \neq \dfrac{\pi}{4} + \dfrac{k\pi}{2}$, and

$\cot^2 x - 3 \neq 0 \Rightarrow x \neq \pm\dfrac{\pi}{6} + k\pi$. We have

$\dfrac{\tan x \cos^2 x}{2\cos^2 x - 1} 2\sin x \cos 2x = \dfrac{2\sin^2 x}{\cos^2 x - 3\sin^2 x}$. Then $\cos x = \dfrac{1}{\cos^2 x - 3\sin^2 x} \Rightarrow$

$\cos x\left(2\cos 2x - 1\right) = 1$, $\cos 3x = 1$ or $x = \dfrac{2k\pi}{3}$, $k \in \mathbf{Z}$;

k) $-\dfrac{1}{2}\cot\dfrac{x}{2} + \dfrac{1}{2}\tan\dfrac{x}{2} + \tan x = 2\sqrt{3} \Rightarrow \tan x - \cot x = 2\sqrt{3}$. Then

$\cot 2x = -\sqrt{3}$. The solutions are $x = -\dfrac{\pi}{12} + \dfrac{k\pi}{2}$, $k \in \mathbf{Z}$;

l) $\sin x \cos^2 x\left(3 - \tan^2 x\right) = 2 + \cos\dfrac{6x}{5}$,

$2\sin x \cos^2 x + \sin x\left(\cos^2 x - \sin^2 x\right) = 2 + \cos\dfrac{6x}{5}$. Then $\sin 3x - \cos\dfrac{6x}{5} = 2$.

Therefore $\sin 3x = 1$ and $\cos\dfrac{6x}{5} = -1$. The solutions are

$x = \dfrac{5\pi}{6} + \dfrac{10k\pi}{3}$, $k \in \mathbf{Z}$, $k \neq 3p + 2$, $p \in \mathbf{Z}$. **V. 16. a)** $\alpha \neq k\pi$, $k \in \mathbf{Z}$.

$$\frac{\sin\alpha}{1+\cos\alpha}=\frac{2\sin\alpha-\cos\alpha}{\sin\alpha} \text{ or } \frac{\sin^2\alpha}{1+\cos\alpha}=2\sin\alpha-\cos\alpha \text{ or }$$

$1-\cos\alpha=2\sin\alpha-\cos\alpha$. The solutions are $\alpha=(-1)^k\dfrac{\pi}{6}+k\pi,\ k\in\mathbf{Z}$.

b) $x\neq k\pi$, $\dfrac{2\sin^2 x}{\sin x}=1$ or $\sin x=\dfrac{1}{2}$. Therefore the solutions

are $x=(-1)^k\dfrac{\pi}{6}+k\pi$; **c)** $\dfrac{2\sin^2 x\cos x}{2\cos^2 x(1+\cos x)}=\dfrac{1-\sin x}{\sin x}\Rightarrow$

$\dfrac{1-\cos x}{\cos x}=\dfrac{1-\sin x}{\sin x}\Rightarrow\tan x=1$. Then the solutions are $x=\dfrac{\pi}{4}+k\pi,\ k'\in\mathbf{Z}$;

d) $\dfrac{\cos x}{\sin x}-\dfrac{\dfrac{\sin\dfrac{x}{2}\sin x}{\cos\dfrac{x}{2}\cos x}+1}{\dfrac{\sin\dfrac{x}{2}}{\cos\dfrac{x}{2}}+\dfrac{\cos x}{\sin x}}=2\sqrt{3}\Rightarrow\dfrac{\cos x}{\sin x}-\dfrac{\dfrac{\cos\dfrac{x}{2}}{\cos\dfrac{x}{2}\cos x}}{\cos\dfrac{x}{2}\sin x}=2\sqrt{3}$, or $\cot 2x=\sqrt{3}$.

Therefore the solutions are $x=\dfrac{\pi}{12}+\dfrac{k\pi}{2}$;

e) $\dfrac{\sin^2 x}{\cos^2 x}=3$ or $\tan x=\pm\dfrac{1}{\sqrt{3}}$. Therefore the solutions are $x=\pm\dfrac{\pi}{6}+k\pi$;

f) $-\dfrac{\cos x}{\sin x}-\dfrac{\sin x}{1+\cos x}=2$. Then $\sin x=-\dfrac{1}{2}$. Therefore the solutions are

$x=(-1)^k\dfrac{\pi}{6}+k\pi,\ k\in\mathbf{Z}$;

g) $\dfrac{2\tan x}{3-\tan^2 x}=\sin 2x\Rightarrow(3\cos^2 x-\sin^2 x-1)\sin x=0$. Then $\cos 2x\sin x=0$.

Therefore the solutions are $x_1=\dfrac{\pi}{4}+\dfrac{k\pi}{2}$ and $x_2=k'\pi,\ k,k'\in\mathbf{Z}$;

h) The set of permissible values of x can be found from conditions:

1) $\sin x\neq 0$ $x\neq k\pi,\ k\in\mathbf{Z}$; 2) $\cos x\neq 0$ $x\neq\dfrac{\pi}{2}+k\pi,\ k\in\mathbf{Z}$;

3) $\tan x\neq -1$ $x\neq-\dfrac{\pi}{4}+k\pi,\ k\in\mathbf{Z}$ and

4) $\tan x \neq 1$ $x \neq \dfrac{\pi}{4} + k\pi$, $k \in \mathbf{Z}$. Finally $x \in \mathbf{R} \setminus \left\{ \dfrac{k\pi}{4} \right\}_{k \in \mathbf{Z}}$.

$$\frac{\cos x}{\sin x} - \frac{\sin x}{\cos x} + 2\left(\frac{\tan x - 1 + \tan x - 1}{\tan^2 x - 1} \right) = 4 \Rightarrow$$

$$\frac{\cos^2 x - \sin^2 x}{\sin x} + \frac{4 \tan x}{\tan^2 x - 1} = 4 \Rightarrow \frac{2\cos 2x}{\sin 2x} - 2\tan 2x = 4 \Rightarrow$$

$\cot 2x - \tan 2x = 2 \Rightarrow \dfrac{\cos^2 2x - \sin^2 2x}{\sin 2x \cos 2x} = 2 \Rightarrow \dfrac{\cos 4x}{\sin 4x} = 1 \Rightarrow \cot 4x = 1$. The

solutions of the initial equation are $x = \dfrac{\pi}{16} + \dfrac{k\pi}{4}$, $k \in \mathbf{Z}$; **i)** Domain

$x \neq k\pi$ and $x \neq \dfrac{\pi}{4} + \dfrac{k\pi}{2}$.

$$\frac{\cos x}{\sin x \cos 2x} = \frac{\sin x + \sin 3x}{\sin x} \Rightarrow \frac{\cos x}{\sin x \cos 2x} = 2\sin 2x \cos x \Rightarrow \cos x = 0 \text{ and}$$

$\cos 4x = 1$. The solutions are $x_1 = \dfrac{\pi}{2} + k\pi$ and $x_2 = \dfrac{\pi}{8} + \dfrac{k'\pi}{2}$, $k, k' \in \mathbf{Z}$;

j) $x \neq \dfrac{k'\pi}{4}$, $k' \in \mathbf{Z}$.

$$2\left(3 - \tan^2 x\right)\cos 2x \cos x = \frac{4\cos 3x \cos 2x}{\sin 2x} \Rightarrow \sin 2x \cos x \left(3 - \tan^2 x\right) = 2\cos 3x,$$

$6\sin x \cos^2 x - 2\sin^3 x = 2\cos 3x$. Then $\tan 3x = 1$. Therefore the solutions

are $x = \dfrac{\pi}{12} + \dfrac{k\pi}{3} = 15° + 60°k$. The images of the solutions (terminal

points) on the unit circle are...

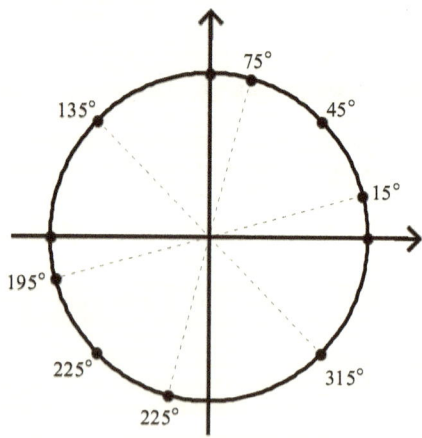

The terminal points of the dashed lines are the solutions for the initial equation, except 135 and 315 terminal lines, which are not valid

because $x \neq \dfrac{k'\pi}{4}$, $k' \in \mathbf{Z}$. Therefore the solutions are $x_1 = \dfrac{\pi}{12} + k\pi$ and

$x_2 = \dfrac{5\pi}{12} + k'\pi$.

Since the domain of the equation is $x \neq \dfrac{k\pi}{4}$, then $\dfrac{\pi}{12} + \dfrac{k\pi}{3} \neq \dfrac{3\pi}{4} + q\pi \Rightarrow$ $k \neq 3q + 2$, $k, q \in \mathbf{Z}$.

V. 17. a) $\dfrac{\sin x}{\cos x \cos^2 5x} - \dfrac{\sin 5x}{\cos 5x \cos^2 x} = 0$ or $\sin 2x - \sin 10x = 0$,

$\sin 4x \cos 6x = 0$. Then $\cos 6x = 0$ and $\sin 4x = 0$. Therefore $x_1 = \dfrac{\pi}{12} + \dfrac{k\pi}{6}$

and $x_2 = k'\pi$, $k, k' \in \mathbf{Z}$; **b)** $x = \dfrac{\pi}{4} + \dfrac{k\pi}{2}$, $k \in \mathbf{Z}$.

c) $\dfrac{\cos 4x}{\sin 4x \sin^2 x} + \dfrac{\cos x}{\sin x \sin^2 4x} = 0$, $\dfrac{\sin 2x + \sin 8x}{2\sin^2 x \sin^2 4x} = 0$ or $\cos 3x \sin 5x = 0$.

Therefore the solutions are $x_1 = \dfrac{\pi}{6} + \dfrac{2k\pi}{3}$ $(k \neq 3p + 1)$ and $x_2 = \dfrac{k'\pi}{5}$ $(k' \neq 5q)$, $k, k' \in \mathbf{Z}$;

d) $\cos x \neq 0 \Rightarrow x \neq \dfrac{\pi}{2} + k\pi$ $k \in \mathbf{Z}$ (1) and $\cos 2x \neq 0 \Rightarrow x \neq \dfrac{\pi}{4} + \dfrac{k\pi}{2}$ $k \in \mathbf{Z}$ (2).

We have $\cos^2 2x \tan 2x = \cos^2 x \tan x \Rightarrow \cos 2x \sin 2x = \cos x \sin x \Rightarrow$

$\sin 4x = \sin 2x \Rightarrow 4x = (-1)^k 2x + k\pi$, $k \in \mathbf{Z}$. There are two situations

I) $k = 2n \Rightarrow 4x = 2x + 2n\pi \Rightarrow x_1 = n\pi$ and

II) $k = 2n + 1 \Rightarrow 4x = -2x + (2n+1)\pi \Rightarrow 6x = (2n+1)\pi \Rightarrow x = \dfrac{\pi}{6} + \dfrac{n\pi}{3}$.

In this situation we eliminate $x = \dfrac{\pi}{2} + k\pi$ see you (1) . Finally

$x_2 = \pm\dfrac{\pi}{6} + k\pi$ $k \in \mathbf{Z}$ is the second set of solutions for the initial equation;

e) Denote $\dfrac{\tan 4x}{\tan 2x} = t$. Then $t + \dfrac{1}{t} + \dfrac{10}{3} = 0$ or $3t^2 + 10t + 3 = 0$.

The roots are $t = -\dfrac{1}{3}$ and $t = -3$. Therefore the solutions are

$$x_1 = \pm\frac{1}{2}\arctan\sqrt{7} + \frac{k\pi}{2} \text{ and } x_2 = \pm\frac{1}{2}\arctan\sqrt{\frac{5}{3}} + \frac{k'\pi}{2}, \ k,k' \in \mathbf{Z};$$

f) Denote $\dfrac{\cot x}{\cot 2x} = t = \dfrac{\tan 2x}{\tan x}$. Then $t + \dfrac{1}{t} + 2 = 0$ or $t = -1$. Therefore

$\dfrac{\tan 2x}{\tan x} = -1$ or $\tan 2x = \tan(-x)$, with the solution $x = \dfrac{k\pi}{3}, \ k \in \mathbf{Z}$, where

$k \neq 3p, \ k, p \in \mathbf{Z}$. Therefore the final set of solutions is $x = \pm\dfrac{\pi}{3} + k\pi, \ k \in \mathbf{Z}$;

g) $x \neq \dfrac{k\pi}{2}, \ k \in \mathbf{Z}$ and $x = -\dfrac{\pi}{4} + k\pi, \ k \in \mathbf{Z}$. After transformation,

we obtain $\sin x - \cos x = \sqrt{2}$. With the solution $x = \dfrac{3\pi}{4} + 2k\pi, \ k \in \mathbf{Z}$

(impossible). Finally the equation does not have any solution;

h) $\dfrac{\cos^2 x}{1 + \sin^2 x} + \dfrac{\sin^2 x}{1 + \cos^2 x} = \dfrac{2}{3} \Rightarrow \dfrac{2(1 - \sin^2 x \cos^2 x)}{2 + \sin^2 x \cos^2 x} = \dfrac{2}{3}$ or $\sin 2x = \pm 1$. The

solutions are $x = \dfrac{\pi}{4} + \dfrac{k\pi}{2}, \ k \in \mathbf{Z}$;

i) $\dfrac{\cos 5x \cos 2x}{\sin 7x} - \dfrac{\sin 5x \sin 2x}{\sin 7x} = \tan 3x, \ \cot 7x = \cot\left(\dfrac{\pi}{2} - 3x\right)$. Therefore the

solutions are $x = \dfrac{\pi}{20} + \dfrac{k\pi}{10}, \ k \neq 5p + 2, \ k, p \in \mathbf{Z}$;

j) $\dfrac{\cos 3x \cos 4x - \sin 3x \sin 4x}{\sin 7x} = -\cot^2 7x, \ \cot 7x = -\cot^2 7x$. Then

$\cot 7x = 0$ and $\cot 7x = -1$. Therefore the solutions are $x_1 = \dfrac{\pi}{14} + \dfrac{k\pi}{7}$,

$k \neq 7p + 3, \ k, p \in \mathbf{Z}$ and $x_2 = \dfrac{3\pi}{28} + \dfrac{k'\pi}{7}, \ k' \neq 7p' + 1, \ k', p' \in \mathbf{Z}$;

k) $\dfrac{\tan x}{1 + \tan^2 x} + \dfrac{\tan 2x}{1 + \tan^2 2x} = \dfrac{\tan 3x}{1 + \tan^2 3x} + \dfrac{\tan 4x}{1 + \tan^2 4x} \Rightarrow$

$\sin 2x + \sin 4x = \sin 6x + \sin 8x \Rightarrow \sin 2x - \sin 8x = \sin 6x - \sin 4x \Rightarrow$

$-2\sin 3x \cos 5x - 2\sin x \sin 5x = 0 \Rightarrow \cos 5x \sin 2x \cos 5x = 0$.

The solutions are $x_1 = \dfrac{\pi}{10} + \dfrac{k\pi}{5}, \ k \in \mathbf{Z}, \ x_2 = \dfrac{p\pi}{2}, \ p \in \mathbf{Z}$, and $x_3 = \dfrac{\pi}{2} + l\pi$,
$l \in \mathbf{Z}$.

V. 18. a) $x \neq \dfrac{\pi}{8} + \dfrac{k\pi}{4}$ and $x \neq \dfrac{k\pi}{2}$. We have $\dfrac{1}{\sin 2x} = \dfrac{1}{\cos 4x}$,

$\cos(90 - 2x) = \cos 4x.$ $x_1 = \dfrac{\pi}{12} - \dfrac{k\pi}{3}, k \in \mathbf{Z}$;

$x_2 = -\dfrac{\pi}{4} + n\pi, n \in \mathbf{Z}$ **b)** The set of permissible values of x can be found

from conditions $\tan x \neq \pm 1$ and this means $x \neq \dfrac{\pi}{4} + \dfrac{k\pi}{2}$, $k \in \mathbf{Z}$

and $\cos x \neq 0$ this means $x \neq \dfrac{\pi}{2} + k\pi$, $k \in \mathbf{Z}$.

We have $\dfrac{\cos^2 x}{\cos^2 x - \sin^2 x} = 2\cos^2 x \Rightarrow \cos^2 x \left(\dfrac{1}{\cos 2x} - 2 \right) = 0 \Rightarrow$

$\cos 2x = \dfrac{1}{2}$, then $x = \pm\dfrac{\pi}{6} + k\pi$, $k \in \mathbf{Z}$;

c) $\tan x \neq 1$, $\sin x \neq 0$ and $\cos x \neq 0$. We have

$$\dfrac{\sin x + \cos x}{\cos x - \sin x} + \dfrac{\left(\cos x \cos\dfrac{x}{2} + \sin x \sin\dfrac{x}{2} \right) \sin x \cos\dfrac{x}{2}}{\cos x \cos\dfrac{x}{2} \left(\cos x \cos\dfrac{x}{2} + \sin x \sin\dfrac{x}{2} \right)} = 1 \Rightarrow$$

$\dfrac{\sin x + \cos x}{\cos x - \sin x} + \dfrac{\sin x}{\cos x} = 1 \Rightarrow 3\cos x = \sin x \Rightarrow$

$\tan x = 3 \Rightarrow x = \arctan 3 + \dfrac{k\pi}{2}$, $k \in \mathbf{Z}$;

d) $\tan x \neq 1$ and $\sin 2x \neq 0$.

$\dfrac{(\sin x + \cos x)^2}{\sin 2x} = \sqrt{2}\left(\cos^2 x - \sin^2 x \right)\dfrac{\cos x + \sin x}{\cos x - \sin x} \Rightarrow$

$(\sin x + \cos x)^2 \left(\dfrac{1}{\sin 2x} - \sqrt{2} \right) = 0.$ $x_1 = -\dfrac{\pi}{4} + k\pi$ $k \in \mathbf{Z}$,

$x_2 = (-1)^k \dfrac{\pi}{8} + \dfrac{k\pi}{2}$ $k \in \mathbf{Z}$;

e) If $\cos x \neq 0$ we have

$\dfrac{\left(2\sin^2 x \right)^2}{\left(2\cos^2 x \right)^2} - \dfrac{2}{\cos^2 x} = 1 \Rightarrow \sin^4 x - 2\cos^2 x = \cos^4 x \Rightarrow \sin^4 x - \cos^4 x - 2\cos^2 x = 0 \Rightarrow$

211

$$\left(\sin^2 x - \cos^2 x\right)\left(\sin^2 x + \cos^2 x\right) - 2\cos^2 x = 0 \Rightarrow \sin^2 x - \cos^2 x - 2\cos^2 x = 0 \Rightarrow$$

$$\tan^2 x = 3 \Rightarrow \tan x = \pm\sqrt{3}. \text{ So } x = \pm\frac{\pi}{3} + k\pi, \ k \in \mathbf{Z};$$

f) If $\tan x \neq \pm 1$ we have $\dfrac{1}{\cos 2x} = 4\sin 2x + 6\cos 2x \Rightarrow$

$$4\sin 2x \cos 2x + 6\cos^2 2x = 1 \Rightarrow$$

$$2\sin 4x + 3(1 + \cos 4x) = 1 \Rightarrow 2\sin 4x + 3\cos 4x = -2. \text{ But } \sin 4x = \frac{2\tan 2x}{1 + \tan^2 2x}$$

and $\cos 4x = \dfrac{1 - \tan^2 2x}{1 + \tan^2 2x}$. Substituting $\tan 2x = t$ we obtain

$$2\frac{2t}{1+t^2} + 3\frac{1-t^2}{1+t^2} = -2 \Leftrightarrow t^2 - 4t - 5 = 0. \text{ This is a quadric equation}$$

whose roots are $t_1 = -1$ and $t_2 = 5$. Thus the solutions of the initial

equation reduce to the solutions of two trigonometric equations:

$$\tan 2x = -1 \Rightarrow x_1 = -\frac{\pi}{8} + \frac{k\pi}{2} \text{ and } \tan 2x = 5 \Rightarrow x_2 = \frac{1}{2}\arctan 5 + \frac{k\pi}{2}, \ k \in \mathbf{Z};$$

g) $\cot^2 x - \dfrac{2\sin 2x \cos 2x - 2\sin 2x}{2\sin 2x \cos 2x + 2\sin 2x} = \dfrac{10}{3} \Rightarrow$

$$\frac{\cos^2 x}{\sin^2 x} - \frac{\cos 2x - 1}{\cos 2x + 1} = \frac{10}{3} \Rightarrow \frac{\cos^2 x}{\sin^2 x} + \frac{2\sin^2 x}{2\cos^2 x} = \frac{10}{3} \Rightarrow$$

$$\frac{\cos^4 x + \sin^4 x}{\sin^2 x \cos^2 x} = \frac{10}{3} \Rightarrow \frac{\left(\sin^2 x + \cos^2 x\right)^2 - 2\sin^2 x \cos^2 x}{\sin^2 x \cos^2 x} = \frac{10}{3} \Rightarrow$$

$$\frac{1 - 2\sin^2 x \cos^2 x}{\sin^2 x \cos^2 x} = \frac{10}{3} \Rightarrow \frac{4}{4\sin^2 x \cos^2 x} = \frac{16}{3} \Rightarrow \sin 2x = \pm\frac{\sqrt{3}}{2}.$$

$$x = \pm\frac{\pi}{3} + k\pi \ \ k \in Z;$$

h) $\tan x \neq 1$. We have $3\dfrac{\cos x + \sin x}{\cos x - \sin x} + \left(\dfrac{\cos x + \sin x}{\cos x - \sin x}\right)^2 - 4 = 0.$

Denoting $\dfrac{\cos x + \sin x}{\cos x - \sin x} = t$, we obtain $t^2 + 3t - 4 = 0$, $t_1 = 1$, $t_2 = -4$.

Therefore $\dfrac{\cos x + \sin x}{\cos x - \sin x} = 1 \Rightarrow \sin x = 0 \Rightarrow x_1 = k\pi$ and

$$\frac{\cos x + \sin x}{\cos x - \sin x} = -4 \Rightarrow \tan x = \frac{5}{3} \Rightarrow x_2 = \arctan\frac{5}{3} + k\pi, \ \ k \in Z .$$

i) $\tan x \neq \pm 1$, $\cos 4x \neq 0$. We have $\dfrac{\cos 2x}{\sin x \cos x} + \dfrac{4}{\cos 4x} = -\dfrac{4\sin x \cos x}{\cos 2x} \Rightarrow$

$$\frac{2\cos 2x}{\sin 2x} + \frac{2\sin 2x}{\cos 2x} + \frac{4}{\cos 4x} = 0 \Rightarrow 2\frac{\sin^2 2x + \cos^2 2x}{\sin 2x \cos 2x} + \frac{4}{\cos 4x} = 0 \Rightarrow$$

$$\frac{4}{\sin 4x} + \frac{4}{\cos 4x} = 0 \Rightarrow \sin 4x + \cos 4x = 0 \Rightarrow \tan 4x = -1 . \text{ Therefore}$$

$$x = -\frac{\pi}{16} + \frac{k\pi}{4} \ \ k \in \mathbf{Z}.$$

j) $\dfrac{\sin^2 x + \cos^2 x + \sin 2x - 2\sin^2 x}{1 + \dfrac{\cos^2 x}{\sin^2 x}} = \sqrt{2} \sin x \cos\left(\frac{\pi}{4} - 2x\right) \Rightarrow$

$$\sin x(\sin 2x + \cos 2x) = \sqrt{2}\left(\cos\frac{\pi}{4}\cos 2x + \sin\frac{\pi}{4}\sin 2x\right) \Rightarrow$$

$$\sin x(\sin 2x + \cos 2x) = \sin 2x + \cos 2x . \text{ Therefore}$$

$$\sin x = 1 \Rightarrow x_1 = \frac{\pi}{2} + 2k\pi \ \ k \in Z \text{ and } \sin 2x + \cos 2x = 0 \Rightarrow \tan 2x = -1,$$

$$x_2 = -\frac{\pi}{8} + \frac{k\pi}{2} \ \ k \in Z .$$

V. 19. a) We have $\dfrac{\cos(3x - 13°)\cos 20° + \sin(3x - 13°)\sin 20°}{\sin 20°} + \dfrac{1}{2\sin 20°} = 0$

or $\cos(3x - 13° - 20°) = -\dfrac{1}{2} \Rightarrow \cos(3x - 33°) = -\dfrac{1}{2} \Rightarrow$

$3x - 33° = \pm \arccos\left(-\dfrac{1}{2}\right) + 2k\pi, \ \ k \in Z.$ Then

1) $3x - 33° = 120° + 2k\pi, \ \ k \in Z \Rightarrow x = 51° + 120° \cdot k, \ \ k \in \mathbf{Z};$

2) $3x - 33° = -120° + 2k\pi, \ \ k \in Z \Rightarrow x = -29° + 120° \cdot k, \ \ k \in \mathbf{Z};$

b) $x = 10° + 120° \cdot k, \ \ k \in \mathbf{Z}$ also $x = -30° + 120° \cdot k, \ \ k \in \mathbf{Z};$

c) $\sin x\left(\cos 2x + \dfrac{1}{2}\right) = \dfrac{1}{4}$, $2\cos 2x \sin x + \sin x = \dfrac{1}{2}$. Then $\sin 3x = \dfrac{1}{2}$. The set of solutions is $x = (-1)^k \dfrac{\pi}{18} + \dfrac{k\pi}{3}$, $k \in \mathbf{Z}$;

d) $\dfrac{2\sin(x-15^\circ)\cos(x+15^\circ)}{2\cos(x-15^\circ)\sin(x+15^\circ)} = \dfrac{1}{3}$, $\dfrac{2\sin 2x - 1}{2\sin 2x + 1} = \dfrac{1}{3}$. Then

$\sin 2x = 1$. The solutions are $x = \dfrac{\pi}{4} + k\pi$, $k \in \mathbf{Z}$; e) We have

$$\tan 2x \dfrac{\sin\left(2x+\dfrac{\pi}{3}\right)\sin\left(2x+\dfrac{2\pi}{3}\right)}{\cos\left(2x+\dfrac{\pi}{3}\right)\cos\left(2x+\dfrac{2\pi}{3}\right)} = \sqrt{3} \Rightarrow \tan 2x \dfrac{\cos\dfrac{\pi}{3} - \cos(4x+\pi)}{\cos\dfrac{\pi}{3} + \cos(4x+\pi)} = \sqrt{3} \Rightarrow$$

$$\tan 2x \dfrac{1 + 2\cos 4x}{1 - 2\cos 4x} = \sqrt{3} \Rightarrow \tan 2x \dfrac{1 + 2(2\cos^2 2x - 1)}{1 - 2(1 - 2\sin^2 2x)} = \sqrt{3} \Rightarrow$$

$$\dfrac{\sin 2x}{\cos 2x} \cdot \dfrac{4\cos^2 2x - 1}{4\sin^2 2x - 1} = \sqrt{3} \Rightarrow \dfrac{2\sin 4x \cos 2x - \sin 2x}{2\sin 4x \sin 2x - \cos 2x} = \sqrt{3} \Rightarrow$$

$$\dfrac{\sin 6x + \sin 2x - \sin 2x}{\cos 2x - \cos 6x - \cos 2x} = \sqrt{3} \Rightarrow \dfrac{\sin 6x}{-\cos 6x} = \sqrt{3} \Rightarrow \tan 6x = -\sqrt{3}. \text{ Therefore}$$

$x = -10^o + k \cdot 30^o$, $k \in \mathbf{Z}$;

f) $-\cot x + 2\cot^2 x = \cot^2 x$ or $\cot x(\cot x - 1) = 0$. Then $\cot x = 0$ and

$\cot x = 1$. Therefore the solutions are $x_1 = \dfrac{\pi}{2} + k\pi$ and $x_2 = \dfrac{\pi}{4} + k'\pi$,

$k, k' \in \mathbf{Z}$;

g) $\sin(x + 20^\circ) + \cos(x - 10^\circ) + \sin(40^\circ - x) = \sqrt{3}$ or

$2\sin 30^\circ \cos(x - 10^\circ) + \cos(x - 10^\circ) = \sqrt{3}$ or $\cos(x - 10^\circ) = \dfrac{\sqrt{3}}{2}$ with the

solutions $x = 10^\circ \pm 30^\circ + 360^\circ k$, $k \in \mathbf{Z}$;

h) $\tan(120^\circ + 3x) + \tan(40^\circ + x) = 2\sin(80^\circ + 2x)$. Denote $x + 40^\circ = t$. Then

$\tan 3t + \tan t = 2\sin 2t$ or $\dfrac{\sin 2t \cos 2t}{\cos t \cos 3t} = \sin 2t$. Therefore $t_1 = k\pi$ and

$t_2 = \dfrac{k'\pi}{3}$. The final solutions are $x = -\dfrac{2\pi}{9} + \dfrac{k\pi}{3}$, $k \in \mathbf{Z}$;

i) $\cot x + \cot 14° = \cot\left(x + 26°\right)\left(\cot x \cot 14° - 1\right)$

$\dfrac{\sin\left(x+14°\right)}{\sin x \sin 14°} = \cot\left(x+26°\right)\dfrac{\cos\left(x+14°\right)}{\sin x \sin 14°}$ or $\tan\left(x+14°\right) = \tan\left(64° - x\right)$;

$x + 14° = 64° - x + k\pi$. Therefore $x = 25° + 90°k$, $k \in \mathbf{Z}$;

j) $\tan x\left(\tan 20° + \tan 40°\right) = 1 - \tan 20° \tan 40°$, $\tan x = \dfrac{1}{\sqrt{3}}$. Therefore the set

of solutions is $x = \dfrac{\pi}{6} + k\pi$, $k \in \mathbf{Z}$;

k) $\tan x\left(\tan 50° \tan 70° - 1\right) = \tan 50° + \tan 70° \Rightarrow \tan x = -\tan 120°$ or

$\tan x = \tan\left(60°\right)$. Therefore the solutions are $x = \dfrac{\pi}{3} + k\pi$, $k \in \mathbf{Z}$;

l) $x = \dfrac{\pi}{4} + k\pi$.

V. 20. a) We have $2 + 2\sin^2 2x = 3 - \sin 4 \Rightarrow \sin 4x = 1 - 2\sin^2 2x \Rightarrow$

$\sin 4x = \cos 4x \Rightarrow \tan 4x = 1$. Therefore $x = \dfrac{\pi}{16} + \dfrac{k\pi}{4}$, $k \in Z$;

b) It is known $\sin a \sin b = \dfrac{1}{2}\left[\cos(a-b) - \cos(a+b)\right]$ and

$\cos a \cos b = \dfrac{1}{2}\left[\cos(a+b) + \cos(a-b)\right]$, also $\cos 2x = 2\cos^2 a - 1 = 1 - 2\sin^2 a$

and $\sin 2a = 2\sin a \cos a$. We have

$\tan 2x \dfrac{\sin\left(2x + \dfrac{\pi}{3}\right)\sin\left(2x + \dfrac{2\pi}{3}\right)}{\cos\left(2x + \dfrac{\pi}{3}\right)\cos\left(2x + \dfrac{2\pi}{3}\right)} = \sqrt{3} \Rightarrow \tan 2x \dfrac{\cos\dfrac{\pi}{3} - \cos(4x + \pi)}{\cos\dfrac{\pi}{3} + \cos(4x + \pi)} = \sqrt{3} \Rightarrow$

$\tan 2x \dfrac{1 + 2\cos 4x}{1 - 2\cos 4x} = \sqrt{3} \Rightarrow$

$\tan 2x \dfrac{1 + 2\left(2\cos^2 2x - 1\right)}{1 - 2\left(1 - 2\sin^2 2x\right)} = \sqrt{3} \Rightarrow \dfrac{\sin 2x}{\cos 2x} \cdot \dfrac{4\cos^2 2x - 1}{4\sin^2 2x - 1} = \sqrt{3}$

$\dfrac{\sin 6x + \sin 2x - \sin 2x}{\cos 2x - \cos 6x - \cos 2x} = \sqrt{3} \Rightarrow \dfrac{\sin 6x}{-\cos 6x} = \sqrt{3} \Rightarrow \tan 6x = -\sqrt{3}$.

So $x = -10° + k \cdot 30°$, $k \in Z$;

c) $-\sin x + \sin\dfrac{x}{2}\sin x - \cos\dfrac{x}{2}\sin^2 x = 0 \Rightarrow$

215

$$\sin x\left(1 - \sin\frac{x}{2} + 2\sin\frac{x}{2}\cos^2\frac{x}{2}\right) = 0 \Rightarrow$$

$$\sin x\left[1 - \sin\frac{x}{2}\left(1 - 2\cos^2\frac{x}{2}\right)\right] = 0 \Rightarrow \sin x\left[1 + \sin\frac{x}{2}\left(1 - 2\sin^2\frac{x}{2}\right)\right] = 0. \text{ Then}$$

$$\sin x = 0 \text{ or } 2\sin^3\frac{x}{2} - \sin\frac{x}{2} - 1 = 0 \Rightarrow \left(\sin\frac{x}{2} - 1\right)\left(2\sin^2\frac{x}{2} + 2\sin\frac{x}{2} + 1\right) = 0.$$

Therefore the solutions are $x = k\pi,\, k \in \mathbf{Z}$;

d) $2\sin^2\left(\dfrac{\pi}{8} + x\right) - 2\sin^2\left(\dfrac{\pi}{8} - x\right) = 2\sin x$,

$\cos\left(\dfrac{\pi}{4} - 2x\right) - \cos\left(\dfrac{\pi}{4} + 2x\right) = 2\sin x$ or $\sqrt{2}\sin x\cos x = \sin x$. Then

$\sin x = 0$ and $\cos x = \dfrac{\sqrt{2}}{2}$. Therefore the solutions are $x_1 = k\pi$ and

$x_2 = \pm\dfrac{\pi}{4} + 2k'\pi,\ k, k' \in \mathbf{Z}$;

e) $x \neq \dfrac{k\pi}{3}$ and $x \neq \dfrac{\pi}{2} + k\pi$.

$\cot 3x\left(\tan^2 x \cot^2 2x - 1\right) = \tan^2 x - \cot^2 2x$,

$\cot 3x \cdot \dfrac{-\sin 3x \sin x}{\cos^2 x \sin^2 2x} = \dfrac{-\cos 3x \cos x}{\cos^2 x \sin^2 2x}$. Then $\cos 3x = 0$ with the solutions

$x_1 = \dfrac{\pi}{6} + \dfrac{k\pi}{3},\ k \in \mathbf{Z}$ where $k \neq 1 + 3l,\ l \in \mathbf{Z}$ and $\tan x = 1$ with the

solutions $x = \dfrac{\pi}{4} + k'\pi,\ k' \in \mathbf{Z},\ k, k' \in \mathbf{Z}$;

f) $\cos 2x \neq 0$, $\cos 3x \neq 0$, $\cos 5x \neq 0$.

$\tan^2 5x - \tan^2 3x = \tan 2x\left(1 - \tan^2 3x \tan^2 5x\right)$,

$\sin^2 5x \cos^2 3x - \sin^2 3x \cos^2 5x = \tan 2x\left(\cos^2 3x \cos^2 5x - \sin^2 3x \sin^2 5x\right)$

or $\sin 8x \sin 2x = \dfrac{\sin 2x}{\cos 2x}\cos 8x \cos 2x$. Then $\sin 2x = 0 \Rightarrow \sin x = 0$ (since

$\cos x \neq 0$) and $\sin 8x = \cos 8x \Rightarrow \tan 8x = 1$. Therefore the solutions are

$x_1 = k\pi$ and $x_2 = \dfrac{\pi}{32} + \dfrac{k\pi}{8}$;

g) $\cos x \neq 0$, $\cos 3x \neq 0$, $\cos 4x \neq 0$.

$$\frac{\sin^2 x \sin^2 \sin 4x}{\cos^2 x \cos^2 3x \cos 4x} = \frac{-\sin 2x}{\cos x \cos 3x} \cdot \frac{\sin 4x}{\cos x \cos 3x} + \frac{\sin 4x}{\cos 4x} \Rightarrow \sin 4x = 0 \text{ or}$$

$$\sin^2 x \sin^2 3x = -\sin 2x \cos 4x + \cos^2 x \cos^2 3x \Rightarrow$$

$(\cos x \cos 3x - \sin x \sin 3x)(\cos x \cos 3x + \sin x \sin 3x) - \sin 2x \cos 4x = 0$, then

$\cos 2x \cos 4x - \sin 2x \cos 4x = 0 \Rightarrow \tan 2x = 1$ (impossible), and $\cos 4x = 0$

(impossible).

From $\sin 4x = 0$ we obtain $2 \sin x \cos x \cos 2x = 0$ or $\cos 2x = 0$ and

$\sin x = 0$.

Therefore the solutions are $x_1 = \dfrac{\pi}{4} + \dfrac{k\pi}{2}$ and $x_2 = k\pi$.

V. 21. a) $4(1 - \cot 2x \tan 3x) = \cot 2x \tan 3x + \tan^2 3x \Rightarrow$

$4(1 - \cot 2x \tan 3x) = \tan 3x(\cot 2x + \tan 3x) \Rightarrow$

$$\frac{-4 \sin x}{\sin 2x \cos 3x} = \frac{\sin 3x}{\cos 3x} \cdot \frac{\cos x}{\sin 2x \cos 3x} \Rightarrow \tan 3x = -4 \tan x. \text{ Then}$$

$\dfrac{3 \tan x - \tan^3 x}{1 - 3 \tan^2 x} = -4 \tan x \Rightarrow \tan x(13 \tan^2 x - 7) = 0$. Therefore the solutions

are $x_1 = k\pi$ (not a solution) and $x_2 = \pm \arctan \sqrt{\dfrac{7}{13}} + k\pi$;

b) Denote $\tan x = t$. Then $t^2 - t + \dfrac{3}{t} - \dfrac{3}{t^2} = 2$, $t^4 - t^3 - 2t^2 + 3t - 3 = 0$ or

$(t^2 - 3)(t^2 - t + 1) = 0$. We obtain $t^2 - 3 = 0$ and $t^2 - t + 1 = 0$ (impossible).

Then $t = \pm \sqrt{3}$. The solutions are $x = \pm \dfrac{\pi}{3} + k\pi$, $k \in \mathbf{Z}$;

c) $\left(\dfrac{1}{\cos^2 3x} - 1\right)^2 = 4 \sin^2 3x \cos^2 3x$. Denote $\cos^2 3x = t$, we get

$\dfrac{(1 - t)^2}{t^2} = 4(1 - t)t$. We obtain $t = 1$, therefore $\cos^2 3x = 1$ or $\sin^2 3x = 0$

with the solution $x = \dfrac{k\pi}{3}$, $k \in \mathbf{Z}$ and $\dfrac{1 - t}{t^2} = 4t$ or $4t^2 + t - 1 = 0$ with the solution $t = \dfrac{1}{2}$ or $\cos^2 3x = \dfrac{1}{2}$ or $\sin 6x = 0$ implies $6x = \dfrac{\pi}{2} + k\pi$ or

$x = \dfrac{\pi}{12} + \dfrac{k\pi}{6}$, $k \in \mathbf{Z}$;

d) $\sin^6 x + 3\sin^2 x \cos^2 x = 1 - 3\sin^2 x \cos^4 x$,

$\sin^4 x + \sin^2 x + 1 = 3\sin^2 x\left(2 - \sin^2 x\right)$ or $4\sin^4 x - 5\sin^2 x + 1 = 0$. Denote

$\sin^2 x = a$. Then $4a^2 - 5a + 1 = 0$. The roots are $a = \pm 1$ (impossible) and

$a = \pm \dfrac{1}{2}$. Therefore the solutions are $x = \pm \dfrac{\pi}{6} + k\pi$, $k \in \mathbf{Z}$;

e) Denote $\sin x + \dfrac{1}{\sin x} = t$. Then $2t^2 + t - 10 = 0$. Therefore the

roots are $t = 2$ and $t = -\dfrac{5}{2}$. The solutions are $x_1 = \dfrac{\pi}{2} + 2k\pi$ and

$x_2 = (-1)^{k+1}\dfrac{\pi}{6} + k\pi$;

f) Denote $\tan x + \dfrac{1}{\tan x} = z$. Then $z^2 - 2 + 3z + 4 = 0$ or $z^2 + 3z + 2 = 0$.

The roots are $z = -1$ and $z = -2$. Therefore $\tan^2 x + \tan + 1 = 0$

(impossible) and $\tan^2 x + 2\tan + 1 = 0$. The solutions are

$x = -\dfrac{\pi}{4} + k\pi$, $k \in \mathbf{Z}$;

g) Denote $\tan^2 x + \dfrac{1}{\tan^2 x} = m$. Then $m^2 + m - 6 = 0$. The roots are $m = 2$

and $m = -3$ (impossible). Therefore the solutions are $x = \dfrac{\pi}{4} + \dfrac{k\pi}{2}$, $k \in \mathbf{Z}$;

h) $x_1 = \pm\dfrac{\pi}{3} + k\pi$, $k \in \mathbf{Z}$, $x_2 = \pm\arctan\sqrt{\dfrac{7 + \sqrt{58}}{3}} + k'\pi$, $k' \in \mathbf{Z}$;

i) $2\left(\tan^2 x + 1\right) + \dfrac{1}{3\tan^2 x} + 10\left(2\tan x + \dfrac{1}{3\tan x}\right) - 1 = 0$, Denote

$2\tan x + \dfrac{1}{3\tan x} = t$. Then $3t^2 + 10t + 7 = 0$. The roots are $t = -1$

(impossible) and $t = -\dfrac{7}{3}$. Therefore the solutions are $x_1 = -\dfrac{\pi}{4} + k\pi$ and

$$x_2 = -\arctan\frac{1}{6} + k\pi \, ;$$

j) $2\left(9\cos^2 x + \dfrac{1}{\cos^2 x}\right) + 5\left(3\cos x + \dfrac{1}{\cos x}\right) + 5 = 0$. Denote $3\cos x + \dfrac{1}{\cos x} = z$.

Then $2t^2 + 5t - 7 = 0$. The roots are $t = -\dfrac{7}{2}$ and $t = 1$ (impossible). Then

$6\cos^2 x + 7\cos x + 2 = 0$. The roots are $\cos x = -\dfrac{1}{2}$ and $\cos x = -\dfrac{2}{3}$. The

solutions are $x_1 = \pm\dfrac{2\pi}{3} + 2k\pi$ and $x_2 = \pm\arccos\left(-\dfrac{2}{3}\right) + 2k'\pi$, $k, k' \in \mathbf{Z}$;

k) Denote $\tan x + \cot x = z$. Then $z^3 + z^2 - 2z - 8 = 0$. We obtain

$z - 2 = 0$ and $z^2 + 3x + 4 = 0$ (impossible). Therefore $z = 2$. The

solutions are $x = \dfrac{\pi}{4} + k\pi$, $k \in \mathbf{Z}$;

l) Denote $\tan x + \cot x = t$. Then $t^3 + t^2 - 3t - 6 = 0$ or

$(t - 2)\left(t^2 + 3y + 3\right) = 0$. We obtain $t - 2 = 0$ and $t^2 + 3y + 3 = 0$

(impossible). Therefore $t = 2$. The solutions are $x = \dfrac{\pi}{4} + k\pi$, $k \in \mathbf{Z}$.

V. 22. a) The initial equations is equivalent to $\cos 3x \cos x = \cos 7x \cos x$.

From $\cos x = 0$, $x_1 = \dfrac{\pi}{2} + k\pi$, $k \in \mathbf{Z}$ and from $\cos 3x = \cos 7x$,

$3x = \pm 7x + 2k\pi$. Which yields to $x_2 = \dfrac{k\pi}{2}$, $k \in \mathbf{Z}$ and $x_3 = \dfrac{k`\pi}{5}$, $k` \in \mathbf{Z}$;

Notice: Since any solution from x_1 is included in the solutions from x_2,

then the solutions of the given equation are those in x_2 and x_3;

b) $x_1 = \dfrac{\pi}{4} + \dfrac{k\pi}{2}$, $k \in \mathbf{Z}$ and $x_2 = \dfrac{\pi}{14} + \dfrac{2k`\pi}{7}$, $k` \in \mathbf{Z}$, $k, k' \in \mathbf{Z}$;

c) $x_1 = -\dfrac{\pi}{2} + 2k`\pi$, $k` \in \mathbf{Z}$, $x_2 = \dfrac{\pi}{4} + \dfrac{k\pi}{2}$, $k \in \mathbf{Z}$;

d) $\cos^3 x + \cos^2 x - 2(1 + \cos x) = 0$, $(1 + \cos x)\left(\cos^2 x - 2\right) = 0$. Then

$1 + \cos x = 0$ and $\cos^2 x - 2 = 0$ (not a valid solution). The solutions are

$x = \pi + 2k\pi$, $k \in \mathbf{Z}$;

e) $x_1 = \dfrac{3\pi}{4} + k`\pi,\ k` \in \mathbf{Z}$, $x_2 = \pm\dfrac{\pi}{6} + k\pi,\ k \in \mathbf{Z}$;

f) $x_1 = -\dfrac{\pi}{4} + k\pi$ and $x_2 = \pm\dfrac{\pi}{6} + k'\pi$;

g) $\sin^3 2x\left(2\sin^2 2x - 1\right) - 3\left(2\sin^2 2x - 1\right) = 0$, $\left(2\sin^2 2x - 1\right)\left(\sin^2 2x - 3\right) = 0$,

$2\sin^2 2x - 1 = 0$ or $\cos 4x = 0$ with the solution $x = \dfrac{\pi}{8} + \dfrac{k\pi}{4},\ k \in \mathbf{Z}$;

h) $\dfrac{(\sin x - \cos x)\left(\sin^2 x + \sin x \cos x + \cos^2 x\right)}{2 + \sin 2x} = \dfrac{1}{3}\left(\cos^2 x - \sin^2 x\right)$ or

$\dfrac{(\sin x - \cos x)\left(1 + \dfrac{\sin 2x}{2}\right)}{2 + \sin 2x} = \dfrac{1}{3}\left(\cos^2 x - \sin^2 x\right)$.

Then $\sin x - \cos x = 0$ or $x = \dfrac{\pi}{4} + k\pi,\ k \in \mathbf{Z}$. The equation $\sin x + \cos x = \dfrac{3}{2}$

is impossible, since $\sin x + \cos x = \cos\left(\dfrac{\pi}{2} - x\right) + \cos x = \sqrt{2}\cos\left(\dfrac{\pi}{4} - x\right) \leq \sqrt{2}$;

i) $x = \arctan\dfrac{4}{5} + 2k\pi,\ k \in \mathbf{Z}$;

j) $\cos^3 x - 2\cos x\sin^2 x - \sin x\cos^2 x = 0$. Then $\cos x = 0$ and

$2\sin^2 x + \sin x \cos x - \cos^2 x = 0$. Therefore the solutions are $x_1 = \dfrac{\pi}{2} + k\pi$,

$x_2 = -\dfrac{\pi}{4} + k\pi$, and $x_3 = \arctan\dfrac{1}{2} + k\pi$;

k) $\dfrac{6\cos^3 x + 2\sin^3 x}{3\cos x - \sin x} = \cos^2 x - \sin^2 x$,

$6\cos^3 x + 2\sin^3 x = 3\cos^3 x - 3\cos x\sin^2 x - \sin x\cos^2 x + \sin^3 x$,

$3\cos^3 x + \sin^3 x = \sin x\cos x(3\sin x + \cos x)$, $3 + \tan^3 x = -\tan x(3\tan x + 1)$.

Denote $\tan x = t$, we get $t^3 + 3t^2 + t + 3 = 0$ or $(t + 3)\left(t^2 + 1\right) = 0$. Therefore

$\tan x = -3$ or $x = -\arctan 3 + k\pi,\ k \in \mathbf{Z}$.

V. 23. $\sin^3 x + \cos^3 x = \sin^2 x + \cos^2 x \Rightarrow$

$\sin^2 x(\sin x - 1) + \cos^2 x(\cos x - 1) = 0 \Rightarrow$

220

$(\sin x - 1)(\cos x - 1)(2 + \sin x + \cos x) = 0 = 0$ Therefore the solutions are

$x_1 = \dfrac{\pi}{2} + 2k\pi$ and $x_2 = 2k'\pi$, $k, k' \in \mathbf{Z}$;

b) $2\sin x \cos x(\sin^2 x - \cos^2 x) = \dfrac{\sqrt{2}}{4}$, $-2\sin 2x \cos 2x = \dfrac{\sqrt{2}}{2}$. Then

$\sin 4x = -\dfrac{\sqrt{2}}{2}$. The solutions are $x = (-1)^{k+1}\dfrac{\pi}{16} + \dfrac{k\pi}{4}$;

c) *Solution* 1. $2\sin^2 x(\sin x + \cos x) - \cos^2 x(\sin x + \cos x) = 0$,

$(3\sin^2 x - 1)(\sin x + \cos x) = 0$ or $(1 - 3\cos 2x)(\sin x + \cos x) = 0$ The

roots are $\sin x = \pm\dfrac{1}{\sqrt{3}}$ and $\tan x = -1$. Therefore the solutions are

$x_1 = \pm\arccos\dfrac{1}{3} + 2k\pi$ and $x_2 = \dfrac{3\pi}{4} + k'\pi$.

Solution 2. Dividing LHS by $\cos^3 x$ and denoting $\tan^3 x = t$, we obtain
the equation $2t^3 + 2t^2 - t - 1 = 0$ and so on;

d) $(1 - \cos^2 x)\sin x \sin 3x + (1 - \sin^2 x)\cos x \cos 3x = \cos^3 4x$,

$\cos 2x - \sin x \cos x \sin 4x = \cos^3 4x$, $\cos 2x(1 - \sin^2 2x) = \cos^3 2x$ or

$\cos 2x = \cos 4x$. Then $4x = \pm 2x + 2k\pi$ with the solutions $x = \dfrac{k\pi}{3}$, $k \in \mathbf{Z}$;

e) $(1 - \cos^2 x)\sin x \cos 3x + (1 - \sin^2 x)\cos x \sin 3x = \dfrac{3}{4}\sin 2x$,

$\sin 4x - \sin x \cos x \cos 2x = \dfrac{3}{4}\sin 2x$, $\sin 4x - \dfrac{\sin 4x}{4} = \dfrac{3}{4}\sin 2x$ or

$\sin 4x = \sin 2x$. Then $4x = (-1)^k 2x + k\pi$ with the solutions $x_1 = k\pi$ and

$x_2 = \dfrac{\pi}{6} + \dfrac{k\pi}{3}$; f) $(1 - \cos^2 x)\sin x \cos 3x + (1 - \sin^2 x)\cos x \sin 3x = -\dfrac{3}{8}$,

$\sin 4x - \sin x \cos x \cos 2x = -\dfrac{3}{8}$, $\sin 4x = -\dfrac{1}{2}$. Then the solutions are

$x = (-1)^{k+1}\dfrac{\pi}{24} + \dfrac{k\pi}{4}$, $k \in \mathbf{Z}$;

g) $x \neq \dfrac{k\pi}{2}$. $\sin^3 x + \cos^3 x - (\sin^2 x \cos x + \cos^2 x \sin x) = \sqrt{2}\cos 2x \Rightarrow$

$$\left(\sin x + \cos x\right)\left(\sin^2 x + \cos^2 x - 2\sin x \cos x\right) = \sqrt{2}\cos 2x \Rightarrow$$

$$\left(\sin x + \cos x\right)\left(\cos x - \sin x\right)^2 = \sqrt{2}\cos 2x \Rightarrow \cos 2x\left(\cos x - \sin x - \sqrt{2}\right) = 0.$$

The solutions are $x = \dfrac{\pi}{4} + \dfrac{k\pi}{2}$, $k \in \mathbf{Z}$;

h) $x = \dfrac{k\pi}{12}$, $k \in \mathbf{Z}$; i) $x = \pm\dfrac{\pi}{30} + \dfrac{k\pi}{5}$, $k \in \mathbf{Z}$;

j) $4\left(\cot^2 x - 3\right)\left(\cot x + 1\right) = 0$. Then $\cot^2 x - 3 = 0$ and $\cot x = -1$. Therefore

the solutions are $x_1 = \pm\dfrac{\pi}{6} + k\pi$ and $x_2 = -\dfrac{\pi}{4} + k'\pi$, $k, k' \in \mathbf{Z}$;

k) $\dfrac{1}{\sin^3 x \sin^3 x} - \dfrac{6}{\cos 2x} = \dfrac{\left(\sin^2 x + \cos^2 x\right)\left(\sin^4 x + \cos^4 x - \sin^2 x \cos^2 x\right)}{\sin^3 x \cos^3 x} \Rightarrow$

$\dfrac{1}{\sin^3 x \sin^3 x} - \dfrac{6}{\cos 2x} = \dfrac{1 - 3\sin^2 x \cos^2 x}{\sin^3 x \cos^3 x} \Rightarrow$

$-\dfrac{6}{\cos 2x} = \dfrac{-3\sin^2 x \cos^2 x}{\sin^3 x \cos^3 x} \Rightarrow \dfrac{2}{\cos 2x} = \dfrac{1}{\sin x \cos x} \Rightarrow \cos 2x = \sin 2x \Rightarrow$

$x = \dfrac{\pi}{8} + \dfrac{k\pi}{2}$ $k \in \mathbf{Z}$.

V. 24. a) $1 - \dfrac{\sin^2 2x}{2} = 1.25 - \sin^2 2x$, $2\sin^2 2x - 1 = 0$ or $\cos 4x = 0$.

Therefore the solutions are $x = \dfrac{\pi}{8} + \dfrac{k\pi}{4}$, $k \in \mathbf{Z}$;

b) $1 - \dfrac{\sin^2 2x}{2} = \sin 2x - \dfrac{1}{2}$, $\sin^2 2x + 2\sin 2x - 3 = 0$. Denote $\sin 2x = t$.

Then $t^2 + 2t - 3 = 0$. The roots are $t = 1$ and $t = -3$ (invalid). The

solutions are $x = \dfrac{\pi}{4} + k\pi$, $k \in \mathbf{Z}$;

c) $5 + 5\cos x = 2 + 1 - 2\cos^2 x$, $2\cos^2 x + 5\cos x + 2 = 0$. Denote $\cos x = t$.

Therefore $2t^2 + 5t + 2 = 0$. The roots are $t = -\dfrac{1}{2}$ and $t = -2$ (invalid).

Therefore the solutions are $x = \pm\dfrac{2\pi}{3} + 2k\pi$;

d) $\sin 2x = \cos x$ or $\cos x = \cos\left(2x - \dfrac{\pi}{2}\right)$. Therefore $2x - \dfrac{\pi}{2} = \pm x + 2k\pi$.

The solutions are $x_1 = \dfrac{\pi}{6} + \dfrac{2k\pi}{3}$ and $x_2 = \dfrac{\pi}{2} + 2k'\pi$, $k, k' \in \mathbf{Z}$;

e) $2 - 3\sin x = \cos 2x$, $2\sin^2 x - 3\sin x + 1 = 0$. Denote $\sin x = t$. Therefore

$2t^2 - 3t + 1 = 0$. The roots are $t = 1$ and $t = \dfrac{1}{2}$. The solutions are

$x_1 = \dfrac{\pi}{2} + 2k\pi$ and $x_2 = (-1)^{k'} \dfrac{\pi}{6} + k'\pi$, $k, k' \in \mathbf{Z}$;

f) $2\sin^4 x - 4\sin^2 x \cos^2 x - 6\cos^4 x = 0$ or $\tan^4 x - 2\tan^2 x - 3 = 0$. Denote

$\tan^2 x = t$. Then $t^2 - 2t - 3 = 0$. The roots are $t = 3$ and $t = -1$ (invalid).

Therefore the solutions are $x = \pm\dfrac{\pi}{3} + k\pi$, $k \in \mathbf{Z}$;

g) $\dfrac{(1 - \cos 2x)^2}{2} + \dfrac{5(1 - \cos^2 2x)}{4} - \dfrac{(1 + \cos 2x)^2}{4} = \cos 2x$. Denote $\cos 2x = t$.

Then $\dfrac{(1-t)^2}{2} + \dfrac{5(1-t^2)}{4} - \dfrac{(1+t)^2}{4} = t$, $2t^2 + 5t - 3 = 0$. The roots are $t = \dfrac{1}{2}$

and $t = -3$. Therefore the solutions are $x = \pm\dfrac{\pi}{6} + k\pi$, $k \in \mathbf{Z}$;

h) $(\sin x + \cos x)(\sin x - 2\cos x)(\sin^2 x - 2\sin x \cos x + 4\cos^2 x) = 0$. Then

$\tan x = -1$ and $\tan x = 2$. The solutions are $x_1 = -\dfrac{\pi}{4} + k\pi$ and

$x_2 = \arctan 2 + k\pi$, $k \in \mathbf{Z}$;

i) We have $(\cos^2 x - \sin^2 x) + 1 = 3(1 - \sin x) \Rightarrow 1 - 2\sin^2 x + 1 = 3(1 - \sin x)$

Denoting $\sin x = t$ we obtain $2t^2 - 3t + 1 = 0$, then $t_1 = 1$ and $t_2 = \dfrac{1}{2}$.

$\sin x = 1$, $x_1 = \dfrac{\pi}{2} + 2k\pi$, $k \in \mathbf{Z}$ and $\sin x = \dfrac{1}{2}$, $x_2 = (-1)^k \dfrac{\pi}{6} + k\pi$, $k \in \mathbf{Z}$;

j) $(2\cos 2x + 3)(\cos^2 x - \sin^2 x)(\cos^2 x + \sin^2 x) = 2$, $2\cos^2 2x + 3\cos 2x - 2 = 0$.

Denote $\cos 2x = t$. Then $2t^2 + 3t - 2 = 0$. The roots are $t = \dfrac{1}{2}$ and $t = -2$.

The solutions are $x = \pm\dfrac{\pi}{6} + k\pi,\ k \in \mathbf{Z}$;

k) $\sqrt{2}(\cos 2x + \sin 2x)(\cos 2x - \sin 2x) = \cos 2x + \sin 2x$. Then

$\cos 2x + \sin 2x = 0$. with the solutions $x_1 = -\dfrac{\pi}{8} + \dfrac{k\pi}{2}$. Also

$\cos 2x - \sin 2x = \dfrac{\sqrt{2}}{2}$, $\sin\left(2x - \dfrac{\pi}{4}\right) = -\dfrac{1}{2}$. Then the other solutions are

$x_2 = -\dfrac{7\pi}{24} + \dfrac{k\pi}{2}$ and $x_3 = \dfrac{\pi}{24} + \dfrac{k\pi}{2}$;

l) $\sin 2x \cdot \left(-2\sin^2 x \cos^2 x\right) = \sin^2 2x$. Then $\sin 2x = 0$ and $\sin 2x = -2$

(impossible). The solutions are $x = \dfrac{k\pi}{2},\ k \in \mathbf{Z}$;

m) $\cos x \neq 0$, $-\cos 2x = 1 - 2\cos^2 x\left(\cos 3x - \dfrac{\sin x}{\cos x}\sin 3x\right) \Rightarrow$

$\dfrac{\cos 3x \cos x - \sin x \sin 3x}{\cos x} = 1 \Rightarrow \cos 4x = \cos x$. Then $4x = \pm x + 2k\pi,\ k \in \mathbf{Z}$;

n) $1 - \sin^2 2x = \cos 2x(1 + \cos 4x),\ \cos^2 2x = \cos 2x\left(2\cos^2 2x\right)$.

Since $\cos 2x \neq 0$ then $\cos 2x = \dfrac{1}{2}$. Therefore the solutions are

$x = \pm\dfrac{\pi}{6} + k\pi,\ k \in \mathbf{Z}$.

V. 25. a) Denote $\tan^2 3x = t$. Then $12t^2 - 7t + 1 = 0$.

The roots are $t = \dfrac{1}{3}$ and $t = \dfrac{1}{4}$. The solutions are $x = \pm\dfrac{\pi}{18} + \dfrac{k\pi}{3},\ k \in \mathbf{Z}$

and $x = \pm\dfrac{1}{3}\arctan\dfrac{1}{2} + \dfrac{k'\pi}{3},\ k' \in \mathbf{Z},\ k, k' \in \mathbf{Z}$;

b) $\dfrac{\sin^4 x + \cos^4 x}{\sin^2 x \cos^2 x} - 2 = 4\tan 2x \Rightarrow \dfrac{1 - 2\sin^2 x \cos^2 x}{\sin^2 x \cos^2 x} - 2 = 4\tan 2x \Rightarrow$

$\dfrac{1 - \dfrac{\sin^2 2x}{2}}{\dfrac{\sin^2 2x}{4}} - 2 = 4\tan 2x \Rightarrow \dfrac{4\cos^2 2x}{\sin^2 2x} = 4\tan 2x \Rightarrow \tan^3 2x = 1$. The solutions are $x = \dfrac{\pi}{8} + \dfrac{k\pi}{2}$, $k \in \mathbf{Z}$;

c) Denote $\sin^2 2x = t \geq 0$. Then $\dfrac{(1-t)^2}{t^2} + \dfrac{1}{t^2} = 25$ or $12t^2 - t - 1 = 0$

. The roots are $t = \dfrac{1}{4}$ and $t = -\dfrac{1}{3}$ (impossible). The solution is

$x = \pm\dfrac{\pi}{12} + \dfrac{k\pi}{2}$, $k \in \mathbf{Z}$.

d) $x = -\dfrac{\pi}{8} + \dfrac{k\pi}{2}$, $k \in \mathbf{Z}$; e) $\dfrac{1 - 2(\cos x \sin x)^2}{\cos 2x} - \dfrac{\cos 2x}{2} = \dfrac{1}{2\sqrt{3}}\dfrac{1}{\sin 2x} \Rightarrow$

$\dfrac{2 - \sin^2 2x}{2\cos 2x} - \dfrac{\cos 2x}{2} = \dfrac{1}{2\sqrt{3}}\dfrac{1}{\sin 2x} \Rightarrow \dfrac{1}{2\cos 2x} = \dfrac{1}{2\sqrt{3}}\dfrac{1}{\sin 2x} \Rightarrow \tan 2x = \dfrac{1}{\sqrt{3}}$.

Then $x = \dfrac{\pi}{12} + \dfrac{k\pi}{2}$, $k \in \mathbf{Z}$;

f) $\dfrac{\sin^8 x + \cos^8 x}{(\sin x \cos x)^4} = \dfrac{17}{4}\dfrac{\sin x \sin 2x + \cos x \cos 2x}{\cos x \cos 2x}\cos 2x \Rightarrow$

$\dfrac{(\sin^4 x + \cos^4 x)^2 - 2(\sin x \cos x)^4}{(\sin x \cos x)^4} = \dfrac{17}{4} \Rightarrow \left(\dfrac{\sin^4 x + \cos^4 x}{(\sin x \cos x)^2}\right)^2 - 2 = \dfrac{17}{4} \Rightarrow$

$\dfrac{\sin^4 x + \cos^4 x}{(\sin x \cos x)^2} = \dfrac{5}{2} \Rightarrow \dfrac{(\sin^2 x + \cos^2 x)^2 - 2(\sin x \cos x)^2}{(\sin x \cos x)^2} = \dfrac{5}{2} \Rightarrow$

$\sin 2x = \pm\dfrac{2\sqrt{2}}{3}$. Then $x = \pm\dfrac{1}{2}\arcsin\dfrac{2\sqrt{2}}{3} + \dfrac{k\pi}{2}$;

g) $\dfrac{\cos^4 \dfrac{x}{2} + \sin^4 \dfrac{x}{2}}{2\left(\sin\dfrac{x}{2}\cos\dfrac{x}{2}\right)^2} = 1 + \dfrac{2}{\sqrt{3}}\cot x \Rightarrow \dfrac{1 - 2\left(\sin\dfrac{x}{2}\cos\dfrac{x}{2}\right)^2}{2\left(\sin\dfrac{x}{2}\cos\dfrac{x}{2}\right)^2} = 1 + \dfrac{2}{\sqrt{3}}\cot x \Rightarrow$

$\dfrac{2 - \sin^2 x}{\sin^2 x} = 1 + \dfrac{2}{\sqrt{3}}\cot x \Rightarrow \dfrac{\cos^2 x}{\sin^2 x} = \dfrac{1}{\sqrt{3}}\dfrac{\cos x}{\sin x}$. Then $\cos x = 0$ and

$\tan x = \sqrt{3}$. Then the solutions are $x_1 = \dfrac{\pi}{2} + k\pi$ and $x_2 = \dfrac{\pi}{3} + k'\pi$,

$k, k' \in \mathbf{Z}$;

h) $4\sin x \cos x\left(1 - 2\sin^2 x \cos^2 x\right) + \sin^3 2x = 1$ or $\sin 2x = \dfrac{1}{2}$. The solutions

are $x = (-1)^k \dfrac{\pi}{12} + \dfrac{k\pi}{2}$, $k \in \mathbf{Z}$;

i) The equation becomes $(\sin x + \cos x)\left(1 + \dfrac{2}{\sin 2x} - \dfrac{\sin 2x}{2} - \dfrac{\sin^2 2x}{4}\right) = 0$.

$\sin x + \cos x = 0$ has the solution $x = -\dfrac{\pi}{4} + k\pi$, $k \in \mathbf{Z}$ and

$\sin 2x = \pm 2$ is impossible; **j)** $\cos x \neq 0$, $\sin x \neq 0$ or $x \neq \dfrac{k\pi}{2}$.

$\dfrac{\sin^2 x\left(1 - \dfrac{1}{\cos^2 x}\right)}{\cos^2 x\left(1 - \dfrac{1}{\sin^2 x}\right)} - \tan^6 x + \tan^4 x - \tan^2 x = 0$, $\tan^2 x\left(\tan^2 x - 1\right) = 0$. Then

$\tan x = 0$ and $\tan x = \pm 1$. Therefore the solutions are $x = \dfrac{\pi}{4} + \dfrac{k\pi}{2}$, $k \in \mathbf{Z}$.

V. 26. a) $1 - 3\sin^2 x \cos^2 x = \dfrac{7}{16}$ or $\sin 2x = \pm\dfrac{\sqrt{3}}{2}$. Therefore the solutions

are $x = \pm\dfrac{\pi}{6} + \dfrac{k\pi}{2}$, $k \in \mathbf{Z}$;

b) $\cos^4 x + 2\sin^2 x \cos^2 x + \sin^4 x - 3\sin^2 x \cos^2 x - \cos^2 2x = \dfrac{1}{16}$,

$1 - \cos^2 2x - \dfrac{3}{4}\sin^2 2x = \dfrac{1}{16}$ or $\sin 2x = \pm\dfrac{1}{2}$. Therefore the solutions are

$x = \pm\dfrac{\pi}{12} + \dfrac{k\pi}{2}$, $k \in \mathbf{Z}$;

c) $-\left(1 - 2\sin^2 2x \cos^2 2x\right) - 2\sin^2 2x \cos^2 2x = \sin x + \cos x \Rightarrow$

$\sin x + \cos x = -1$. Then $\cos\left(x - \dfrac{\pi}{4}\right) = -\dfrac{1}{\sqrt{2}}$ with the solutions

$x_1 = -\dfrac{\pi}{2} + 2k\pi$

and $x_2 = \pi + 2k\pi$.

d) We have $2\left(\sin^2 x + \cos^2 x\right)\left(\sin^4 x - \sin^2 x \cos^2 x + \cos^4 x\right) -$

$$-3\left[\left(\sin^2 x + \cos^2 x\right)^2 - 2\sin^2 x \cos^2 x\right] = \sin 2x \Rightarrow$$

$$2\left[\left(\sin^2 x + \cos^2 x\right)^2 - 3\sin^2 x \cos^2 x\right] - 3\left[1 - 2\sin^2 x \cos^2 x\right] = \sin 2x \Rightarrow$$

$$2\left[1 - 3\frac{\sin^2 2x}{4}\right] - 3\left[1 - \frac{\sin^2 2x}{2}\right] = \sin 2x \Rightarrow \sin 2x = -1.$$

Therefore $x = -\dfrac{\pi}{4} + k\pi$, $k \in \mathbf{Z}$;

e) $2\left(1 - 3\sin^2 x \cos^2 x\right) - 3\left(1 - 2\sin^2 x \cos^2 x\right) = \cos 2x$, $\cos 2x = -1$. Therefore

the solutions are $x = \dfrac{\pi}{2} + k\pi$, $k \in \mathbf{Z}$;

f) $\dfrac{\left(1 - \sin^2 x\right)\left(1 + \sin^2 x + \sin^4 x\right) - \cos^6 x}{\left(1 - \sin^2 x\right)\left(1 + \sin^2 x\right) - \cos^4 x} = 2\cos^2 4x$

$\dfrac{1 + \sin^2 x + \sin^4 x - \cos^4 x}{1 + \sin^2 x - \cos^2 x} = 2\cos^2 4x$; $\dfrac{1 + \sin^2 x - \cos 2x}{1 - \cos 2x} = 2\cos^2 4x$;

$\dfrac{2\sin^2 x + \sin^2 x}{2\sin^2 x} = 2\cos^2 4x$; $\dfrac{3}{2} = 2\cos^2 4x$; $\cos^2 4x = \dfrac{3}{4}$. Therefore

$x = \pm\dfrac{\pi}{24} + \dfrac{k\pi}{4}$, $k \in \mathbf{Z}$;

g) $\dfrac{\sin^3 x}{\cos^3 x} + \dfrac{\cos^3 x}{\sin^3 x} - \dfrac{8}{8\sin^3 x \cos^3 x} = 12$, $\dfrac{\sin^6 x + \cos^6 x - 1}{\sin^3 x \cos^3 x} = 12$,

$-\dfrac{6}{\sin 2x} = 12$. Then $\sin 2x = -\dfrac{1}{2}$. The solutions are $x = (-1)^k \dfrac{7\pi}{12} + \dfrac{k\pi}{2}$, $k \in \mathbf{Z}$.

h) $\dfrac{\sin^6 x + \cos^6 x}{\left(\cos x \sin x\right)^3} - \dfrac{1}{\left(\cos x \sin x\right)^3} = -\dfrac{2\sqrt{3}}{\cos 2x}$,

$\dfrac{\left(\sin^2 x + \cos^2 x\right)\left(\sin^4 x + \cos^4 x - \sin^2 x \cos^2 x\right) - 1}{\left(\cos x \sin x\right)^3} = -\dfrac{2\sqrt{3}}{\cos 2x}$,

$\dfrac{\left(\sin^2 x + \cos^2 x\right)^2 - 3\sin^2 x \cos^2 x - 1}{\left(\cos x \sin x\right)^3} = -\dfrac{2\sqrt{3}}{\cos 2x}$, $\tan 2x = \sqrt{3}$.

Then $x = \dfrac{\pi}{6} + \dfrac{k\pi}{2}$, $k \in \mathbf{Z}$;

i) $\sin^{10} x + \cos^{10} x = \dfrac{1}{4} \sin^2 x \cos^2 x \Rightarrow \dfrac{\sin^8 x}{\cos^2 x} + \dfrac{\cos^8 x}{\sin^2 x} = \dfrac{1}{4} \Rightarrow$

$\dfrac{(1-\cos 2x)^4}{8(1+\cos 2x)} + \dfrac{(1+\cos 2x)^4}{8(1-\cos 2x)} = \dfrac{1}{4} \Rightarrow$

$(1-\cos 2x)^5 + (1-\cos 2x)^5 = 2\big(1-\cos^2 2x\big) \Rightarrow 5\cos^4 2x + 1 \ \cos^2 2x = 0$ so

$\cos^2 2x = 0$ then $x = \dfrac{\pi}{4} + \dfrac{k\pi}{2}, \ \ k \in \mathbf{Z}$.

V. 27. a) $\sin^3 x \dfrac{\sin x + \cos x}{\sin x} + \cos^3 x \dfrac{\sin x + \cos x}{\cos x} = 2\sqrt{\sin x \cos x}$,

$\sin x + \cos x - 2\sqrt{\sin x \cos x} = 0$ or $\left(\sqrt{\sin x} - \sqrt{\cos x}\right)^2 = 0$, where

$\sin x > 0$ and $\cos x > 0$. Then $\tan x = 1$. Therefore the solutions are

$x = \dfrac{\pi}{4} + 2k\pi, \ k \in \mathbf{Z}$;

b) $\cos 2x \le 0$, $x \in \left[\dfrac{\pi}{4} + k\pi, \ \dfrac{3\pi}{4} + k\pi\right] \cup \left[\dfrac{5\pi}{4} + k\pi, \ \dfrac{7\pi}{4} + k\pi\right]$. Squaring on

both sides, we have $1 + \sin 2x = 1 - 2\cos^2 x$, $2\sin x \cos x = -2\cos^2 x$. Then

$\cos x = 0$ and $\cos x = \cos\left(\dfrac{\pi}{2} + x\right)$. Therefore the solutions are $x_1 = \dfrac{\pi}{2} + k\pi$

and $x_2 = \dfrac{7\pi}{4} + k'\pi$, $k, k' \in \mathbf{Z}$;

c) Squaring in both sides, we have

$(1-\cos x)^2 = 1 - \sqrt{4\cos^2 x - 7\cos^4 x}$, $\cos x(\cos x - 2) = -\cos x\sqrt{4 - 7\cos^2 x}$.

Then $\cos x = 0$ and $2 - \cos x = \sqrt{4 - 7\cos^2 x}$. Therefore the solutions are

$x_1 = \dfrac{\pi}{2} + k\pi$ and $x_2 = \pm\dfrac{\pi}{3} + 2k'\pi$, $k, k' \in \mathbf{Z}$;

d) $\tan\dfrac{x}{2} \ge 0$, then $0 \le \dfrac{x}{2} < \dfrac{\pi}{2}$ or in general $x \in [2k\pi, \ \pi + 2k\pi)$. Then

$(1 + \cos x)\sqrt{\tan\dfrac{x}{2}} - 2(1 + \cos x) + \sin x = 0$,

228

$2\cos^2\dfrac{x}{2}\left(\sqrt{\tan\dfrac{x}{2}}-2\right)+2\sin\dfrac{x}{2}\cos\dfrac{x}{2}=0$. Therefore $\cos\dfrac{x}{2}=0$ (impossible)

and $\cos\dfrac{x}{2}\left(\sqrt{\tan\dfrac{x}{2}}-2\right)+\sin\dfrac{x}{2}=0\Rightarrow\tan\dfrac{x}{2}+\sqrt{\tan\dfrac{x}{2}}-2=0$. Then

$\sqrt{\tan\dfrac{x}{2}}=1$ or $\sqrt{\tan\dfrac{x}{2}}=-2$ (impossible). Finally $x=\dfrac{\pi}{2}+2k\pi,\ k\in\mathbf{Z}$;

e) Squaring in both sides, we have

$\cos^2 x+0.5+2\sqrt{\cos^2 x+0.5}\sqrt{\sin^2 x+0.5}+\sin^2 x+0.5=4\Rightarrow$

$\sqrt{\cos^2 x\sin^2 x+0.5(\cos^2 x+\sin^2 x)+0.25}=1\Rightarrow\cos^2 x\sin^2 x=0.25\Rightarrow$

$\sin^2 2x=1$. Finally $x=\dfrac{\pi}{4}+\dfrac{k\pi}{2},\ k\in\mathbf{Z}$; f) $x=\dfrac{\pi}{2}+2k\pi,\ k\in\mathbf{Z}$; g) With

the condition of $x\neq(2k+1)\pi$, $x\neq-\dfrac{\pi}{2}+2k\pi,k\in\mathbf{Z}$, the equation becomes

successively $\sqrt{\dfrac{1-\cos\left(\dfrac{\pi}{2}-x\right)}{1+\cos\left(\dfrac{\pi}{2}-x\right)}}+\tan\dfrac{x}{2}=2\Leftrightarrow\left|\tan\left(\dfrac{\pi}{4}-\dfrac{x}{2}\right)\right|+\tan\dfrac{x}{2}=2$ or

$\left|\dfrac{1-\tan\dfrac{x}{2}}{1+\tan\dfrac{x}{2}}\right|+\tan\dfrac{x}{2}=2$. We denote $y=\tan\dfrac{x}{2}$. We have $\dfrac{1-y}{1+y}+y=2$, the

convenient solution is $y=1-\sqrt{2}\Leftrightarrow x=2\arctan\left(1-\sqrt{2}\right)+2k\pi,k\in\mathbf{Z}$ and

from $\dfrac{y-1}{1+y}+y=2\Leftrightarrow y=\pm\sqrt{3}$,

the convenient solution is $x=-\dfrac{2\pi}{3}+2k\pi,k\in\mathbf{Z}$;

h) For the domain $\tan x\geq 0$ and $\cos x>0$. Therefore $x\in\left[0,\dfrac{\pi}{2}\right)$, in

general $x\in\left[2k\pi,\dfrac{\pi}{2}+2k\pi\right)_{k\in\mathbf{Z}}$.

We have $\sqrt{\tan x}\left[\left(\sqrt{1+\cos x}+\sqrt{1-\cos x}\right)-2\cos x\right]=0$. From $\tan x=0$,

$x=2k\pi,\ k\in\mathbf{Z}$.

$\left|\cos\dfrac{x}{2}\right|+\left|\sin\dfrac{x}{2}\right|=\sqrt{2}\cos x$. According to the domain, we have

$\cos\dfrac{x}{2}+\sin\dfrac{x}{2}=\sqrt{2}\left(\cos^2\dfrac{x}{2}-\sin^2\dfrac{x}{2}\right)\Rightarrow$

$\cos\dfrac{x}{2}-\sin\dfrac{x}{2}=\dfrac{1}{\sqrt{2}}\Rightarrow\cos\dfrac{x}{2}+\cos\left(\dfrac{\pi}{2}+\dfrac{x}{2}\right)=\dfrac{1}{\sqrt{2}}\Rightarrow\cos\left(\dfrac{\pi}{4}+\dfrac{x}{2}\right)=\dfrac{1}{2}$. The

only solution, in this case, is $x=\dfrac{(12k+1)\pi}{6},k\in\mathbf{Z}$;

i) We have $x\in\left[0,\dfrac{\pi}{2}\right)\cup\left[\pi,\dfrac{3\pi}{2}\right)$ and $\sqrt{2}\left(\left|\sin\dfrac{x}{2}\right|+\left|\cos\dfrac{x}{2}\right|\right)=4\sin x\cos x$.

Squaring in both sides, we obtain $2(1+|\sin x|)=16\sin^2 x(1-\sin^2 x)$.

Denoting $|\sin x|=y$, it yields to $8y^4-8y^2+y+1=0$ and then

$(2y-1)(y+1)(4y^2-2y-1)=0$ $|\sin x|=\dfrac{1}{2}$, etc.

j) $y=\cos 2x\in\left[-\dfrac{1}{2},\dfrac{1}{2}\right]$. Squaring the given equation, we obtain

$\sqrt{\left(\dfrac{1}{2}-y\right)\left(\dfrac{1}{2}+y\right)}=0\Leftrightarrow\cos 2x=\pm\dfrac{1}{2}$;

k) $\left(\sqrt[3]{\sin^2 x}-\sqrt[3]{\cos^2 x}\right)^3=\left(\sqrt[3]{\cos 2x}\right)^3\Rightarrow$

$-\cos 2x-3\sqrt[3]{\sin x}\sqrt[3]{\cos x}\left(\sqrt[3]{\sin x}-\sqrt[3]{\cos x}\right)=\cos 2x\Rightarrow$

$-3\sqrt[3]{\sin^2 x}\sqrt[3]{\cos^2 x}\sqrt[3]{\cos 2x}=2\cos 2x\Rightarrow-27\sin^2 x\cos^2 x\cos 2x=8\cos^3 2x\Rightarrow$

$\cos 2x=0$. Therefore $x=\dfrac{\pi}{4}+\dfrac{k\pi}{2},k\in\mathbf{Z}$;

l) We have $2\cos x\geq 1$, also $4\cos^2 x=2+\sin 3x\Rightarrow$

$4\left(1-\sin^2 x\right)=2+3\sin x-4\sin^3 x\Rightarrow 4\sin^3 x-4\sin^2 x-3\sin x+2=0$.

The convenient solution is only $\sin x=\dfrac{1}{2}$;

230

m) $\cos 2x \geq 0$.

$$\sqrt{\sin x + \cos x}\left(\sqrt{\sin x - \cos x} + \sqrt{\sin x + \cos x} - 2\right) = 0.$$

I) $\sqrt{\sin x + \cos x} = 0 \Rightarrow x_1 = -\dfrac{\pi}{4} + k\pi, k \in \mathbf{Z}$.

II $\sqrt{\sin x - \cos x} + \sqrt{\sin x + \cos x} - 2 = 0 \Leftrightarrow \cos^2 x + 4\cos x - 5 = 0$,

$\cos x = 1$, $x_2 = 2n\pi, n \in \mathbf{Z}$;   **n)** $x = k\pi, \ k \in \mathbf{Z}$.

V. 28. a) $\sin x \neq 0$. Obviously $x = \dfrac{k\pi}{2}, k \in \mathbf{Z}$ are not solution for the equation.

I) $\cos x > 0 \Rightarrow x \in \left(0, \dfrac{\pi}{2}\right) \cup \left(\dfrac{3\pi}{2}, 2\pi\right)$ or

$x \in \left(2k\pi, \dfrac{\pi}{2} + 2k\pi\right) \cup \left(\dfrac{3\pi}{2} + 2k\pi, 2\pi + 2k\pi\right)$, the equation becomes

$\sqrt{2}\cos x = 1 + \cot x \Rightarrow \dfrac{\sqrt{2}}{2}\sin 2x = \sin x + \cos x$.

Squaring on both sides, we obtain $\dfrac{1}{2}\sin^2 2x = 1 + \sin 2x$. Then

$\sin 2x = 1 - \sqrt{3} \Rightarrow x = \dfrac{(-1)^k}{2}\arcsin\left(1 - \sqrt{3}\right) + \dfrac{k\pi}{2}, k \in \mathbf{Z}$. The solutions on the

interval $(0, 2\pi)$ are $x_1 = -\dfrac{1}{2}\arcsin\left(1 - \sqrt{3}\right) + \dfrac{\pi}{2}$, $x_2 = \dfrac{1}{2}\arcsin\left(1 - \sqrt{3}\right) + \pi$,

$x_3 = -\dfrac{1}{2}\arcsin\left(1 - \sqrt{3}\right) + \dfrac{3\pi}{2}$, and $x_4 = \dfrac{1}{2}\arcsin\left(1 - \sqrt{3}\right) + 2\pi$.

Only $x_3, x_4 \in \left(0, \dfrac{\pi}{2}\right) \cup \left(\dfrac{3\pi}{2}, 2\pi\right)$ *are solutions.*

I) $\cos x < 0 \Rightarrow x \in \left(\dfrac{\pi}{2}, \dfrac{3\pi}{2}\right)$ or $x \in \left(\dfrac{\pi}{2} + 2k\pi, \dfrac{3\pi}{2} + 2k\pi\right)$, the equation

becomes $-\sqrt{2}\cos x = 1 + \cot x \Rightarrow -\dfrac{\sqrt{2}}{2}\sin 2x = \sin x + \cos x$. Squaring on

both sides, we obtain $\dfrac{1}{2}\sin^2 2x = 1 + \sin 2x$. Then $\sin 2x = 1 - \sqrt{3} \Rightarrow$

$2x = (-1)^k \arcsin\left(1 - \sqrt{3}\right) + k\pi \Rightarrow x = \dfrac{(-1)^k}{2}\arcsin\left(1 - \sqrt{3}\right) + \dfrac{k\pi}{2}$. The

solutions in the interval $(0, 2\pi)$ are $x_1 = -\dfrac{1}{2}\arcsin\left(1 - \sqrt{3}\right) + \dfrac{\pi}{2}$,

$x_2 = \dfrac{1}{2}\arcsin\!\left(1-\sqrt{3}\right)+\pi,$

$x_3 = -\dfrac{1}{2}\arcsin\!\left(1-\sqrt{3}\right)+\dfrac{3\pi}{2}$, and $x_4 = \dfrac{1}{2}\arcsin\!\left(1-\sqrt{3}\right)+2\pi$. Only

$x_2, x_3 \in \left(\dfrac{\pi}{2}, \dfrac{3\pi}{2}\right)$ *are solutions.* Therefore the general solutions are

$x = -\dfrac{1}{2}\arcsin\!\left(1-\sqrt{3}\right)+\dfrac{\pi}{2}+k\pi$ and $x = \dfrac{1}{2}\arcsin\!\left(1-\sqrt{3}\right)+\pi+k\pi,\ k \in \mathbf{Z}$.

b) I) $\sin x \geq 0$ and $\cos x \geq 0$, $x \in \left[0, \dfrac{\pi}{2}\right]$ or in general $x \in \left[2k\pi, \dfrac{\pi}{2}+2k\pi\right]$.

The equation is $\cos\!\left(x - \dfrac{\pi}{4}\right) = \dfrac{3\sqrt{2}}{5}$. The solutions are

$x = \pm\arccos\dfrac{3\sqrt{2}}{5} + \dfrac{\pi}{4} + 2k\pi$.

II) $\sin x > 0$ and $\cos x < 0$, $x \in \left(\dfrac{\pi}{2}, \pi\right)$ or in general

$x \in \left(\dfrac{\pi}{2}+2k\pi,\ \pi+2k\pi\right)$, $k \in \mathbf{Z}$. The equation is $\cos\!\left(\dfrac{\pi}{4}+x\right) = -\dfrac{3\sqrt{2}}{5}$. The

solutions are $x = \pm\arccos\!\left(-\dfrac{3\sqrt{2}}{5}\right) - \dfrac{\pi}{4} + 2k\pi, k \in \mathbf{Z}$.

III) $\sin x < 0$ and $\cos x < 0$, $x \in \left(\pi, \dfrac{3\pi}{2}\right)$ or in general

$x \in \left(\pi+2k\pi, \dfrac{3\pi}{2}+2k\pi\right), k \in \mathbf{Z}$. The equation is $\cos\!\left(x - \dfrac{\pi}{4}\right) = -\dfrac{3\sqrt{2}}{5}$. The

solutions are $x = \pm\arccos\!\left(-\dfrac{3\sqrt{2}}{5}\right) + \dfrac{\pi}{4} + 2k\pi, k \in \mathbf{Z}$.

IV) $\sin x < 0$ and $\cos x > 0$, $x \in \left(\dfrac{3\pi}{2}, 2\pi\right)$ or in general

$x \in \left(\dfrac{3\pi}{2}+2k\pi, 2\pi+2k\pi\right)$. The equation is $\cos\!\left(\dfrac{\pi}{4}+x\right) = \dfrac{3\sqrt{2}}{5}$. The

solutions are $x = \pm\arccos\dfrac{3\sqrt{2}}{5} - \dfrac{\pi}{4} + 2k\pi, k \in \mathbf{Z}$.

Finally, the sets of solutions of the equation are

$$x = -\arccos\frac{3\sqrt{2}}{5} + \frac{\pi}{4} + \frac{k\pi}{2}, k \in \mathbf{Z} \text{ and } x = \arccos\frac{3\sqrt{2}}{5} - \frac{\pi}{4} + \frac{k\pi}{2}, k \in \mathbf{Z}.$$

c) The equation becomes $\left|\cos\left(\frac{\pi}{4} - x\right)\right| = 1$. Therefore the solutions are

$$x = \frac{\pi}{4} + k\pi, k \in \mathbf{Z}.$$

d) The equation becomes $\left|\dfrac{1}{\sin 2x \cos 2x}\right| = \dfrac{4}{\sqrt{3}}$ or $|\sin 4x| = \dfrac{\sqrt{3}}{2}$. Then:

I) $\sin 4x = \dfrac{\sqrt{3}}{2}$. The solutions are $x_1 = \dfrac{\pi}{12} + \dfrac{k\pi}{4}$, $k \in \mathbf{Z}$;

II) $\sin 4x = -\dfrac{\sqrt{3}}{2}$. The solutions are $x_2 = -\dfrac{\pi}{12} + \dfrac{k\pi}{4}$, $k \in \mathbf{Z}$.

V. 29. We have $\dfrac{1}{2}\cos x + \dfrac{\sqrt{3}}{2}\sin x = \dfrac{m}{2} \Rightarrow \cos\left(\dfrac{\pi}{3} - x\right) = \dfrac{m}{2}$. For $m \in [-2, 2]$,

$$x = \pm\arccos\frac{m}{2} + \frac{\pi}{3} + 2k\pi, \ k \in \mathbf{Z}.$$

b) I) $\cos x - \sin x = 0$, $x = \dfrac{\pi}{4} + k\pi$, $k \in \mathbf{Z}$ (This solution does not depend on m).

II) $\cos x + \sin x = \dfrac{1}{m} \Rightarrow \cos\left(x - \dfrac{\pi}{4}\right) = \dfrac{\sqrt{2}}{2m}$. For $m \in \left(-\infty, -\dfrac{\sqrt{2}}{2}\right] \cup \left[\dfrac{\sqrt{2}}{2}, \infty\right)$,

$$x = \pm\arccos\frac{\sqrt{2}}{2m} + \frac{\pi}{4} + 2k\pi, \ k \in \mathbf{Z}.$$

c) Since $\sin x = \dfrac{2\tan\dfrac{x}{2}}{1 + \tan^2\dfrac{x}{2}}$ and $\cos t = \dfrac{1 - \tan^2\dfrac{x}{2}}{1 + \tan^2\dfrac{x}{2}}$, taking $\tan\dfrac{x}{2} = t$, we

obtain $m\dfrac{2t}{1+t^2} + (2-m)\dfrac{1-t^2}{1+t^2} = m+1 \Rightarrow 3t^2 - 2m + 2m - 1 = 0$. Therefore

the solutions are $x = 2\arctan\dfrac{m \pm \sqrt{m^2 - 6m + 3}}{3} + 2k\pi$, $k \in \mathbf{Z}$, where

$$m \in \left(-\infty, 3 - \sqrt{6}\right] \cup \left[3 + \sqrt{6}, \infty\right).$$

233

d) $\cos\left(2x - \dfrac{\pi}{4}\right) = \dfrac{m\sqrt{2}}{2}$. For $m \in \left[-\sqrt{2}, \sqrt{2}\right]$, $x = \pm\dfrac{1}{2}\arccos\dfrac{m\sqrt{2}}{2} + \dfrac{\pi}{8} + k\pi$, $k \in \mathbf{Z}$.

e) I) $\sin x = \dfrac{2m}{m+1}$, $x = (-1)^k \arcsin\left(\dfrac{2m}{m+1}\right) + k\pi$, $k \in \mathbf{Z}$, when $m \in \left[-\dfrac{1}{3}, 1\right]$.

II) $\sin x = \dfrac{1}{2}$, $x = (-1)^k \dfrac{\pi}{6} + k\pi$, $k \in \mathbf{Z}$ (This solution does not depend on m).

f) $\cos 2x - \sqrt{3}\sin 2x = m - 2 \Rightarrow \cos\left(\dfrac{\pi}{3} - 2x\right) = \dfrac{m-2}{2}$. For $m \in [0, 4]$,

$x = \pm\dfrac{1}{2}\arccos\left(\dfrac{m-2}{2}\right) + \dfrac{\pi}{6} + k\pi$, $k \in \mathbf{Z}$.

g) I) $\sin x = \dfrac{1}{2}$, $x = (-1)^k \dfrac{\pi}{6} + k\pi$, $k \in \mathbf{Z}$ (This solution does not depend on m).

II) $\sin x = -m$. For $m \in [-1, 1]$, $x = \pm\arcsin(-m) + 2k\pi$, $k \in \mathbf{Z}$.

h) $m\cos 2x + 2(m^2 + 3)\sin x - 7m = 0$;

Solution I) $\sin x = m$. For $m \in [-1, 1]$, $x = \pm\arcsin m + 2k\pi$, $k \in \mathbf{Z}$.

II) $\sin x = \dfrac{3}{m}$. For $m \in (-\infty, -3] \cup [3, \infty)$, $x = \pm\arcsin\dfrac{3}{m} + 2k\pi$, $k \in \mathbf{Z}$.

i) We have $m(1 - \sin 2x) - (1 - 2\sin^2 2x) - 1 = 0 \Rightarrow 2\sin^2 2x - m\sin 2x + m - 2 = 0 \Rightarrow$

$(\sin 2x - 1)(2\sin 2x - m + 2) = 0$.

I) $\sin 2x = \dfrac{m-2}{2}$, $x = (-1)^k \dfrac{1}{2}\arcsin\left(\dfrac{m-2}{2}\right) + \dfrac{k\pi}{2}$, $k \in \mathbf{Z}$, when $m \in [0, 4]$.

II) $\sin 2x = 1$, $x = \dfrac{\pi}{4} + k\pi$, $k \in \mathbf{Z}$ (This solution does not depend on m);

j) We have $\left(\dfrac{1 - \cos 2x}{2}\right)^2 + \left(\dfrac{1 + \cos 2x}{2}\right)^2 = m \Rightarrow$

$\cos^2 2x = 2m - 1$. Since $0 \le \cos^2 2x \le 1$, then $m \in \left[\dfrac{1}{2}, 1\right]$. Therefore

$x = \pm\dfrac{1}{2}\arccos\left(\pm\sqrt{2m-1}\right) + k\pi$, $k \in \mathbf{Z}$;

234

k) We have $\sin^4 x + \cos^4 x = m\left(\sin^2 x + \cos^2 x\right)\left(\sin^4 x + \cos^4 x - \sin^2 x \cos^2 x\right) \Rightarrow$

$$\frac{1+\cos^2 2x}{2} = m\left(\frac{1+\cos^2 2x}{2} - \frac{\sin^2 2x}{2}\right) \Rightarrow \cos^2 2x = \frac{1}{2m-1}.$$

For $m \in (-\infty, 0] \cup [1, \infty)$, $x = \pm\frac{1}{2}\arccos\left(\pm\sqrt{\frac{1}{2m-1}}\right) + k\pi$, $k \in \mathbf{Z}$;

V. 30. The roots are $x = \frac{1}{3}\arctan a + k\frac{\pi}{3}$, $k = 1, 2, 3$.

V. 31. Since $\sin 2x = \frac{1}{2}$, then $x = \frac{\pi}{12}$. The equation becomes

$$\cos x \cos 2x \left(2\sin 4x - 2\sqrt{2-\sqrt{3}}\right) = 0, \text{ then } x = (-1)^k \frac{\pi}{48} + \frac{k\pi}{4}, \ k \in \mathbf{Z}.$$

V. 32. $\sqrt{2+\sqrt{2}} = 2\cos\alpha \Rightarrow 2+\sqrt{2} = 4\cos^2\alpha$ or $\cos 2\alpha = \frac{\sqrt{2}}{2}$ then $\alpha = \frac{\pi}{8}$.

$$\sqrt{2+\sqrt{2+\sqrt{2}}} = \sqrt{2+2\cos\frac{\pi}{8}} = \sqrt{4\cos^2\frac{\pi}{16}} = 2\cos\frac{\pi}{16}.$$ The equation becomes

$$2^n\left(\cos\frac{n\pi}{16} + i\sin\frac{n\pi}{16}\right) + 2^n = 0; \text{ or } \cos\frac{n\pi}{16} + i\sin\frac{n\pi}{16} = -1 \text{ with the solution}$$

$$n = 16(2k+1), \ k \in \mathbf{N}.$$

Unit VI

Inverses of Trigonometric Functions

VI. 1. a) Let $\arcsin x = \theta$, $x \in [-1,1]$ and $\theta \in \left[-\dfrac{\pi}{2}, \dfrac{\pi}{2}\right]$, $\sin \theta = x$.

Since $-x \in [-1,1]$ and $-\theta \in \left[-\dfrac{\pi}{2}, \dfrac{\pi}{2}\right]$, $\sin(-\theta) = -x$, therefore

$\arcsin(-x) = -\theta = -\arcsin x$;

b) Let $\arccos x = \theta$, $x \in [-1,1]$ and $\theta \in [0, \pi]$, $\cos \theta = x$.

Since $-x \in [-1,1]$ and $\pi - \theta \in [0, \pi]$, $\cos(\pi - \theta) = -x$, therefore

$\arccos(-x) = \pi - \theta = \pi - \arccos x$.

VI. 3. a) For $\arcsin(\sin 1)$, $1 \in \left[-\dfrac{\pi}{2}, \dfrac{\pi}{2}\right]$, $\arcsin(\sin 1) = 1$;

b) Let $\arcsin(\sin 3) = \alpha$, $\alpha \in \left[-\dfrac{\pi}{2}, \dfrac{\pi}{2}\right]$. Since $3 \notin \left[-\dfrac{\pi}{2}, \dfrac{\pi}{2}\right]$ and

$\sin 3 = \sin(\pi - 3)$, $\pi - 3 \in \left[-\dfrac{\pi}{2}, \dfrac{\pi}{2}\right]$, we obtain $\arcsin(\sin 3) = \pi - 3$;

c) Let $\arcsin(\sin 10) = \beta$, $\beta \in \left[-\dfrac{\pi}{2}, \dfrac{\pi}{2}\right]$. Since $10 \notin \left[-\dfrac{\pi}{2}, \dfrac{\pi}{2}\right]$ and

$\sin 10 = \sin(3\pi - 10)$, $3\pi - 10 \in \left[-\dfrac{\pi}{2}, \dfrac{\pi}{2}\right]$, we obtain $\arcsin(\sin 10) = 3\pi - 10$;

d) Let $\arcsin(\cos 1) = \alpha$, $\alpha \in \left[-\dfrac{\pi}{2}, \dfrac{\pi}{2} \right]$, $\cos 1 = \sin \alpha \Rightarrow$

$\cos 1 = \cos\left(\dfrac{\pi}{2} - \alpha \right) \Rightarrow \dfrac{\pi}{2} - \alpha = \pm 1 + 2k\pi$, $k \in \mathbf{Z}$.

Since $\alpha \in \left[-\dfrac{\pi}{2}, \dfrac{\pi}{2} \right]$, $\alpha = \dfrac{\pi}{2} - 1$;

e) Let $\arcsin(\cos 3) = \alpha$, $\alpha \in \left[-\dfrac{\pi}{2}, \dfrac{\pi}{2} \right]$, $\cos 3 = \sin \alpha \Rightarrow$

$\cos 3 = \cos\left(\dfrac{\pi}{2} - \alpha \right) \Rightarrow \dfrac{\pi}{2} - \alpha = \pm 3 + 2k\pi$, $k \in \mathbf{Z}$. Since $\alpha \in \left[-\dfrac{\pi}{2}, \dfrac{\pi}{2} \right]$,

$\dfrac{\pi}{2} - \alpha \in [0, \pi]$, then $\alpha = \dfrac{\pi - 6}{2}$;

f) Let $\arcsin(\cos 10) = \alpha$, $\alpha \in \left[-\dfrac{\pi}{2}, \dfrac{\pi}{2} \right]$, $\cos 10 = \sin \alpha \Rightarrow$

$\cos 10 = \cos\left(\dfrac{\pi}{2} - \alpha \right) \Rightarrow \dfrac{\pi}{2} - \alpha = \pm 10 + 2k\pi$, $k \in \mathbf{Z}$. Since $\alpha \in \left[-\dfrac{\pi}{2}, \dfrac{\pi}{2} \right]$

or $\dfrac{\pi}{2} - \alpha \in [0, \pi]$, a solution could be $\dfrac{\pi}{2} - \alpha = \pm 10$ or $\dfrac{\pi}{2} - \alpha = \pm 10 \pm 2\pi$

or $\dfrac{\pi}{2} - \alpha = \pm 10 \pm 4\pi$. The only solution is $\dfrac{\pi}{2} - \alpha = -10 + 4\pi$. Therefore

$\alpha = \dfrac{20 - 7\pi}{2}$; **g)** $\arccos(\cos 10) = 4\pi - 10$, $4\pi - 10 \in [0, \pi]$;

h) $\dfrac{\pi}{2} - \alpha = -2014 + 642\pi$.

II. 4. I) $x \in \left[0, \dfrac{\pi}{2} \right]$, then $f(x) = x$, for any $x \in \left[0, \dfrac{\pi}{2} \right]$;

II) $x \in \left(\dfrac{\pi}{2}, \pi \right]$, then $\pi - x \in \left[0, \dfrac{\pi}{2} \right) \Rightarrow f(x) = \arcsin(\sin(\pi - x)) = \pi - x$;

III) $x \in \left(\pi, \dfrac{3\pi}{2} \right]$, then $x - \pi \in \left[0, \dfrac{\pi}{2} \right] \Rightarrow$

$f(x) = \arcsin(-\sin(x - \pi)) = -\arcsin(\sin(x - \pi)) = -x + \pi$

IV) $x \in \left(\dfrac{3\pi}{2}, 2\pi \right]$, then $2\pi - x \in \left[0, \dfrac{\pi}{2} \right) \Rightarrow f(x) = \arcsin(-\sin(2\pi - x)) = -2\pi + x$.

Therefore

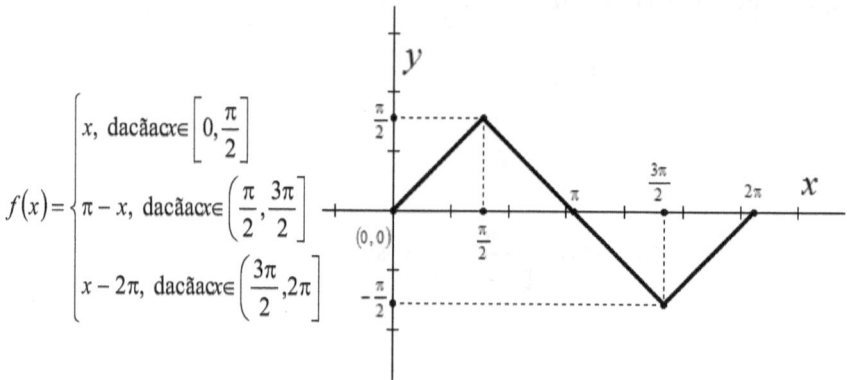

$$f(x)=\begin{cases} x, & \text{dacăacx}\in\left[0,\dfrac{\pi}{2}\right] \\[2mm] \pi-x, & \text{dacăacx}\in\left(\dfrac{\pi}{2},\dfrac{3\pi}{2}\right] \\[2mm] x-2\pi, & \text{dacăacx}\in\left(\dfrac{3\pi}{2},2\pi\right] \end{cases}$$

VI. 5. f is an increasing function, therefore, it is one to one (injective) and $f\left(\left[0,\dfrac{\pi}{2}\right]\right)=[-1,1]$, that is, f is onto (surjective).

$f(x)=\sqrt{2}\sin\left(x-\dfrac{\pi}{4}\right)$. We obtain $f^{-1}:[-1,1]\to\left[0,\dfrac{\pi}{2}\right]$,

$f^{-1}(x)=\dfrac{\pi}{4}+\arcsin\dfrac{x}{\sqrt{2}}$.

VI. 6. a) Taking $\sin x=y$, then $\sin(\pi-x)=y$ and $\pi-x\in\left[-\dfrac{\pi}{2},\dfrac{\pi}{2}\right]$.

Therefore $\pi-x=\arcsin y$. The inverse of f is $f^{-1}:[-1,1]\to\left[\dfrac{\pi}{2},\dfrac{3\pi}{2}\right]$,

$f^{-1}(x)=\pi-\arcsin x$. **b)** $g^{-1}:[-1,1]\to[\pi,2\pi]$, $g^{-1}(x)=\pi-\arccos x$;

c) $h^{-1}:[-1,1]\to[3\pi,4\pi]$, $h^{-1}(x)=4\pi-\arccos x$;

d) $f^{-1}:\mathbf{R}\to\left[\dfrac{3\pi}{2},\dfrac{5\pi}{2}\right]$, $f^{-1}(x)=2\pi+\arctan x$;

VI. 7. a) $x\in[-3,3]$; **b)** $x\in\left[-\dfrac{7}{3},1\right]$; **c)** $x\in[0,\infty)$;

d) $x\in\left[-1,-\dfrac{\sqrt{2}}{2}\right]\cup\left[\dfrac{\sqrt{2}}{2},1\right]$; **e)** $x\in(-\infty,-2]\cup[0,\infty)$;

f) $-1 \le \dfrac{x^2 - 6x + 8}{x^2 - 9} \le 1$, $x \in \left[\dfrac{3 - \sqrt{11}}{2}, \dfrac{17}{6}\right] \cup \left[\dfrac{3 + \sqrt{11}}{2}, \infty\right)$.

VI. 8. $x \in [0, 1]$, then $\arcsin\sqrt{1 - x^2} = \arccos\sqrt{1 - \left(\sqrt{1 - x^2}\right)^2} = \arccos\sqrt{x}$.

Since $f(x) = \arcsin\sqrt{x} + \arccos\sqrt{x} = \dfrac{\pi}{2}$ the graph is a closed segment

connecting the points $\left(0, \dfrac{\pi}{2}\right)$ and $\left(1, \dfrac{\pi}{2}\right)$.

VI. 9.

Since $\cos : \mathbf{R} \to [-1, 1]$ and

$\arccos : [-1, 1] \to [0, \pi]$,

then $0 \le 2\arccos x \le 2\pi$,

that is, $f : [-1, 1] \to [-1, 1]$.

$f(x) = 2\cos^2(\arccos x) - 1 = 2x^2 - 1$.

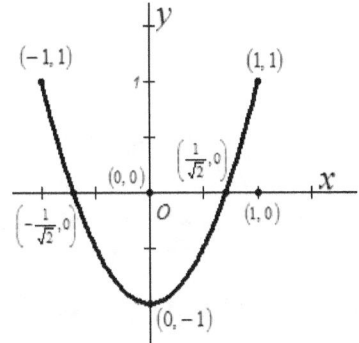

VI. 10. $f(x) = \arcsin x\left(\dfrac{\pi}{2} - \arcsin x\right) = -\arcsin^2 x + \dfrac{\pi}{2}\arcsin x - \left(\dfrac{\pi}{2}\right)^2 + \left(\dfrac{\pi}{2}\right)^2$.

Then $f(x) = \left(\dfrac{\pi}{2}\right)^2 - \left(\dfrac{\pi}{2} - \arcsin x\right)^2$. Therefore $-\dfrac{3\pi^2}{4} \le f(x) \le \dfrac{\pi^2}{4}$.

VI. 11. a) Let $\arctan\dfrac{1}{4} = \alpha$, $\alpha > 0$.

$$\sin\alpha = \dfrac{\tan\alpha}{\sqrt{1 + \tan^2\alpha}} = \dfrac{\tan\left(\arctan\dfrac{1}{4}\right)}{\sqrt{1 + \tan^2\left(\arctan\dfrac{1}{4}\right)}} = \dfrac{1}{\sqrt{17}}.$$

$$\cos\alpha = \dfrac{1}{\sqrt{1 + \tan^2\alpha}} = \dfrac{1}{\sqrt{1 + \tan^2\left(\arctan\dfrac{1}{4}\right)}} = \dfrac{4}{\sqrt{17}}\text{Therefore}$$

$$\arctan\frac{1}{5} = \arcsin\frac{1}{\sqrt{17}} \text{ and } \arctan\frac{1}{5} = \arccos\frac{4}{\sqrt{17}};$$

b) $\operatorname{arc\,cot}3 = \arcsin\dfrac{1}{\sqrt{10}}, \ \operatorname{arc\,cot}3 = \dfrac{3}{\sqrt{10}}, \ \operatorname{arc\,cot}3 = \arctan\dfrac{1}{3}.$

VI. 12. If $x = 2\arctan\dfrac{1}{3} \Rightarrow \tan\dfrac{x}{2} = \dfrac{1}{3}$. Using $\sin x = \dfrac{2\tan\dfrac{x}{2}}{1+\tan^2\dfrac{x}{2}}$ and

$\cos x = \dfrac{1-\tan^2\dfrac{x}{2}}{1+\tan^2\dfrac{x}{2}}$, we obtain $\sin x = \dfrac{3}{5}$ and $\cos x = \dfrac{4}{5}$.

VI. 13. a) $y = \arcsin x$ is an increasing function, therefore, the order is

$\arcsin\dfrac{1}{2}, \ \arcsin\dfrac{1}{3}, \ \arcsin 1;$ **b)** $y = \arccos x$ is a decreasing function,

therefore, the order is $\arccos\dfrac{2}{3}, \ \arccos\dfrac{1}{\sqrt{3}}, \ \arccos\dfrac{1}{\sqrt{2}};$

c) Let $\alpha = \arccos\dfrac{1}{3}, \beta = \arcsin\dfrac{3}{5}, \gamma = \arctan\dfrac{11}{5}.$

$OC = \dfrac{1}{3} = \cos\left(\overset{\wedge}{NOA}\right) = \cos\beta$

$OD = \dfrac{3}{5} = \sin\left(\overset{\wedge}{MOA}\right) = \sin\alpha$

$AE = \dfrac{11}{5} = \tan\left(\overset{\wedge}{EOA}\right) = \tan\gamma$

$\alpha < \gamma < \beta.$

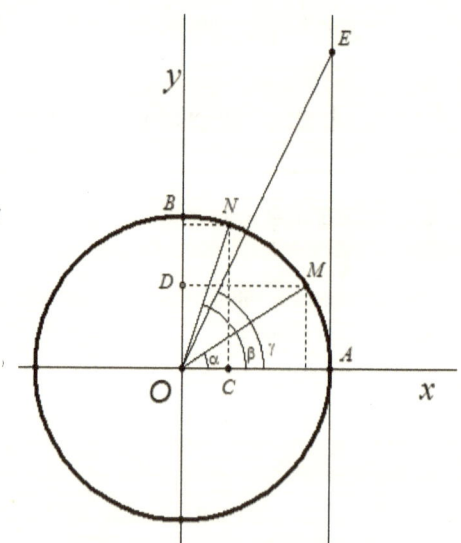

240

VI. 14. a) $\dfrac{2\sqrt{2}+3}{6}$;   **b)** $\dfrac{1-\sqrt{2}}{\sqrt{6}}$;   **c)** Let $\arctan\dfrac{1}{2}=\alpha$ and $\arccos\dfrac{4}{5}=\beta$.

We have $\alpha,\beta\in\left(0,\dfrac{\pi}{2}\right)$, $\tan\alpha=\dfrac{1}{2}$ and $\cos\beta=\dfrac{4}{5}$.

Then $\sin\left(2\arctan\dfrac{1}{2}+\arccos\dfrac{4}{5}\right)=\sin(2\alpha+\beta)=$

$=\sin 2\alpha\cos\beta+\sin\beta\cos 2\alpha=\dfrac{2\tan\alpha}{1+\tan^2\alpha}\cos\beta+\sqrt{1-\cos^2\beta}\,\dfrac{1-\tan^2\alpha}{1+\tan^2\alpha}=1$;

d) Let $\arcsin\dfrac{4}{5}=\alpha\in\left[-\dfrac{\pi}{2},\dfrac{\pi}{2}\right]$, then $\sin\alpha=\dfrac{4}{5}$ and $\cos\alpha=\dfrac{3}{5}$ or

$\sin\dfrac{\alpha}{4}=\sqrt{\dfrac{\sqrt{5}-2}{2\sqrt{5}}}$ and $\cos\dfrac{\alpha}{4}=\sqrt{\dfrac{\sqrt{5}+2}{2\sqrt{5}}}$. Therefore

$\tan\left(-\dfrac{\pi}{4}+\dfrac{\alpha}{4}\right)=\dfrac{\tan\left(-\dfrac{\pi}{4}\right)+\tan\dfrac{\alpha}{4}}{1-\tan\left(-\dfrac{\pi}{4}\right)\tan\dfrac{\alpha}{4}}=\dfrac{1-\sqrt{5}}{2}$;   **e)** $\dfrac{8+6\sqrt{2}}{15}$;   **f)** $\dfrac{\sqrt{21}}{5}$;   **g)** $\dfrac{2}{\sqrt{5}}$;

h) Denote $\arcsin\dfrac{12}{13}=\alpha\in\left[-\dfrac{\pi}{2},\dfrac{\pi}{2}\right]$. Then $\sin\alpha=\dfrac{12}{13}$ and $\cos\alpha=\dfrac{5}{13}$.

Also, denote $\arccos\dfrac{5}{\sqrt{26}}=\beta\in[0,\pi]$. Then $\sin\beta=\dfrac{1}{\sqrt{26}}$ and $\cos\beta=\dfrac{5}{\sqrt{26}}$,

also $\sin 2\beta=\dfrac{5}{13}$ and $\cos 2\beta=\dfrac{12}{13}$;

Therefore $\tan(2\beta-\alpha)=\dfrac{\cos 2\beta\sin\alpha-\sin 2\beta\cos\alpha}{\cos 2\beta\cos\alpha+\sin 2\beta\sin\alpha}=\dfrac{119}{120}$.

VI. 15. a) Let $\arctan\dfrac{1}{3}=\alpha$ and $\arctan\dfrac{1}{7}=\beta$. We have $\alpha,\beta\in\left(0,\dfrac{\pi}{2}\right)$ and

$\tan\alpha=\dfrac{1}{3}$ and $\tan\beta=\dfrac{1}{7}$. Then $\sin\left(4\arctan\dfrac{1}{3}\right)=\sin 2\alpha\cos 2\alpha=$

$=2\dfrac{2\tan\alpha}{1+\tan^2\alpha}\dfrac{1-\tan^2\alpha}{1+\tan^2\alpha}=\dfrac{24}{25}$ and $\cos\left(2\arctan\dfrac{1}{7}\right)=\cos 2\beta=\dfrac{1-\tan^2\beta}{1+\tan^2\beta}=\dfrac{24}{25}$;

c) Using the formulae $\sin 2\alpha=\dfrac{2\tan\alpha}{1+\tan^2\alpha}$, $\tan\dfrac{\alpha}{2}=\dfrac{\sin\alpha}{1+\cos\alpha}$ we have

$$\sin\left(2\arctan\frac{1}{3}\right)+\tan\left(\frac{1}{2}\arcsin\frac{3}{5}\right)=\frac{2\tan\left(\arctan\frac{1}{3}\right)}{1+\tan^2\left(\arctan\frac{1}{3}\right)}+\frac{\sin\left(\arcsin\frac{3}{5}\right)}{1+\cos\left(\arcsin\frac{3}{5}\right)}=$$

$$=\frac{2\cdot\frac{1}{3}}{1+\left(\frac{1}{3}\right)^2}+\frac{\frac{3}{5}}{1+\sqrt{1-\left(\frac{3}{5}\right)^2}}=\frac{14}{15};$$

d) $\sin\left(2\arctan\frac{1}{2}\right)-\tan\left(\arccos\frac{3}{5}\right)=\dfrac{2\tan\left(\arctan\frac{1}{2}\right)}{1+\tan^2\left(\arctan\frac{1}{2}\right)}-\dfrac{\sin\left(\arccos\frac{3}{5}\right)}{\cos\left(\arccos\frac{3}{5}\right)}=$

$$=\frac{2\cdot\frac{1}{2}}{1+\left(\frac{1}{2}\right)^2}-\frac{\sqrt{1-\left(\frac{3}{5}\right)^2}}{\frac{3}{5}}=\frac{4}{5}-\frac{4}{3}=-\frac{8}{15};$$

e) $\cos\left(2\arctan\frac{1}{2}\right)-\sin\left(3\arcsin\frac{1}{2}\right)=\dfrac{1-\tan^2\left(\arctan\frac{1}{2}\right)}{1+\tan^2\left(\arctan\frac{1}{2}\right)}-3\sin\left(\arcsin\frac{1}{2}\right)+$

$$+4\sin^3\left(\arcsin\frac{1}{2}\right)=\frac{1-\left(\frac{1}{2}\right)^2}{1+\left(\frac{1}{2}\right)^2}-3\cdot\frac{1}{2}+4\cdot\left(\frac{1}{2}\right)^3=-\frac{2}{5}.$$

VI. 16. a) $-\dfrac{1}{k}$;   **b)** $-\dfrac{1}{k}$;   **c)** 0.02;   **d)** -0.007;   **e)** 0.009.

VI. 17. a) $\arccos\left(\cos\left(2\arctan\left(\sqrt{2}-1\right)\right)\right)=\arccos\left(2\cos^2\left(\arctan\left(\sqrt{2}-1\right)\right)-1\right)=$

$$=\arccos\left(2\frac{1}{1+\tan^2\left(\arctan\left(\sqrt{2}-1\right)\right)}-1\right)=\arccos\left(2\frac{1}{1+\left(\sqrt{2}-1\right)^2}-1\right)=$$

$=\arccos\dfrac{\sqrt{2}}{2}=\dfrac{\pi}{4}$;   **b)** Denoting $\operatorname{arccot}\left(\sqrt{2}-1\right)=\alpha$, then $\cot\alpha=\sqrt{2}-1$

and $\tan\alpha = \sqrt{2}+1$. Thus $\cos 2\alpha = -\dfrac{1}{\sqrt{2}}$ and finally $\arccos\left(-\dfrac{1}{\sqrt{2}}\right) = \dfrac{3\pi}{4}$;

c) $\arcsin\left(\cos\left(2\arctan\left(\sqrt{2}-1\right)\right)\right) = \arcsin\left(2\cos^2\left(\arctan\left(\sqrt{2}-1\right)\right)-1\right) =$

$= \arcsin\left(2\dfrac{1}{1+\tan^2\left(\arctan\left(\sqrt{2}-1\right)\right)}-1\right) = \arcsin\left(2\dfrac{1}{1+\left(\sqrt{2}-1\right)^2}-1\right) =$

$= \arcsin\dfrac{\sqrt{2}}{2} = \dfrac{\pi}{4}$;   d) $-\dfrac{\pi}{4}$;   e) $-\dfrac{\pi}{4}$.

VI. 18. a) 0.98;   **b)** Denote $\arcsin\dfrac{4}{5} = \alpha \in \left[-\dfrac{\pi}{2}, \dfrac{\pi}{2}\right]$, then $\sin\alpha = \dfrac{4}{5}$

and $\cos\alpha = \dfrac{3}{5}$, Also, denote $\arctan 2 = \beta \in \left(-\dfrac{\pi}{2}, \dfrac{\pi}{2}\right)$ then $\tan\beta = 2$ and

$\cos\beta = \dfrac{1}{\sqrt{5}}$, $\sin\beta = \dfrac{2}{\sqrt{5}}$. Moreover $\sin 2\beta = \dfrac{4}{5}$, $\cos 2\beta = 1 - 2\sin^2\beta = -\dfrac{3}{5}$,

$\sin 4\beta = -\dfrac{24}{25}$, and $\cos 4\beta = 1 - 2\sin^2 2\beta = -\dfrac{7}{25}$. Therefore

$\sin^2\left(\dfrac{1}{2}\alpha + 2\beta\right) = \dfrac{1}{2}\left[1 - \cos(\alpha + 4\beta)\right] = \dfrac{1}{2}\left[1 - \cos\alpha\cos 4\beta + \sin\alpha\sin 4\beta\right] =$

$= \dfrac{1}{2}\left[1 - \dfrac{3}{5}\cdot\left(-\dfrac{7}{25}\right) + \dfrac{4}{5}\cdot\left(-\dfrac{24}{25}\right)\right] = \dfrac{1}{5}$;   **c)** Denote $\arccos\dfrac{3}{5} = \alpha \in [0, \pi]$, then

$\cos\alpha = \dfrac{3}{5}$, $\sin\alpha = \dfrac{4}{5}$. Since $\tan\alpha = \dfrac{2\tan\dfrac{\alpha}{2}}{1 - \tan^2\dfrac{\alpha}{2}}$, we have $\dfrac{4}{3} = \dfrac{2\tan\dfrac{\alpha}{2}}{1 - \tan^2\dfrac{\alpha}{2}}$ or

$\tan\dfrac{\alpha}{2} = \dfrac{1}{2}$ because $\dfrac{\alpha}{2} \in \left(0, \dfrac{\pi}{2}\right)$.

Also, denote $\text{arccot}\left(-\dfrac{1}{2}\right) = \beta \in (0, \pi)$ then $\cot\beta = -\dfrac{1}{2}$ or $\tan\beta = -2$.

$\tan 2\beta = \dfrac{2\tan\beta}{1 - \tan^2\beta} = \dfrac{2(-2)}{1 - (-2)^2} = \dfrac{4}{3}$.

Therefore $\cot\left(\dfrac{1}{2}\alpha - 2\beta\right) = \dfrac{1 + \tan\left(\dfrac{1}{2}\alpha\right)\tan(2\beta)}{\tan\left(\dfrac{1}{2}\alpha\right) - \tan(2\beta)} = \dfrac{1 + \dfrac{1}{2}\cdot\dfrac{4}{3}}{\dfrac{1}{2} - \dfrac{4}{3}} = -2$;

d) Let $\arccos\dfrac{3}{5} = \alpha \in [0, \pi]$, then $\cos\alpha = \dfrac{3}{5}$ or $\sin\dfrac{\alpha}{2} = \dfrac{1}{\sqrt{5}}$ and $\cos\dfrac{\alpha}{2} = \dfrac{2}{\sqrt{5}}$.

Also, denote $\text{arccot}(-2) = \beta \in (0, \pi)$, then $\cot\beta = -2$ or $\sin 3\beta = \dfrac{11}{5\sqrt{5}}$ and

$\cos 3\beta = \dfrac{2}{5\sqrt{5}}$. Therefore $\tan\left(\dfrac{1}{2}\alpha - 3\beta\right) = \dfrac{\cos\dfrac{\alpha}{2}\sin 3\beta - \sin\dfrac{\alpha}{2}\cos 3\beta}{\cos\dfrac{\alpha}{2}\cos 3\beta + \sin\dfrac{\alpha}{2}\sin 3\beta} = \dfrac{4}{3}$;

e) Let $\arcsin\dfrac{4}{5} = \alpha \in \left[-\dfrac{\pi}{2}, \dfrac{\pi}{2}\right]$, then $\sin\alpha = \dfrac{4}{5}$ and $\cos\alpha = \dfrac{3}{5}$ or

$\sin\dfrac{\alpha}{2} = \dfrac{1}{\sqrt{5}}$ and $\cos\dfrac{\alpha}{2} = \dfrac{2}{\sqrt{5}}$. Also, denote $\text{arccot}\left(-\dfrac{1}{2}\right) = \beta \in (0, \pi)$,

then $\cot\beta = -\dfrac{1}{2}$ or $\sin\beta = \dfrac{2}{\sqrt{5}}$ and $\cos\beta = -\dfrac{1}{\sqrt{5}}$ or $\sin 2\beta = -\dfrac{4}{5}$ and

$\cos 2\beta = -\dfrac{3}{5}$.

Therefore $\cos\left(\dfrac{1}{2}\alpha - 2\beta\right) = \cos\dfrac{\alpha}{2}\cos 2\beta + \sin\dfrac{\alpha}{2}\sin 2\beta = -\dfrac{2}{\sqrt{5}}$;

f) Let $\arccos\dfrac{3}{5} = \alpha \in [0, \pi]$, then $\cos\alpha = \dfrac{3}{5}$ or $\sin\dfrac{\alpha}{2} = \dfrac{1}{\sqrt{5}}$ and $\cos\dfrac{\alpha}{2} = \dfrac{2}{\sqrt{5}}$.

Also, denote $\arctan(-2) = \beta \in \left(-\dfrac{\pi}{2}, \dfrac{\pi}{2}\right)$, then $\tan\beta = -2$ or $\sin 2\beta = \dfrac{4}{5}$

and $\cos 2\beta = \dfrac{3}{5}$. Therefore $\tan\left(\dfrac{1}{2}\alpha - 2\beta\right) = -\dfrac{1}{2}$; **g)** $-\dfrac{2}{\sqrt{5}}$.

VI. 19. a) Let $\arcsin\dfrac{3}{5} = \alpha \in \left[-\dfrac{\pi}{2}, \dfrac{\pi}{2}\right]$, then $\sin\alpha = \dfrac{3}{5}$ and $\cos\alpha = \dfrac{4}{5}$ or

$\sin\dfrac{\alpha}{2} = \dfrac{1}{\sqrt{10}}$. Therefore $\dfrac{1}{4} - \cos^4\left(\dfrac{3\pi}{2} - \dfrac{\alpha}{2}\right) = \dfrac{1}{4} - \sin^4\dfrac{\alpha}{2} = \dfrac{6}{25}$;

b) $-\dfrac{2\sqrt[4]{2}}{3}$; **c)** $\dfrac{2a}{b}$; **d)** $(x + y)(x^2 + x^2)$.

VI. 20. $\arccos\alpha + \arccos\beta = \pi - \arccos\gamma$,

or $\cos(\arccos\alpha + \arccos\beta) = \cos(\pi - \arccos\gamma)$. Then

$\alpha\beta - \sqrt{(1-\alpha^2)(1-\beta^2)} = -\gamma$ or $\sqrt{(1-\alpha^2)(1-\beta^2)} = \alpha\beta + \gamma$. Squaring in both sides, we obtain $\alpha^2 + \beta^2 + \gamma^2 + 2\alpha\beta\gamma = 1$.

VI. 21. a) Let $\arcsin\dfrac{1}{\sqrt5} = \alpha$ and $\arccos\dfrac{3}{\sqrt{10}} = \beta$.

We have $\alpha, \beta \in \left(0, \dfrac{\pi}{2}\right)$ and $\sin\alpha = \dfrac{1}{\sqrt5}$ and $\cos\beta = \dfrac{3}{\sqrt{10}}$.

Then $\cos\left(\arcsin\dfrac{1}{\sqrt5} + \arccos\dfrac{3}{\sqrt{10}}\right) = \cos(\alpha+\beta) = \dfrac{\sqrt2}{2}$;

c) Let $\alpha = \arctan\dfrac{1}{5}$, $\beta = \arctan\dfrac{1}{4}$, then $\tan\alpha = \dfrac{1}{5}$, $\tan\beta = \dfrac{1}{4}$ $\alpha, \beta \in \left(0, \dfrac{\pi}{2}\right)$.

Then $\tan(2\alpha+\beta) = \dfrac{32}{43}$;

f) Using $\cos 2\alpha = \dfrac{\cot^2\alpha - 1}{1 + \cot^2\alpha}$ we obtain

$\cos(2\operatorname{arccot}7) = \dfrac{\cot^2(\operatorname{arccot}7) - 1}{1 + \cot^2(\operatorname{arccot}7)} = \dfrac{7^2 - 1}{1 + 7^2} = \dfrac{24}{25}$, using $\sin 4\alpha = \dfrac{2\cot 2\alpha}{1 + \cot^2 2\alpha}$

and $\cot 2\alpha = \dfrac{\cot^2\alpha - 1}{2\cot\alpha}$, we obtain $\sin(4\operatorname{arccot}3) = \dfrac{24}{25}$;

g) $\sin\left(\arccos\dfrac{36}{85} + \arcsin\dfrac{4}{5}\right) =$

$= \sin\left(\arccos\dfrac{36}{85}\right)\cos\left(\arcsin\dfrac{4}{5}\right) + \sin\left(\arcsin\dfrac{4}{5}\right)\cos\left(\arccos\dfrac{36}{85}\right) =$

$= \dfrac{3}{5}\cdot\dfrac{77}{85} + \dfrac{4}{5}\cdot\dfrac{36}{85} = \dfrac{15}{17}$ and $\sin\left(\dfrac{\pi}{2} + \arccos\dfrac{15}{17}\right) = \cos\left(\arccos\dfrac{15}{17}\right) = \dfrac{15}{17}$;

i) We have $\arcsin\dfrac{2}{5} = \alpha$, $\arcsin\dfrac{3}{5} = \beta$, $\arcsin\dfrac{4}{5} = \theta$, where $\alpha, \beta, \theta \in \left(0, \dfrac{\pi}{2}\right)$.

Then $\sin\alpha = \dfrac{2}{5}$, $\sin\beta = \dfrac{3}{5}$, $\sin\theta = \dfrac{4}{5}$.

Calculating $\sin(\alpha + \beta + \theta)$ we obtain the result.

VI. 22. a) Let $\alpha = \arctan\sqrt{\dfrac{1-x}{1+x}}$, $\beta = \arccos x$, then $\tan\alpha = \sqrt{\dfrac{1-x}{1+x}}$,

$\alpha \in \left(0, \dfrac{\pi}{2}\right)$, $\cos\beta = x$ $\beta \in (0, \pi)$. $\tan(2\alpha) = \dfrac{\sqrt{1-x^2}}{x}$ and

$\tan\beta = \dfrac{\sqrt{1-\cos^2\beta}}{\cos\beta} = \dfrac{\sqrt{1-x^2}}{x}$;

b) $\tan\left[2\arctan\left(\dfrac{2\sin^2\dfrac{x}{2}}{2\sin\dfrac{x}{2}\cos\dfrac{x}{2}}\right)\right] = \tan\left[2\arctan\left(\tan\dfrac{x}{2}\right)\right] = \tan x$;

c) Since $\tan(\arccos x) = \dfrac{\sin(\arccos x)}{\cos(\arccos x)} = \dfrac{\sqrt{1-x^2}}{x}$, then

$\tan\left(\arccos\dfrac{1}{\sqrt{1+x^2}} + \arccos\dfrac{x}{\sqrt{1+x^2}}\right) = \dfrac{\tan\left(\arccos\dfrac{1}{\sqrt{1+x^2}}\right) + \tan\left(\arccos\dfrac{x}{\sqrt{1+x^2}}\right)}{1 - \tan\left(\arccos\dfrac{1}{\sqrt{1+x^2}}\right)\tan\left(\arccos\dfrac{x}{\sqrt{1+x^2}}\right)} =$

$= \dfrac{-x + \dfrac{1}{x}}{1-(-x)\dfrac{1}{x}} = \dfrac{1-x^2}{2x}$; **d)** $\tan(\arctan x^2) = x^2$,

$\tan\left(\dfrac{\pi}{2} - \arctan\dfrac{1}{x^2}\right) = \cot\left(\arctan\dfrac{1}{x^2}\right) = x^2$;

f) For $x = 1$ the equality is true.

For $x > 1$, let $\alpha = 2\arctan x$ and $\beta = \arcsin\dfrac{2x}{1+x^2}$. We have

$\tan\alpha = \tan(2\arctan x) = \dfrac{2x}{1-x^2}$ and $\sin\beta = \dfrac{2x}{1+x^2}$. Since $x > 1$, then

$\cos\beta = \dfrac{|x^2-1|}{1+x^2} = \dfrac{x^2-1}{1+x^2}$. Thus $\tan\beta = \dfrac{2x}{x^2-1}$ and

$\tan\alpha = -\tan\beta \Rightarrow \tan\alpha = \tan(\pi-\beta)$. Therefore $\alpha + \beta = \pi$. Similarly for $x < 1$;

k) Use $\arctan x + \operatorname{arccot} x = \dfrac{\pi}{2}$, $\sin(\arccos x) = \sqrt{1-x^2}$, and
$\cos(\arcsin x) = \sqrt{1-x^2}$;

l) $\arctan\dfrac{4x+4}{2-3x} - \arctan\dfrac{5x+2}{5} = \arctan\dfrac{3}{4}$, $x < \dfrac{2}{3}$.

For $x > \dfrac{2}{3}$, $\arctan\dfrac{4x+4}{2-3x} < 0$, $-\arctan\dfrac{5x+2}{5} < 0$, and $\arctan\dfrac{3}{4} > 0$, which
is impossible.

For $x < \dfrac{2}{3}$, we have $\tan\left(\arctan\dfrac{4x+4}{2-3x} - \arctan\dfrac{5x+2}{4}\right) = \dfrac{\dfrac{4x+4}{2-3x} - \dfrac{5x+2}{4}}{1 + \dfrac{4x+4}{2-3x}\dfrac{5x+2}{4}} = \dfrac{3}{4}$.

VI. 23. a) $x \in [-1,1]$. Use $\cos 3\alpha = 4\cos^3\alpha - 3\cos\alpha$ and
$\cos 2\alpha = 2\cos^2\alpha - 1$.
We have $4x^3 - 3x = 1 + 2x^2 - 1 \Rightarrow x(4x^2 - 2x - 3) = 0$ with the convenient

roots $x_1 = 0$ and $x_2 = \dfrac{1-\sqrt{13}}{4}$; **b)** $x_1 = 0$ and $x_2 = \dfrac{1-\sqrt{13}}{4}$;

c) We have $2(2\cos^2(2\arccos x) - 1) - 8 \cdot \dfrac{1 - \cos(2\arccos x)}{2} + 7 = 0 \Rightarrow$

$[2\cos(2\arccos x) + 1]^2 = 0 \Rightarrow \cos(2\arccos x) = -\dfrac{1}{2} \Rightarrow 2x^2 - 1 = -\dfrac{1}{2}$.

Therefore $x = \pm\dfrac{1}{2}$;

d) We have $2(2\cos^2(2\arcsin x) - 1) - 8 \cdot \dfrac{1 + \cos(2\arcsin x)}{2} + 7 = 0$

$[2\cos(2\arcsin x) - 1]^2 = 0 \Rightarrow \cos(2\arcsin x) = \dfrac{1}{2} \Rightarrow 1 - 2x^2 = \dfrac{1}{2}$.

Therefore $x = \pm\dfrac{1}{2}$;

VI. 24. a) $\arcsin x = \dfrac{1}{2}$, $x_1 = \dfrac{\pi}{6}$ and $\arcsin x = \dfrac{1}{3}$, $x_2 = \sin\dfrac{1}{3}$;

b) $-1 \le x - 1 \le 1$, $x \in [0, 2]$.

$\cos[2\arctan(2x-1)] = \dfrac{1-(2x-1)^2}{1+(2x-1)^2} = \dfrac{-4x^2+4x}{4x^2-4x+2}$. Then

$\dfrac{-4x^2+4x}{4x^2-4x+2} = x-1 \Rightarrow x-1 = 0$ or $\dfrac{-4x}{4x^2-4x+2} = 1$. The only solution is $x = 1$;   c) $-1 \le \dfrac{2-x}{3} \le 1 \Rightarrow x \in [-1, 5]$, $-1 \le x+2 \le 1 \Rightarrow x \in [-3, -1]$,

$x = -1$ is a solution;   d) $x \in \left[-\dfrac{1}{2}, \dfrac{1}{2} \right], 0, \pm\dfrac{1}{2}$;

e) $x \in \left[-\dfrac{1}{\sqrt{2}}, \dfrac{1}{\sqrt{2}} \right], 0, \pm\dfrac{1}{\sqrt{2}}$;   f) $x \in [0,1]$;

g) We have: I) $4 \arctan x = \dfrac{\pi}{3} + 2k\pi \Rightarrow \arctan x = \dfrac{\pi}{12} + \dfrac{k\pi}{2}$, $k \in \mathbf{Z}$. Since

$\arctan x \in \left(-\dfrac{\pi}{2}, \dfrac{\pi}{2} \right)$, then $\arctan x = \dfrac{\pi}{12}$ or $\arctan x = -\dfrac{5\pi}{12}$. Therefore,

$x_1 = \tan\dfrac{\pi}{12} = 2 - \sqrt{3}$ or $x_2 = \tan\left(-\dfrac{5\pi}{12} \right) = -2 - \sqrt{3}$;

II) $4 \arctan x = \dfrac{2\pi}{3} + 2k\pi \Rightarrow \arctan x = \dfrac{\pi}{6} + \dfrac{k\pi}{2}$, $k \in \mathbf{Z}$. Since

$\arctan x \in \left(-\dfrac{\pi}{2}, \dfrac{\pi}{2} \right)$, then $\arctan x = \dfrac{\pi}{6}$ or $\arctan x = -\dfrac{\pi}{3}$. Therefore,

$x_3 = \tan\dfrac{\pi}{6} = \dfrac{\sqrt{3}}{3}$ or $x_4 = \tan\left(-\dfrac{\pi}{3} \right) = -\sqrt{3}$;   h) $x \in (0, 1]$;   i) $x = \dfrac{1}{2}$;

j) $x = -2$;   k) $x = 1$.

VI. 25. a) $x \in \left[-\dfrac{1}{2}, \dfrac{1}{2} \right]$, $2x = \sin\left(\dfrac{\pi}{3} - \arcsin x \right) \Rightarrow 2x = \dfrac{\sqrt{3}\sqrt{1-x^2}}{2} - \dfrac{x}{2}$

with the solution $x = \dfrac{\sqrt{3}}{2\sqrt{7}}$;   b) We have

$2x = \sin\left(-\dfrac{\pi}{2} - \arcsin\left(2x\sqrt{3} \right) \right) \Rightarrow 2x = -\cos\left(\arcsin\left(2x\sqrt{3} \right) \right)$. Then

$2x = -\sqrt{1 - 12x^2}$, with the solution $x = -\dfrac{1}{4}$;

c) $x \in \left[-\dfrac{3}{4}, \dfrac{3}{4} \right]$, $\sin(\arccos x - \pi) = \sin\left(\arcsin\dfrac{4x}{3} \right)$, then $-\sqrt{1-x^2} = \dfrac{4x}{3}$.

Finally $x = -\dfrac{3}{5}$;   d) $\dfrac{1}{2}$;   e) $-1, 0$;   f) $[-1,0]$;

g) $x = 1$;   h) We have $\arcsin x + \arccos x = \dfrac{\pi}{2}$ also

$$\arcsin^3 x + \arccos^3 x = \frac{\pi^3}{32}.$$

Then $\arcsin^3 x + \left(\dfrac{\pi}{2} - \arcsin x\right)^3 x = \dfrac{\pi^3}{32} \Rightarrow \dfrac{3\pi}{2}\left(\arcsin x - \dfrac{\pi}{4}\right)^2 = 0$.

Therefore $\arcsin x = \dfrac{\pi}{4}$ and $x = \dfrac{\sqrt{2}}{2}$.

VI. 26. a) $\dfrac{x+1-x}{1+(x+1)x} = 1 \Rightarrow x_1 = 0$ and $x_2 = -1$;

b) $\dfrac{1}{2}$; **c)** 1; **d)** $\dfrac{1+\sqrt{2}}{3}$; **e)** $x = -1$; **f)** 0, $\dfrac{1}{2}$; **g)** $x = \pm\dfrac{1}{3}$;

h) $\operatorname{arc cot} x = \dfrac{\pi}{2} - \arctan x$, denoting $\arctan x = t$, then

$t^2 + \left(\dfrac{\pi}{2} - t\right)^2 = \dfrac{5\pi^2}{8}$, it yields to $x = -1$; **i)** $x = \sqrt{3}$; **j)** $x = \pm 1$; **k)** $\dfrac{1}{3}$.

VI. 27. a) We have $\cos(\arccos x - \arcsin x) = x\sqrt{3} \Rightarrow 2x\sqrt{1-x^2} = x\sqrt{3}$,

with the solutions $0, \pm\dfrac{1}{2}$; **b)** $x \in \left[\dfrac{2}{5}, \dfrac{4}{5}\right], \dfrac{1}{2}$;

c) $x \in [-1,1]$, $\cos[\arccos x + \arccos(1-x)] = x(1-x) - \sqrt{1-x^2}\sqrt{2x-x^2}$.

Then $x(1-x) - \sqrt{1-x^2}\sqrt{2x-x^2} = -x \Rightarrow \sqrt{1-x^2}\sqrt{2x-x^2} = 2x - x^2 \Rightarrow$

I) $2x - x^2 = 0$, $x = 0$

II) $\sqrt{1-x^2}\sqrt{2x-x^2} = 1 \Rightarrow x^4 - 2x^3 - x^2 + 2x - 1 = 0$. Denoting

$f(x) = x^4 - 2x^3 - x^2 + 2x - 1$, since $f(0)f(-1) > 0$ and $f(0)f(1) > 0$, that means, the function f does not have any roots on the interval $[-1,1]$.

d) $x \in [0,1]$, $\dfrac{1}{4}, \dfrac{1}{2}$; **e)** We have

$$\arctan\frac{x}{\sqrt{1-x^2}} + \arctan\sqrt{\frac{1-x}{x}} = \arctan\left(\sqrt{x+1} + \sqrt{x}\right)^2.$$

Applying tan in both sides and after all calculations, we obtain the equation

$x^3 - 2x^2 - 2x + 1 = 0$ with the convenient roots $x_1 = \dfrac{1}{2}$ and $x_2 = \dfrac{\sqrt{2}}{2}$;

f) $x \in \left[-\sqrt{2}, \sqrt{2}\right]$, $2\arctan\left(x^2 - 1\right) = \arccos\left(1 - x^2\right) + \arcsin\left(1 - x^2\right)$.

Since $\arcsin x + \arccos x = \dfrac{\pi}{2}$ for any $x \in [-1, 1]$, we obtain

$\arctan\left(x^2 - 1\right) = \dfrac{\pi}{4} \Rightarrow x^2 - 1 = 1$, Therefore $x = \pm\sqrt{2}$.

VI. 28. a) $x = 0$;   **b)** $x = \pm 1$;   **c)** $x = 0$ and $x = \pm\dfrac{2\sqrt{3}}{3}$;   **d)** We have

$\arctan(x - 1) + \arctan(x + 1) = \arctan 3x - \arctan x$. Applying tan in both sides,

we obtain $\dfrac{x + 1 + x - 1}{1 - \left(x^2 - 1\right)} = \dfrac{3x - x}{1 + 3x^2}$. Therefore, $x = 0$ and $x = \pm\dfrac{1}{2}$;   **e)** 0, 1;

f) We have $\tan(\arcsin x + \arctan x) = 2x \Rightarrow \dfrac{\dfrac{x}{\sqrt{1 - x^2}} + x}{1 - \dfrac{x}{\sqrt{1 - x^2}} \cdot x} = 2x \Rightarrow x = 0$.

VI. 29. a) We have

$\arcsin x \cdot \left(\dfrac{\pi}{2} - \arcsin x\right) = \dfrac{\pi^2}{18} \Rightarrow \left(\arcsin x - \dfrac{\pi}{6}\right)\left(\arcsin x - \dfrac{\pi}{3}\right) = 0$.

Therefore $x_1 = \dfrac{1}{2}$ and $x_2 = \dfrac{\sqrt{3}}{2}$;   **b)** $x = \pm\dfrac{\sqrt{3}}{2}$;

c) We have $\arctan x \cdot \left(\dfrac{\pi}{2} - \arctan x\right) = \dfrac{\pi^2}{16} \Rightarrow \left(\arctan x - \dfrac{\pi}{4}\right)^2 = 0$.

Therefore $x = 1$;

d) We have $(\arcsin x)^2 + \left(\dfrac{\pi}{2} - \arcsin x\right)^2 = (\arctan x)^2 + \left(\dfrac{\pi}{2} - \arctan x\right)^2 \Rightarrow$

$2(\arcsin x)^2 - \pi \arcsin x + \left(\dfrac{\pi}{2}\right)^2 = 2(\arctan x)^2 - \pi \arctan x + \left(\dfrac{\pi}{2}\right)^2 \Rightarrow$

$2(\arcsin x - \arctan x)(\arcsin x + \arctan x - \pi) = 0 \Rightarrow$.

$$2(\arcsin x - \arctan x)\left(\arcsin x - \frac{\pi}{2} + \arctan x - \frac{\pi}{2}\right) = 0 \Rightarrow$$

$$2(\arcsin x - \arctan x)(\arccos x + \text{arccot}) = 0.$$

Since $x \in [-1, 1]$, $\arccos x \in [0, \pi]$, and $\text{arc cot} \in (0, \pi)$, then

$\arccos x + \text{arccot} > 0$. Therefore $\arcsin x = \arctan x \Rightarrow$

$\tan(\arcsin x) = \tan(\arctan x) \Rightarrow \dfrac{x}{\sqrt{1 - x^2}} = x$. The only solution for this

equation is $x = 0$.

VI. 30. We have $\tan(\arctan x_1 + \arctan x_2) = \tan(\arctan x_3) \Rightarrow \dfrac{x_1 + x_2}{1 - x_1 x_2} = x_3$.

Using Vieta's formulae $\begin{cases} x_1 + x_2 + x_3 = 0 \\ x_1 x_2 + x_1 x_3 + x_2 x_3 = -a \\ x_1 x_2 x_3 = -b \end{cases}$, we obtain:

$x_1 = \sqrt{a}$, $x_2 = -\sqrt{a}$, $x_3 = 0$ or $x_{1,2} = \dfrac{a \pm \sqrt{a^2 - 8}}{2}$, $x_3 = -\dfrac{b}{2}$.

UNIT VII

Trigonometric Ratios in a Triangle

VII. 1. a) $165°42' = 9942'$ (1) and $27°37' = 1657'$ (2). Dividing (1) by (2), we obtain 6.
b) Let x and y be the angles. Then, $x - y = 13°14'18"$ and $x + y = 90°$. Finally, $x = 51°37'8"$ and $y = 38°22'52"$.
c) Let x and y be the angles. Then, $x - y = 63°24'12"$ and $x + y = 180°$. Finally, $x = 121°42'6"$ and $y = 58°29'48"$.
d) Two opposite angles are equal. $A = 60°$ and $B = 120°$.
e) $x = 37°30'$ and $y = 52°30'$.

VII. 2. The triangle does exist, $B = 26°$.

VII. 3. a) one: m(C) = 102.8°, m(B) = 33.2°;
b) two: m(A) = 6°, m(C) = 147° or m(A) = 120°, m(C) = 33° ;
c) one: m(B) = 90°, m(C) = 60°;
d) two: m(P) = 102.8°, m(Q) = 39.9° or m(P) = 2.6°, m(Q) = 140.1° ;
e) No triangle is possible.

VII. 4. $3\sqrt{3}$.   **VII. 5.** $S_{\triangle ABC} = 6.2$ cm².   **VII. 6.** $2.4\sqrt{3}$ m.   **VII. 7.** 24°.

VII. 8. $2A = 3B = 6C \Rightarrow A = \dfrac{3B}{2}$ and $C = \dfrac{B}{2}$. Since $A + B + C = 180°$, then $A = 90°$, $C = 30°$, $a = 20\sqrt{3}$.   **VII. 9.** $\dfrac{3\sqrt{2} + \sqrt{3}}{6}$.   **VII. 10.** $\dfrac{2\sqrt{2}}{\sqrt{41 - 20\sqrt{2}}}$.

VII. 11. $\sin A = \sin(B+C) = \dfrac{1}{3} \cdot \dfrac{\sqrt{80}}{9} + \dfrac{\sqrt{2}}{3} \cdot \dfrac{1}{9} = \dfrac{4\sqrt{5}+2\sqrt{2}}{27}$, $AB = \dfrac{24}{4\sqrt{5}+2\sqrt{2}}$.

VII. a) $A = \dfrac{\pi}{3}$, $b = 4\sqrt{6}$, $c = 4\sqrt{6}\left(\sqrt{3}+1\right)$;　**b)** $A = \dfrac{\pi}{4}$, $B = \dfrac{\pi}{6}$, $C = \dfrac{7\pi}{12}$;

c) $a = 8\left(1+\sqrt{3}\right)$ cm, $B = \dfrac{\pi}{3}$, $b = 8\sqrt{6}$;　**d)** $c = \dfrac{4\sqrt{6}}{3}$, $a = \dfrac{2\left(3\sqrt{2}+\sqrt{6}\right)}{3}$;

e) $a = \sqrt{7}$, $\sin B = \dfrac{\sqrt{3}}{\sqrt{7}}$, $\sin C = \dfrac{3\sqrt{3}}{2\sqrt{7}}$.

VII. 13. a) $B = \arcsin\dfrac{24}{25} = 73.74°$, and $A = 76.26°$.

$$\sin A = \sin(B+C) = \sin\left(\arcsin\dfrac{24}{25}\right)\cos 30° + \sin 30° \cos\left(\arcsin\dfrac{24}{25}\right) =$$

$$= \dfrac{24}{25}\dfrac{\sqrt{3}}{2} + \dfrac{1}{2}\dfrac{7}{25} = \dfrac{24\sqrt{3}+7}{50}.$$ From $\dfrac{a}{\sin A} = \dfrac{b}{\sin B} = \dfrac{c}{\sin C}$ we obtain

$$b = \dfrac{36\left(24\sqrt{3}+7\right)}{125}, \quad c = \dfrac{13\sqrt{3}\left(24\sqrt{3}+7\right)}{100};$$ **b)** $C = 67.38°$,

$$\sin B = \sin(A+C) = \sin 45° \cos\left(\arcsin\dfrac{12}{13}\right) + \cos 45° \sin\left(\arcsin\dfrac{12}{13}\right) =$$

$$= \dfrac{\sqrt{2}}{2}\dfrac{5}{13} + \dfrac{\sqrt{2}}{2}\dfrac{12}{13} = \dfrac{17\sqrt{2}}{26}.$$ From $\dfrac{a}{\sin A} = \dfrac{b}{\sin B} = \dfrac{c}{\sin C}$ we obtain

$a = 13\sqrt{2}$, $c = 24$;

c) $\cos A = -\dfrac{1}{2}$, then $A = 120°$, $\cos B = \dfrac{176}{216}$, $\cos C = \dfrac{112}{168}$;

d) $\cos A = \dfrac{1}{2}$, then $A = 60°$, $\sin B = \dfrac{1}{2\sqrt{6}}$, $A_{\Delta ABC} = \dfrac{\sqrt{3}+1}{16}$;

e) $A = \dfrac{\pi}{10}$. We have $\sin 2A = \sin\dfrac{2\pi}{10} = \cos\left(\dfrac{\pi}{2} - \dfrac{2\pi}{10}\right) = \cos\dfrac{3\pi}{10} = \cos 3A$ or

$$2\sin A \cos A = 4\cos A\left(4\cos^2 A - 3\right) \Rightarrow 4\sin^2 A + 2\sin A - 1 = 0$$

$$\sin A = \dfrac{-1\pm\sqrt{5}}{4}.$$ Since $\sin A > 0$, then $\sin A = \dfrac{-1+\sqrt{5}}{4}$;

f) $\cos B = \pm\dfrac{1}{\sqrt{7}}$, $b = 4$, $A = 60°$.

VII. 14. From Law of Cosines $a^2 = (b+c)^2 - 2bc - 2bc\cos A$, we obtain

$bc = 2 + 2\sqrt{3}$. Therefore $S_{\Delta ABC} = \dfrac{bc\sin A}{2} = \dfrac{3+\sqrt{3}}{2}$. **VII. 15.** $\dfrac{\sqrt{2}}{6}$.

VII. 16. $\dfrac{7}{5}$.   **VII. 17.** Since $\cos\dfrac{\pi}{12} = \dfrac{\sqrt{2}+\sqrt{6}}{4}$ and $a = 2b\cos\dfrac{\pi}{12}$, we have

$2b\cos\dfrac{\pi}{12} + b + b = 8 + 2\sqrt{2} + 2\sqrt{6} \Rightarrow b\left(1 + \cos\dfrac{\pi}{12}\right) = 4 + \sqrt{2} + \sqrt{6}$. Therefore

$b = 2$ and $a = \sqrt{2} + \sqrt{6}$, $S = 1$.

VII. 18. I) ΔABC is acute:

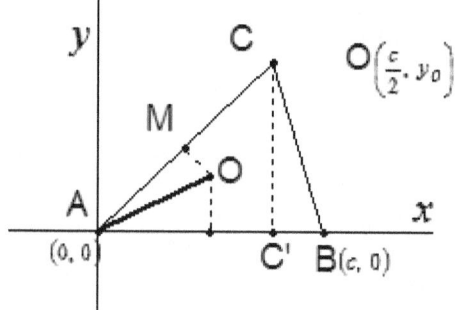

In the right $\Delta ACC'$:

$CC' = b\sin A$,

$C'A = b\cos A$,

the coordinates of C are

$(b\cos A,\ b\sin A)$ and for M the midpoint of CA are

$\left(\dfrac{b\cos A}{2},\ \dfrac{b\sin A}{2}\right)$.

The center of the

circumscribed circle of ΔABC is the point O, the intersection of

the perpendicular bisectors of the sides CA and AB. The slope of

the line MO is $m_{MO} = -\dfrac{1}{m_{CA}} = -\cot A$, then the equation of MO is

$y - y_M = -\cot A \cdot (x - x_M)$, since it passes through the point O, we obtain

$y_O - \dfrac{b\sin A}{2} = -\cot A \cdot \left(\dfrac{c}{2} - \dfrac{b\cos A}{2}\right)$.

Therefore $R = AO = \sqrt{\left(\dfrac{c}{2}\right)^2 + \left(\dfrac{b}{2\sin A} - \dfrac{c}{2}\cot A\right)^2} =$

$$= \sqrt{c^2\left(1+\frac{\cos^2 A}{\sin^2 A}\right)+\frac{1}{\sin^2 A}\left(b^2-2b\,\cos A\right)}=\frac{a}{2\sin A}\,.$$

II) $\triangle ABC$ is obtuse (*a similar proof*):

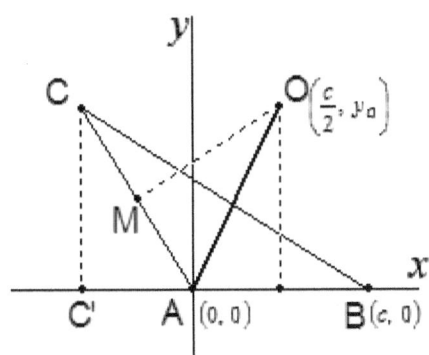

In the right $\triangle ACC'$ (

$\angle CAC'=180°-A$):

$CC'=b\sin A$, $C'A=-b\cos A$,

the coordinates of C are

$(-b\cos A,\ b\sin A)$ and for

M, the midpoint of CA are

$\left(-\dfrac{b\cos A}{2},\dfrac{b\sin A}{2}\right)$. The center

of the circumscribed circle of

$\triangle ABC$ is the point O, the intersection of the perpendicular bisectors of

the sides CA and AB. The slope of the line MO is $m_{MO}=-\dfrac{1}{m_{CA}}=\cot A$,

then the equation of MO is $y-y_M=\cot A\cdot(x-x_M)$, since it passes

through the point O, we obtain $y_O-\dfrac{b\sin A}{2}=\cot A\cdot\left(\dfrac{c}{2}+\dfrac{b\cos A}{2}\right)$.

Therefore $R=AO=\sqrt{\left(\dfrac{c}{2}\right)^2+\left(\dfrac{b}{2\sin A}+\dfrac{c}{2}\cot A\right)^2}=$

$$= \sqrt{c^2\left(1+\frac{\cos^2 A}{\sin^2 A}\right)+\frac{1}{\sin^2 A}\left(b^2+2b\,\cos A\right)}=\frac{a}{2\sin A}\,.$$

VII. 19.

$$BC^2=a^2+\frac{8}{9}a^2-2a\frac{2\sqrt2}{3}a\cos 45°=\frac{a\sqrt5}{3},$$

$$\cos B=\frac{a^2+\left(\dfrac{a\sqrt5}{3}\right)^2-\left(\dfrac{2\sqrt2}{3}a\right)^2}{2(a)\left(\dfrac{a\sqrt5}{3}\right)}=\frac{1}{\sqrt5},$$

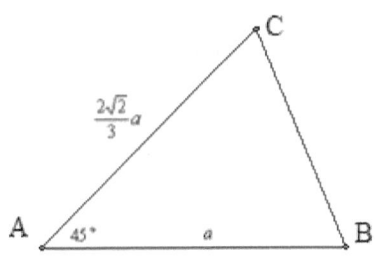

255

then $\sin B = \dfrac{2\sqrt{5}}{5}$. Therefore $\tan B = 2$.

VII. 20. Using *Law of Tangents* $\tan\dfrac{B-C}{2}\tan\dfrac{A}{2} = \dfrac{b-c}{b+c}$ we obtain

$\tan\dfrac{B-C}{2}\tan 30° = \dfrac{1-\dfrac{c}{b}}{1+\dfrac{c}{b}}$. Thus $\tan\dfrac{B-C}{2} = 1$ which yields $B - C = 90°$.

Since $B + C = 120°$, then $B = 105°$ and $C = 15°$.

VII. 21. Use Law of Sines and Cosines.

VII. 22. Use Law of Sines.

VII. 23. From the given relation we obtain $b^2 + c^2 - a^2 = bc$. Using Law of cosine $\cos A = \dfrac{1}{2}$.

VII. 24. a) $BA = 7.5$, $B = 60°$, $AC = \dfrac{15\sqrt{3}}{2}$;

b) $CB = 16\sqrt{3}$, $C = 30°$, $AC = 24$;

c) $BA = 15$, $B = 28°$;   d) $CB = 13$, $B = 22.6°$;

e) $c = 10$, $C = \dfrac{\pi}{6}$, $b = 15\sqrt{3}$;   f) $b = 5$, $B = 22.6°$;

g) $B = \dfrac{\pi}{3}$, $a = \sqrt{3}$;   h) $a = 13$;   i) $a = 6$, $B = \dfrac{\pi}{3}$;

j) $C = 75°$, $\sin 15° = \dfrac{\sqrt{6}-\sqrt{2}}{4}$, $b = \sqrt{6} - \sqrt{2}$.

Also $\sin 75° = \dfrac{\sqrt{6}+\sqrt{2}}{4}$, $c = \sqrt{6} + \sqrt{2}$;   k) $a = 6$, $b = 2\sqrt{\dfrac{7}{3}}$.

VII. 25. $\sin A = -2$ (impossible), $\sin A = 1 \Rightarrow A = 90°$.

VII. 26. $c^2 = a^2 - b^2 = (a-b)(a+b) = (a-b)3c$, then $c = 3(a-b)$.

Since $a + b = 3c$ we obtain $a = \dfrac{5c}{3}$ and $b = \dfrac{4c}{3}$. Therefore

$$\sin 2B = 2\sin B \cos B = 2\frac{b}{a}\frac{c}{a} = \frac{24}{25} \text{ and } \frac{2\tan\dfrac{C}{2}}{1-\tan^2\dfrac{C}{2}} = \tan C = \frac{c}{b} = \frac{3}{4},$$

which yields $\tan\dfrac{C}{2} = \dfrac{1}{3}$.

VII. 27. $2R\sin B\cos B + 2R\sin C\cos C = 4R\sin A\sin B\sin C \Rightarrow$

$\sin 2B + \sin 2C = 2[\cos(A-B) - \cos(A+B)]\sin C \Rightarrow$

$\sin 2B + \sin 2C = 2\cos(A-B)\sin C + \sin 2C \Rightarrow$

$\sin 2B = \sin(A-B+C) + \sin(A-B-C) \Rightarrow \sin(2A-\pi) = 0$. Therefore $A = \dfrac{\pi}{2}$.

VII. 28. $4R\sin A\sin B\cos B = \dfrac{4R^2\sin B\sin C}{R} \Rightarrow 2\sin A\cos B = 2\sin C$ or

$\sin(A-B) = \sin C$, then $A-B = C$.

VII. 29. $2\sin\dfrac{C}{2}\cos\dfrac{C}{2} = 2\cos\dfrac{A+B}{2}\cos\dfrac{A-B}{2} \Rightarrow$

$2\sin\dfrac{C}{2}\left(\cos\dfrac{C}{2} - \cos\dfrac{A-B}{2}\right) = 0$. Since $\sin\dfrac{C}{2} \neq 0$, then $\cos\dfrac{C}{2} - \cos\dfrac{A-B}{2} = 0$.

Finally $A = B+C$.

VII. 30. $2\cos\dfrac{B+C}{2}\cos\dfrac{B-C}{2} = 2\sin\dfrac{B+C}{2}\cos\dfrac{B-C}{2}$. Since $\cos\dfrac{B-C}{2} \neq 0$,

then $\tan\dfrac{B+C}{2} = 1$, $B+C = 90°$.

VII. 31. a) Use Law of Sines: $\dfrac{1+\cos B}{\sin B} = \dfrac{2R\sin A + 2R\sin C}{2R\sin B}$,

$\dfrac{2\cos^2\dfrac{B}{2}}{\sin B} = \dfrac{\sin\dfrac{A+C}{2}\cos\dfrac{A-C}{2}}{\sin B}$. Since $\sin\dfrac{A+C}{2} = \cos\dfrac{B}{2} \neq 0$, we obtain

$\cos\dfrac{B}{2} = \cos\dfrac{A-C}{2}$ which yields $A = C+B$.

b) We have $\dfrac{1+\sin 2B}{\cos 2B} = \dfrac{\sin C + \sin B}{\sin C - \sin B} \Rightarrow \dfrac{(\sin B + \cos B)^2}{\cos^2 B - \sin^2 B} = \dfrac{\sin C + \sin B}{\sin C - \sin B} \Rightarrow$

$\dfrac{\sin B + \cos B}{\cos B - \sin B} = \dfrac{\sin C + \sin B}{\sin C - \sin B} \Rightarrow \cos B = \sin C \Rightarrow \cos B = \cos(90° - C)$, which yields $B + C = 90°$.

VII. 32.

$\dfrac{1 + \cos 2A}{2} + \dfrac{1 + \cos 2B}{2} + \dfrac{1 + \cos 2C}{2} = 1 \Rightarrow \cos 2A + \cos 2B + \cos 2C + 1 = 0 \Rightarrow$

$2\cos(A + B)\cos(A - B) + 2\cos^2 C = 0 \Rightarrow 2\cos C[\cos(A - B) - \cos C] = 0$.

Therefore $C = 90°$ or $A = C + B$.

VII. 33.
$1 - \cos 2A + 1 - \cos 2B + 1 - \cos 2C = 4$ or

$1 + \cos 2A + \cos 2B + \cos 2C = 0$ or

$2\cos^2 A + 2\cos(B + C)\cos(B - C) = 0$ or

$2\cos A[\cos A - \cos(B - C)] = 0$. It yields $\cos A = 0 \Rightarrow A = 90°$ or

$\cos A = \cos(B - C) \Rightarrow A = B - C$. $A + C = B$ that yields the triangle is right.

VII. 34.

$\dfrac{\sin B + \sin C}{\cos B + \cos C} = \dfrac{2\sin\dfrac{B + C}{2}\cos\dfrac{B - C}{2}}{2\cos\dfrac{B + C}{2}\cos\dfrac{B - C}{2}} = \tan\dfrac{B + C}{2} = \tan\left(90° - \dfrac{A}{2}\right) = \cot\dfrac{A}{2}$.

Then $2\sin\dfrac{A}{2}\cos\dfrac{A}{2} = \cot\dfrac{A}{2} \Rightarrow 2\sin^2\dfrac{A}{2} = 1 \Rightarrow \cos A = 0$. Finally $A = 90°$.

VII. 35. a) $\sin B + \tan B\sin C = \sin B + \cos C \Rightarrow \tan B\sin C = \cos C \Rightarrow$
$\sin B\sin C = \cos B\cos C \Rightarrow \cos(B + C) = 0$. Therefore $B + C = 90°$.

VII. 36. $\dfrac{1 + \tan B}{1 - \tan B} = \dfrac{1 + \tan C}{-1 + \tan C} \Rightarrow \tan B = \cot C \Rightarrow \tan B = \tan\left(\dfrac{\pi}{2} - C\right)$.

Therefore $B + C = 90°$.

VII. 37. a)

In $\triangle ABD (AD \perp BD)$: $\cos B = \dfrac{BD}{c}$,

In $\triangle ACD (AD \perp CD)$: $\cos C = \dfrac{CD}{b}$.

Therefore $a = CD + DB = b\cos C + c\cos B$.

b) From $\dfrac{a}{\sin A}=\dfrac{b}{\sin B}=\dfrac{c}{\sin C}=2R$, $b=2R\sin B$ and $c=2R\sin C$. Therefore

$b\cos B+c\cos C=2R\sin B\cos B+2R\sin C\cos C=R(\sin 2B+\sin 2C)=$

$=\dfrac{a}{2\sin A}(\sin 2B+\sin 2C)=\dfrac{a}{\sin A}\sin(B+C)\cos(B-C)=a\cos(B-C).$

VII. 38. b) $2a^2=8R^2\Rightarrow a=2R$

c) We have $a^2+2bc\cos A=b^2+c^2$ (Law of Cosines). Using Law of Sines, we obtain $\sin^2 B+\sin^2 C=\sin^2 A+2\cos A\sin B\sin C$. Since $\sin^2 B+\sin^2 C=1+2\cos A\sin B\sin C$, then $\sin^2 A=1\Rightarrow A=90^\circ.$

VII. 39. From Law of Sines $a=2R\sin A$, $b=2R\sin B$, and $c=2R\sin C$,

therefore $\dfrac{a+c}{b}=\dfrac{2R\sin A+2R\sin C}{2R\sin B}=\dfrac{2\sin\dfrac{A+C}{2}\cos\dfrac{A-C}{2}}{\sin B}=$

$=\dfrac{2\sin\dfrac{\pi-B}{2}\cos\dfrac{A-C}{2}}{2\sin\dfrac{B}{2}\cos\dfrac{B}{2}}=\dfrac{\cos\dfrac{A-C}{2}}{\sin\dfrac{B}{2}}.$ Since $\dfrac{\cos\dfrac{A-C}{2}}{\sin\dfrac{B}{2}}=\cot\dfrac{B}{2}$, we obtain

$\cos\dfrac{A-C}{2}=\cos\dfrac{B}{2}.$ Thus $\dfrac{A-C}{2}=\pm\dfrac{B}{2}.$ Finally $A=90^\circ$ or $C=90^\circ.$

VII. 40. $\tan 2C=\dfrac{2\tan C}{1-\tan^2 C}=\dfrac{2\dfrac{c}{b}}{1-\dfrac{c^2}{b^2}}=\dfrac{2bc}{b^2-c^2}.$

VII. 41. $(1+\cos B)(1+\cos C)=\dfrac{a+b}{a}\dfrac{a+c}{a}=\dfrac{a^2+ab+bc+ca}{a^2}=$

$=\dfrac{a^2+b^2+c^2+2ab+2bc+2ca}{2a^2}=\dfrac{2p^2}{a^2}.$

VII. 42. $\sqrt{\dfrac{p(p-a)}{b}}+\sqrt{\dfrac{(p-b)(p-c)}{b}}=\sqrt{2}\Rightarrow\cos\dfrac{A}{2}+\sin\dfrac{A}{2}=\sqrt{2}\Rightarrow$

$\cos\dfrac{A}{2}+\cos\left(\dfrac{\pi}{2}-\dfrac{A}{2}\right)=\sqrt{2}\Rightarrow 2\cos\dfrac{\pi}{2}\cos\left(\dfrac{\pi}{4}-\dfrac{A}{2}\right)=\sqrt{2}\Rightarrow$

$\cos\left(\dfrac{\pi}{4}-\dfrac{A}{2}\right)=1$. Therefore $A=\dfrac{\pi}{2}$.

VII. 43. $\tan\dfrac{x}{2}\left(\tan\dfrac{y}{2}+\tan\dfrac{z}{2}\right)+\tan\dfrac{y}{2}\tan\dfrac{z}{2}-1=0$ or

$$\tan\dfrac{x}{2}\left(\dfrac{\sin\dfrac{y+z}{2}}{\cos\dfrac{y}{2}\cos\dfrac{z}{2}}\right)-\dfrac{\sin\dfrac{y+z}{2}}{\cos\dfrac{y}{2}\cos\dfrac{z}{2}}=0.$$ Since $x+y+z=180°$ then

$\cos\dfrac{y}{2}\cos\dfrac{z}{2}\neq 0$ $\sin\dfrac{y+z}{2}\neq 0\Rightarrow\tan\dfrac{x}{2}-1=0$. Finally $x=90°$.

VII. 44. Use $r=(p-a)\tan\dfrac{A}{2}$ and $r_a=p\tan\dfrac{A}{2}$. $A=90°$.

VII. 45. Using Law of Cosines

$$\dfrac{a+b+c}{2}=\dfrac{b^2+c^2-a^2}{2c}+\dfrac{a^2+c^2-b^2}{2a}+\dfrac{a^2+b^2-c^2}{2b}\Rightarrow$$
$$(b-a)(c-b)(c-a)(a+b+c)=0.$$

VII. 47. $2\cos\dfrac{A+B}{2}\cos\dfrac{A-B}{2}+1-2\sin^2\dfrac{C}{2}=\dfrac{3}{2}\Rightarrow$

$$4\cos\dfrac{A+B}{2}\cos\dfrac{A-B}{2}-4\cos^2\dfrac{A+B}{2}=1\Rightarrow$$

$$4\cos\dfrac{A+B}{2}\cos\dfrac{A-B}{2}-4\cos^2\dfrac{A+B}{2}=\cos^2\dfrac{A-B}{2}+\sin^2\dfrac{A-B}{2}\Rightarrow$$

$$\left(2\cos\dfrac{A+B}{2}-\cos\dfrac{A-B}{2}\right)^2+\sin^2\dfrac{A-B}{2}=0.$$ Finally

$$2\cos\dfrac{A+B}{2}-\cos\dfrac{A-B}{2}=0 \text{ and } \sin\dfrac{A-B}{2}=0.$$

VII. 48. $\sqrt{a^2+b^2+c^2}=\dfrac{3abc}{4S}\Rightarrow\sqrt{a^2+b^2+c^2}=3R\Rightarrow$

$a^2+b^2+c^2=9R^2$. Now, using Law of Sines

$4R^2\sin^2 A+4R^2\sin^2 B+4R^2\sin^2 C=9R^2\Rightarrow$

$$\sin^2 A + \sin^2 B + \sin^2 C = \frac{9}{4} \Rightarrow A = B = C.$$

VII. 49. a) $4\cos A[\cos(B+C)+\cos(B-C)]-1=0 \Rightarrow$

$4\cos A[-\cos A + \cos(B-C)]-1=0 \Rightarrow 4\cos^2 A - 4\cos A\cos(B-C)+1=0 \Rightarrow$

$4\cos^2 A - 4\cos A\cos(B-C)+\cos^2(B-C)+\sin^2(B-C)=0.$

Then $2\cos A - \cos(B-C)=0$ and $\sin(B-C)=0$. This yields $B = C$ and

$\cos A = \dfrac{1}{2}$.

VII. 50. From the first condition, we obtain $b^2 + c^2 - bc = a^2$. Using

Law of Cosine, then $\cos A = \dfrac{1}{2}$. Therefore $A = 60°$. From the second

relation we obtain $\cos(B-C)-\cos(B+C)=\dfrac{3}{2}$, thus $\cos(B-C)=1$ or, $B =$

C, since $B + C = 120°$ it follows that the triangle is equilateral.

VII. 51. We have

$$\frac{\sqrt{(p-b)(p-c)}}{\sqrt{p(p-a)}}+\frac{\sqrt{(p-c)(p-a)}}{\sqrt{p(p-b)}}+\frac{\sqrt{(p-a)(p-b)}}{\sqrt{p(p-c)}}=\frac{1}{4S}\left(a^2+b^2+c^2\right).$$

Using Heron's formula, we obtain

$$\frac{1}{S}[(p-b)(p-c)+(p-c)(p-a)+(p-a)(p-b)]=\frac{1}{4S}\left(a^2+b^2+c^2\right)\Rightarrow$$

$$-p^2+ab+bc+ca=\frac{1}{4}\left(a^2+b^2+c^2\right)\Rightarrow (a-b)^2+(b-c)^2+(c-a)^2=0.$$

Then $a=b=c$.

VII. 52. a) We have $\dfrac{1}{\sqrt{r_a}\sqrt{r_b}}+\dfrac{1}{\sqrt{r_c}\sqrt{r_a}}+\dfrac{1}{\sqrt{r_b}\sqrt{r_c}}=\dfrac{1}{r}$. Using

$\dfrac{1}{r_a}+\dfrac{1}{r_b}+\dfrac{1}{r_c}=\dfrac{1}{r}$, we obtain $\dfrac{1}{\sqrt{r_a}\sqrt{r_b}}+\dfrac{1}{\sqrt{r_b}\sqrt{r_c}}+\dfrac{1}{\sqrt{r_c}\sqrt{r_a}}=\dfrac{1}{r_a}+\dfrac{1}{r_b}+\dfrac{1}{r_c}\Rightarrow$

$$\left(\frac{1}{\sqrt{r_a}}-\frac{1}{\sqrt{r_b}}\right)^2+\left(\frac{1}{\sqrt{r_b}}-\frac{1}{\sqrt{r_c}}\right)^2+\left(\frac{1}{\sqrt{r_c}}-\frac{1}{\sqrt{r_a}}\right)^2=0\,;$$

b) Squaring on both sides, we have $r_a^2+r_b^2+r_c^2+2(r_ar_b+r_br_c+r_ar_c)=3p^2$,

since $r_ar_b+r_br_c+r_ar_c=p^2$ (see the problem **VII. 80. h)**), we obtain

(1) $r_a^2 + r_b^2 + r_c^2 = p^2$. Using $\dfrac{r_a}{\tan\dfrac{A}{2}} = \dfrac{r_b}{\tan\dfrac{B}{2}} = \dfrac{r_c}{\tan\dfrac{C}{2}} = p$ (see the problem

VII. 80. d)), then (1) becomes (2) $\tan^2\dfrac{A}{2} + \tan^2\dfrac{B}{2} + \tan^2\dfrac{C}{2} = 1$.

Now, using $\tan\dfrac{A}{2}\tan\dfrac{B}{2} + \tan\dfrac{B}{2}\tan\dfrac{C}{2} + \tan\dfrac{C}{2}\tan\dfrac{A}{2} = 1$

(see the problem **VII. 71. u)**) we obtain

$$\left(\tan\dfrac{A}{2} - \tan\dfrac{B}{2}\right)^2 + \left(\tan\dfrac{B}{2} - \tan\dfrac{C}{2}\right)^2 + \left(\tan\dfrac{C}{2} - \tan\dfrac{A}{2}\right)^2 = 0.$$

Isosceles Triangle

VII. 53. $2R(\sin B + \sin C) = 4R\cos\dfrac{A}{2}$, $2\sin\dfrac{B+C}{2}\cos\dfrac{B-C}{2} = 2\cos\dfrac{A}{2}$.

Since $\cos\dfrac{A}{2} \neq 0$ in a triangle, then $\cos\dfrac{B-C}{2} = 1$, which yields $B = C$.

VII. 54. $2a = b+c \Rightarrow$

$2\sin A = \sin B + \sin C \Rightarrow 2\sin A = 2\sin\dfrac{B+C}{2}\cos\dfrac{B-C}{2} \Rightarrow$

$2\sin A = 2\sin A\cos\dfrac{B-C}{2}$. Since $\sin A \neq 0$, then $\cos\dfrac{B-C}{2} = 1$.

Therefore $B - C = 0$.

VII. 55. Using Law of Sines, we obtain $2\sin A\cos B = \sin C$ or
$\sin(A+B) + \sin(A-B) = \sin C$ or $\sin(A+B) + \sin(A-B) = \sin C$.
This yields $\sin(A-B) = 0$, then $A = B$.

VII. 56. From $2a = b+c$, using Law of Sines we obtain
$4R\sin A = 2R\sin B + 2R\sin C$ or $2\sin A = \sin B + \sin C$ or
$2\sin A = 2\sin\dfrac{B+C}{2}\cos\dfrac{B-C}{2}$ or $2\sin A = 2\sin A\cos\dfrac{B-C}{2}$. Since

$0 < \sin A < \pi$, $\cos\dfrac{B-C}{2} = 1$ or $B = C$.

VII. 57. a) $\sin A = \sin(B+C) + \sin(B-C) \Rightarrow \sin(B-C) = 0 \Rightarrow B = C$

262

Or $\dfrac{a}{2R} = 2\dfrac{b}{2R}\dfrac{a^2+b^2-c^2}{2ab} \Rightarrow b=c$;

b) $2R\sin A = 4R\sin B\cos C \Rightarrow \sin A = \sin(B+C)+\sin(B-C) \Rightarrow$
$\sin(B-C)=0 \Rightarrow B=C$;

c) $2R\sin A = 4R\sin B\sin\dfrac{A}{2} \Rightarrow 2\sin\dfrac{A}{2}\cos\dfrac{A}{2} = 2\sin B\sin\dfrac{A}{2}$. Since $\sin\dfrac{A}{2} \neq 0$,

then $\cos\dfrac{A}{2} = \sin B \Rightarrow \sin\dfrac{\pi-A}{2} = \sin B \Rightarrow \pi-A=2B$, using $A+B+C=\pi$

we obtain $B=C$; **d)** $B=A$;

e) $a = 2\dfrac{S}{P}\dfrac{\cos\dfrac{B}{2}}{\sin\dfrac{B}{2}} \Rightarrow ap\sin\dfrac{B}{2} = 2S\cos\dfrac{B}{2} \Rightarrow ap\sqrt{\dfrac{(p-a)(p-c)}{ac}} = 2S\sqrt{\dfrac{(p-b)p}{ac}}$.

Since $S = \sqrt{p(p-a)(p-b)(p-c)}$, we obtain $b=c$;

f) Using $h_a = c\sin B$, we obtain $2c\sin B = a\tan B \Rightarrow$

$4R\sin C\sin B = 2R\sin A\tan B \Rightarrow 2\sin C\cos B = \sin A$ and see b);

g) $2R\sin A\cos B = 2R\sin B\cos A \Rightarrow \sin(A-B)=0 \Rightarrow A=B$;

h) Since $r_a = p\tan\dfrac{A}{2}$, $r_c = p\tan\dfrac{C}{2}$, the given relation becomes

$\tan^2\dfrac{A}{2} + 2\tan\dfrac{A}{2}\tan\dfrac{C}{2} = 1$, and using **VII. 71. u)** we obtain $A=B$.

VII. 58. a) $\sin\dfrac{A}{2}\cos\dfrac{B}{2} - \sin\dfrac{B}{2}\cos\dfrac{A}{2} = 0$ or $\sin\left(\dfrac{A}{2}-\dfrac{B}{2}\right) = 0$ or $A=B$.

VII. 59. a) $A=B$; **b)** We have $\dfrac{\sin(A+B)\sin C}{\sin(B+C)\sin A} = \dfrac{a}{c} \Rightarrow$

$\dfrac{\sin(A+B)}{\sin(B+C)} = \dfrac{a\sin A}{c\sin C} = 1 \Rightarrow \dfrac{\sin(A+B)}{\sin(B+C)} = 1 \Rightarrow A=C$.

VII. 60. a) $(p-c)\sqrt{\dfrac{p(p-a)}{(p-b)(p-c)}} = p\sqrt{\dfrac{(p-b)(p-a)}{p(p-c)}} \Rightarrow b=c$;

b) $(b-c)\sqrt{\dfrac{p(p-a)}{(p-b)(p-c)}} = (a-c)\sqrt{\dfrac{p(p-b)}{(p-a)(p-c)}} \Rightarrow$

$(b-c)(p-a)=(a-c)(p-b) \Rightarrow a=b$;   c) Use Law of coSines.

VII. 61. We have $a\left(\tan A - \tan \dfrac{A+B}{2} \right) = b\left(\tan \dfrac{A+B}{2} - \tan B \right)$ or

$a\left(\dfrac{\sin \dfrac{A-B}{2}}{\cos A \cos \dfrac{A+B}{2}} \right) = b\left(\dfrac{\sin \dfrac{A-B}{2}}{\cos B \cos \dfrac{A+B}{2}} \right)$ and using Law of Sines we

obtain $\tan A \sin \dfrac{A-B}{2} = \tan B \sin \dfrac{A-B}{2}$. Finally $A = B$.

VII. 62. Using $\sin A + \sin B + \sin C = 4\cos \dfrac{A}{2} \cos \dfrac{B}{2} \cos \dfrac{C}{2}$,

$h_a = c \sin B$ and Law of Sines, after simplifications, we obtain

$2\sin \dfrac{B}{2} \sin \dfrac{C}{2} = \dfrac{\cos^2 \dfrac{A}{2}}{1+\sin \dfrac{A}{2}} \Rightarrow 2\sin \dfrac{B}{2} \sin \dfrac{C}{2} = 1 - \sin \dfrac{A}{2} \Rightarrow$

$2\sin \dfrac{B}{2} \sin \dfrac{C}{2} = 1 - \cos \dfrac{B+C}{2} \Rightarrow \cos \dfrac{B-C}{2} = 1$. Therefore the triangle is

isosceles $B = C$.

VII. 63. $\sin B - \sin C = \cos C - \cos B$, then

$2\sin\left(\dfrac{B-C}{2} \right)\cos\left(\dfrac{B+C}{2} \right) = 2\sin\left(\dfrac{B+C}{2} \right)\sin\left(\dfrac{B-C}{2} \right)$,

From $\sin\left(\dfrac{B-C}{2} \right) = 0$, we obtain $B = C$ and from it follows

$\cos\left(\dfrac{B+C}{2} \right) = \sin\left(\dfrac{B+C}{2} \right)$ or $\tan\left(\dfrac{B+C}{2} \right) = 1$ we obtain $B+C = \dfrac{\pi}{2}$.

Therefore the triangle is isosceles or right.

VII. 64. $2R \sin B\left(\dfrac{\sin \dfrac{B}{2} + \cos \dfrac{B}{2}}{\sin \dfrac{B}{2}} \right) = 2R \sin C\left(\dfrac{\sin \dfrac{C}{2} + \cos \dfrac{C}{2}}{\sin \dfrac{C}{2}} \right)$, then

$2\cos \dfrac{B}{2}\left(\sin \dfrac{B}{2} + \cos \dfrac{B}{2} \right) = 2\cos \dfrac{C}{2}\left(\sin \dfrac{C}{2} + \cos \dfrac{C}{2} \right)$,

$\sin B + \cos B = \sin C + \cos C$, see the previous problem.

VII. 65. a) $2R\sin A\cos A = 2R\sin B\cos B \Rightarrow \sin 2B = \sin 2C$ we obtain
$2B = (-1)^k 2C + k\pi,\ k \in \mathbf{Z}$. For $k = 0$ and $k = 1$ the result follows.

b) Similarly to the previous one.

c) Using Law of Sines and Cosines we, we obtain $\dfrac{b^2 - c^2}{b^2 + c^2} = \dfrac{b^2 - c^2}{a^2}$.

VII. 66. Use the Law of Cosines.

VII. 67. From Law of Sines $a = 2R\sin A$, $b = 2R\sin B$, and $c = 2R\sin C$.
Then $S = \dfrac{ab\sin C}{2} = 2R^2 \sin A \sin B \sin C$.

VII. 68. Use Law of Sines.

VII. 69. Using Law of Sines: $\dfrac{a-b}{a+b} = \dfrac{2R\sin A - 2R\sin B}{2R\sin A + 2R\sin B} = \dfrac{\sin A - \sin B}{\sin A + \sin B} =$

$$= \frac{2\sin\dfrac{A-B}{2}\cos\dfrac{A+B}{2}}{2\sin\dfrac{A+B}{2}\cos\dfrac{A-B}{2}} = \frac{\tan\dfrac{A-B}{2}}{\tan\dfrac{A+B}{2}}.$$

VII. 70. Use Law of Sines.

VII. 72. $\dfrac{\sin(A+B)}{\cos A\cos B} + \tan C = \dfrac{\sin(\pi - C)}{\cos A\cos B} + \dfrac{\sin C}{\cos C} =$

$$= \sin C\left[\frac{1}{\cos A\cos B} - \frac{1}{\cos(A+B)}\right] = \sin C\left[\frac{\cos(A+B) - \cos A\cos B}{\cos A\cos B\cos(A+B)}\right] =$$

$$\sin C\left[\frac{\cos(A+B) - \cos A\cos B}{\cos A\cos B\cos(A+B)}\right] = \tan A\tan B\tan C .$$

VII. 74. a) $\dfrac{\sin A\cos C - \sin B\cos B}{\sin A\cos B - \sin B\cos A} + \cos C = \dfrac{\sin 2A - \sin 2B}{2\sin(A-B)} + \cos C =$

$$= \frac{2\sin(A-B)\cos(A+B)}{2\sin(A-B)} + \cos C = \cos(A+B) + \cos C =$$

$$= \cos(\pi - C) + \cos C = 0 ;$$

265

b) $\dfrac{\sin(A-B)\sin C}{1+\cos(A-B)\cos C} = \dfrac{\cos(A-B-C)-\cos(A-B+C)}{2+\cos(A-B+C)+\cos(A-B-C)} =$

$= \dfrac{\cos(2A-\pi)-\cos(\pi-2B)}{2+\cos(\pi-2B)+\cos(2A-\pi)} = \dfrac{-\cos 2A+\cos 2B}{2-\cos 2B-\cos 2A} =$

$\cos 2x = 1-2\sin^2 x$

$= \dfrac{-\cos 2A+\cos 2B}{2-\cos 2B-\cos 2A} = \dfrac{\sin^2 A-\sin^2 B}{\sin^2 B+\sin^2 A} = \dfrac{a^2-b^2}{a^2+b^2} ,$

Where $a=2R\sin A$ and $b=2R\sin B$;

c) $2R\left(\sin A\cos\dfrac{B}{4}+\sin C\cos\dfrac{B}{4}\right)+2R\sin B\cos\left(A+B-\dfrac{B}{4}\right) =$

$= R\sin\left(A+\dfrac{B}{4}\right)+R\sin\left(A-\dfrac{B}{4}\right)+R\sin\left(C+\dfrac{B}{4}\right)+R\sin\left(C-\dfrac{B}{4}\right)+$

$+R\sin\left(A+2B-\dfrac{B}{4}\right)-R\sin\left(A-\dfrac{B}{4}\right) =$

$= R\sin\left(A+\dfrac{B}{4}\right)+R\sin\left(C+\dfrac{B}{4}\right)+R\sin\left(C-\dfrac{B}{4}\right)+R\sin\left(C-\dfrac{3B}{4}\right) =$

$= R\left[\sin\left(A+\dfrac{B}{4}\right)+\sin\left(C-\dfrac{3B}{4}\right)\right]+R\left[\sin\left(C+\dfrac{B}{4}\right)+\sin\left(C-\dfrac{B}{4}\right)\right] =$

$= 2R\cos\dfrac{3B}{4}\sin C+2R\sin C\cos\dfrac{B}{4} = c\left(\cos\dfrac{3B}{4}+\cos\dfrac{B}{4}\right) =$

$= 2c\cos\dfrac{B}{2}\cos\dfrac{B}{4} .$

VII. 75. *Solution* 1: From Law of Sines $b=\dfrac{a\sin B}{\sin A}$, $c=\dfrac{a\sin C}{\sin A}$ then

$b\cos B+c\cos C = \dfrac{a(\sin B\cos B+\sin C\cos C)}{\sin A} =$

$= \dfrac{a(\sin 2B+\sin 2C)}{2\sin A} = \dfrac{2a\sin(B+C)\cos(B-C)}{2\sin(B+C)} = a\cos(B-C).$

Solution 2 : $b\cos B+c\cos C =$

$= 2R(\sin B\sin B+\sin C\cos C) = R(\sin 2B+\sin 2C) =$

$= 2R\sin(B+C)\cos(B-C) = a\cos(B-C).$

VII. 76. a) $1 + \cos A \cos(B - C) = 1 + \dfrac{1}{2}\left[\cos(A + B - C) + \cos(A - B + C)\right] =$

$= 1 - \dfrac{1}{2}\left[\cos 2C + \cos 2B\right] = $ using $\cos 2x = 1 - 2\sin^2 x$

$= \sin^2 B + \sin^2 A = \dfrac{b^2 + c^2}{4R^2}$; $\quad \sin \dfrac{A}{2} = \sqrt{\dfrac{(p - b)(p - c)}{bc}}$

b) $\tan A = \dfrac{\sin A}{\cos A} = \dfrac{2 \sin \dfrac{A}{2} \cos \dfrac{A}{2}}{2 \cos^2 \dfrac{A}{2} - 1} = \dfrac{2 \sqrt{\dfrac{(p - b)(p - c)}{bc}} \sqrt{\dfrac{p(p - a)}{bc}}}{2 \dfrac{p(p - a)}{bc} - 1} =$

$= \dfrac{2S}{2p(p - a) - b} = \dfrac{4S}{b^2 + c^2 - a^2}$;

c) $\dfrac{b + c}{2c \cos \dfrac{A}{2}} = \dfrac{2R(\sin B + \sin C)}{4R \sin C \cos \dfrac{A}{2}} = \dfrac{\sin \dfrac{B + C}{2} \cos \dfrac{B - C}{2}}{\sin(B + C) \cos \dfrac{A}{2}} = \dfrac{\cos \dfrac{B - C}{2}}{\sin(B + C)} =$

$= \dfrac{\cos \dfrac{\pi - A - C - C}{2}}{\sin(B + C)} = \dfrac{\sin\left(\dfrac{A}{2} + C\right)}{\sin(A + B)}$;

d) $\cot \dfrac{A}{2} \cdot \cot \dfrac{B}{2} \cdot \cot \dfrac{C}{2} = \sqrt{\dfrac{p(p - a)}{(p - b)(p - c)}} \cdot \sqrt{\dfrac{p(p - b)}{(p - a)(p - c)}} \cdot \sqrt{\dfrac{p(p - c)}{(p - a)(p - b)}} =$

$= \sqrt{\dfrac{p^3}{(p - a)(p - b)(p - c)}} = \dfrac{p^2}{S} = \dfrac{p}{r}$, since $S = pr$.

VII. 77. Let AM be the median of the triangle ABC.

In $\triangle ABM$: $c^2 = \left(\dfrac{a}{2}\right)^2 + m_a^2 - 2\dfrac{a}{2} m_a \cos BMA$,

In $\triangle ACM$: $b^2 = \left(\dfrac{a}{2}\right)^2 + m_a^2 - 2\dfrac{a}{2} m_a \cos(180° - BMA)$. Since

$\cos(180° - BMA) = -\cos BMA$, adding (1) and (2) we get the result.

VII. 78. a) $\sin\dfrac{A}{2} = \sqrt{\dfrac{(p-b)(p-c)}{bc}}$, $\cos\dfrac{A}{2} = \sqrt{\dfrac{p(p-a)}{bc}}$, then

$\tan\dfrac{A}{2} = \sqrt{\dfrac{(p-b)(p-c)}{p(p-a)}}$. Therefore $(p-a)\tan\dfrac{A}{2} = (p-a)\sqrt{\dfrac{(p-b)(p-c)}{p(p-a)}} =$

$= \sqrt{\dfrac{p(p-a)(p-b)(p-c)}{p^2}} = \dfrac{S}{p} = r$;   **b)** $\dfrac{S}{p} = (p-a)\tan\dfrac{A}{2}$;

c) $4R\cos\dfrac{A}{2}\cos\dfrac{B}{2}\cos\dfrac{C}{2} = 4\dfrac{abc}{4S}\sqrt{\dfrac{p(p-a)}{bc}}\sqrt{\dfrac{p(p-b)}{ac}}\sqrt{\dfrac{p(p-c)}{ab}} = p$;

e) $4R\cos\dfrac{A}{2}\sin\dfrac{B}{2}\sin\dfrac{C}{2} =$

$= 4\dfrac{abc}{4S}\sqrt{\dfrac{p(p-a)}{bc}}\sqrt{\dfrac{(p-a)(p-c)}{ac}}\sqrt{\dfrac{(p-a)(p-b)}{ab}} = p-a$;

f) $m_a^2 = \dfrac{b^2+c^2}{2} - \dfrac{a^2}{4} = \dfrac{a^2+2bc\cos A}{2} - \dfrac{a^2}{4} = \dfrac{a^2+4bc\cos A}{4} =$

$= \dfrac{4R^2\sin^2 A + 4(2R\sin B)(2R\sin C)\cos A}{4} =$

$= R^2\left(\sin^2 A + 4\cos A\sin B\sin C\right)$.

g) $\left(4m_a^2 - a^2\right)\tan A = \left(2\left(b^2+c^2\right) - a^2 - a^2\right)\tan A =$

$= 2\left(b^2+c^2-a^2\right)\tan A = 2(2bc\cos A)\tan A = 8\dfrac{bc\sin A}{2} = 8S$.

VII. 79. From the bisector theorem

$\dfrac{BD}{DC} = \dfrac{AB}{AC}$ and Law of Sines in $\triangle ABD$

$\dfrac{l_a}{\sin B} = \dfrac{BD}{\sin\dfrac{A}{2}}$.

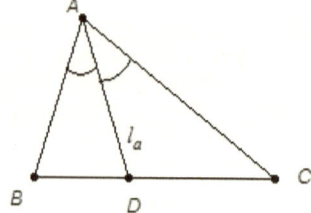

VII. 80.

a) $r_a \cdot r_b \cdot r_c = \dfrac{S^3}{(p-a)(p-b)(p-c)} = \dfrac{pS^3}{p(p-a)(p-b)(p-c)} = pS = p^2 r;$

b) $bc\sin^2 \dfrac{A}{2} = (p-b)(p-c) = \dfrac{S}{r_b} \cdot \dfrac{S}{r_c} = \dfrac{S^2 r_a}{r_a \cdot r_b \cdot r_c} = \dfrac{S^2 r_a}{p^2 r} = \dfrac{(pr)^2 r_a}{p^2 r} = r \cdot r_a;$

c) $r_b \cdot r_c - r \cdot r_a = \dfrac{S}{p-b}\dfrac{S}{p-c} - \dfrac{S}{p}\dfrac{S}{p-a} = S^2 \dfrac{p(p-a)-(p-b)(p-c)}{p(p-a)(p-b)(p-c)} =$

$p(-a+b+c)-bc = \dfrac{b^2+c^2-a^2}{2} = b\ \cos A;$

d) $\dfrac{r_a}{\tan\dfrac{A}{2}} = \dfrac{S}{p-a}\sqrt{\dfrac{p(p-a)}{(p-b)(p-c)}} = p$, since $\tan\dfrac{A}{2} = \sqrt{\dfrac{(p-b)(p-c)}{p(p-a)}}$;

e) $r_a + r_b + r_c - r = \left(\dfrac{S}{p-a}+\dfrac{S}{p-b}\right) + \left(\dfrac{S}{p-c}-\dfrac{S}{p}\right) =$

$= \dfrac{S(p-a+p-b)}{(p-a)(p-b)} + \dfrac{S(p-p+c)}{p(p-c)} = Sc\left(\dfrac{1}{(p-a)(p-b)}+\dfrac{1}{p(p-c)}\right) = \dfrac{abc}{S} = 4R;$

f) $\dfrac{1}{r_a}+\dfrac{1}{r_b}+\dfrac{1}{r_c} = \dfrac{p-a}{S}+\dfrac{p-b}{S}+\dfrac{p-c}{S} = \dfrac{p}{S} = \dfrac{1}{r};$

g) $\sqrt{r \cdot r_a \cdot r_b \cdot r_c} = \sqrt{\dfrac{S}{p}\dfrac{S}{p-a}\dfrac{S}{p-b}\dfrac{S}{p-c}} = \sqrt{\dfrac{S^4}{S^2}} = S.$

h) $r_a r_b + r_b r_c + r_a r_c = \dfrac{S}{p-a}\dfrac{S}{p-b} + \dfrac{S}{p-b}\dfrac{S}{p-c} + \dfrac{S}{p-a}\dfrac{S}{p-c} =$

$= \dfrac{S^2(p-c+p-a+p-b)}{(p-a)(p-b)(p-c)} = p^2.$

VII. 81. a) Using $S = \dfrac{ah_a}{2}$ and $S = r_a(p-a)$, we have

$\dfrac{h_a - h_b}{r_a - r_b} = \dfrac{2S(b-a)}{ab} \cdot \dfrac{(p-a)(p-b)}{S(a-b)} = -\dfrac{2(p-a)(p-b)}{ab} = -2\sin^2\dfrac{C}{2}$ and

$\dfrac{h_a h_b}{r\, r_c} = \dfrac{4S(p-c)}{rab} = \dfrac{4p(p-c)}{ab} = 4\cos^2\dfrac{C}{2}$. From the last two identities the

given identity is obvious;

b) $\dfrac{h_b + h_c}{r_a} = \dfrac{a\sin C + a\sin B}{\dfrac{S}{p-a}} = \dfrac{p-a}{S}\left(\dfrac{ab\sin C}{b} + \dfrac{ac\sin B}{c}\right) =$

$= \dfrac{p-a}{S}\left(\dfrac{2S}{b} + \dfrac{2S}{c}\right) = \dfrac{2(p-a)}{b} + \dfrac{2(p-a)}{c} = \dfrac{-a+b+c}{b} + \dfrac{-a+b+c}{c}.$

Similarly for the other two relations and finally add them;

c) $\dfrac{1}{l_a \sin\dfrac{A}{2}} = \dfrac{1}{\dfrac{2bc}{b+c}\cos\dfrac{A}{2}\sin\dfrac{A}{2}} = \dfrac{b+c}{bc\sin A} = \dfrac{b+c}{2S}$ and similarly.

Therefore $\dfrac{(b+c)^2}{bc}l_a^2 + \dfrac{(c+a)^2}{ca}l_b^2 + \dfrac{(a+b)^2}{ab}l_c^2 = \dfrac{b+c}{2S} + \dfrac{c+a}{2S} + \dfrac{a+b}{2S} =$

$= \dfrac{2(a+b+c)}{2S} = \dfrac{2p}{S};$

d) Since $\cos\dfrac{A}{2} = \sqrt{\dfrac{p(p-a)}{bc}}$, we have $\dfrac{(b+c)^2}{bc}l_a^2 = \dfrac{(b+c)^2}{bc}\left(\dfrac{2b}{b+c}\cos\dfrac{A}{2}\right)^2 =$

$= 4p(p-a)$ and so on.

VII. 82. $S_n = \dfrac{n}{2}R^2 \cdot \sin\dfrac{2\pi}{n}$ is the formula for the area of a regular polygon.

$n = 3 \Rightarrow S_3 = \dfrac{3}{2}R^2 \cdot \sin\dfrac{2\pi}{3} = \dfrac{3R^2\sqrt{3}}{4};$

$n = 4 \Rightarrow S_4 = \dfrac{4}{2}R^2 \cdot \sin\dfrac{2\pi}{4} = 2R^2;$

$n = 6 \Rightarrow S_6 = \dfrac{6}{2}R^2 \cdot \sin\dfrac{2\pi}{6} = \dfrac{3R^2\sqrt{3}}{2};$

$n = 8 \Rightarrow S_3 = \dfrac{8}{2}R^2 \cdot \sin\dfrac{2\pi}{8} = 2R^2\sqrt{2};$

$n = 12 \Rightarrow S_{12} = \dfrac{12}{2}R^2 \cdot \sin\dfrac{2\pi}{12} = 3R^2;$

$n = 20 \Rightarrow S_{20} = \dfrac{20}{2}R^2 \cdot \sin\dfrac{2\pi}{20} = 10R^2\dfrac{\sqrt{5}-1}{4} = \dfrac{5}{2}\left(\sqrt{5}-1\right)R^2.$

VII. 83.

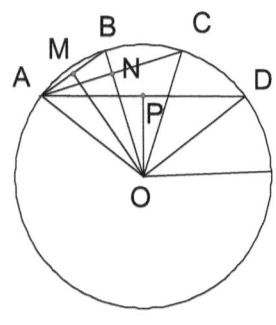

Let M, N, P be the midpoints of AB, AC, AD respectively.

Denote
$arcAB = 2\alpha \Rightarrow \angle AOB = 2\alpha$.
In $\triangle BOM\,(\angle OMB = 90°)$,

$$\sin\alpha = \frac{BM}{BO} \Rightarrow BM = R\cdot\sin\alpha,\text{ then}$$

$AB = 2R\cdot\sin\alpha$ (1).

In $\triangle NOC$, $\sin 2\alpha = \dfrac{NC}{OC} \Rightarrow NC = R\cdot\sin 2\alpha \Rightarrow AC = 2R\sin 2\alpha$ (2).

In $\triangle POD$, $\sin 3\alpha = \dfrac{DP}{OD} \Rightarrow DP = 2R\cdot\sin 3\alpha$ (3).

Replacing (1), (2), and (3) in the given relation, we get

$$\frac{1}{2R\cdot\sin\alpha} = \frac{1}{2R\cdot\sin 2\alpha} + \frac{1}{2R\cdot\sin 3\alpha} \Rightarrow \frac{1}{\sin\alpha} = \frac{1}{\sin 2\alpha} + \frac{1}{\sin 3\alpha} \Rightarrow$$

$$\frac{1}{\sin 2\alpha} = \frac{\sin 3\alpha - \sin\alpha}{\sin\alpha\cdot\sin 3\alpha} = \frac{2\sin\alpha\cdot\cos 2\alpha}{\sin\alpha\cdot\sin 3\alpha} \Rightarrow 2\sin 2\alpha\cdot\cos 2\alpha = \sin 3\alpha$$

$\Rightarrow \sin 4\alpha = \sin 3\alpha \Rightarrow 4\alpha = \pi - 3\alpha$. Then $\alpha = \dfrac{\pi}{7} \Rightarrow arcAB = \dfrac{2\pi}{7}$.

$n = 2\pi \div \dfrac{2\pi}{7} = 7$. Therefore the polygon has 7 sides. $S_7 = \dfrac{7}{2}R^2\sin\dfrac{2\pi}{7}$.

VII. 84.

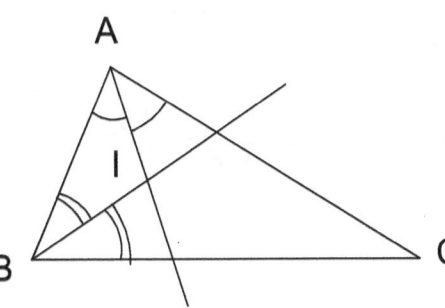

Applying Law of Sines in $\triangle ABI$, we get

$$\frac{AI}{\sin\dfrac{B}{2}} = \frac{BI}{\sin\dfrac{A}{2}} = \frac{AB}{\sin\angle BIA}\ (1).$$

$$\angle BIA = 180° - \frac{A+B}{2} = 90° + \frac{C}{2}.$$

$\sin \angle BIA = \sin\left(90° + \dfrac{C}{2}\right) = \cos\dfrac{C}{2}$. Applying Law of Sines

in $\triangle ABC$, we obtain $\dfrac{AB}{\sin C} = 2R$. Therefore from (1)

$$AI = AB \cdot \dfrac{\sin \dfrac{B}{2}}{\cos \dfrac{C}{2}} = 2R \cdot \sin C \cdot \dfrac{\sin \dfrac{B}{2}}{\cos \dfrac{C}{2}} = 4R \cdot \sin\dfrac{C}{2} \cdot \sin\dfrac{B}{2}.$$

VII. 85.

Obviously $IA' = IB' = IC' = r$ and
$IA' \perp BC$,
$IB' \perp AC$, and $IC' \perp AB$.

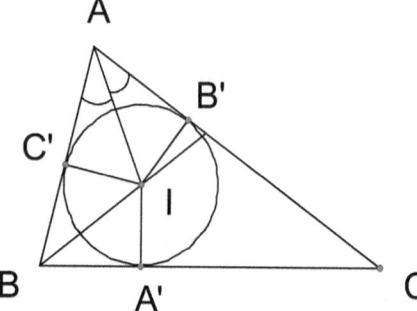

Therefore $AC'IB'$ is an inscribed
quadrilateral and
$\sin(C'IB') = \sin(\pi - A) = \sin A \cdot$

Then

$$A_{C'IB'} = \dfrac{IC' \cdot IB' \sin(C'IB')}{2} = \dfrac{r^2 \sin A}{2} = \dfrac{r^2 \dfrac{a}{2R}}{2} = \dfrac{r^2 a}{4R}.$$ Similarly for the

triangles $C'IB$ and $A'IB'$. Thus $\dfrac{S_{\triangle A'B'C'}}{S_{\triangle ABC}} = \dfrac{\dfrac{r^2 a}{4R}(a+b+c)}{rp} = \dfrac{r}{2R}.$

VII. 86. Since $0 < \dfrac{A}{2} < \dfrac{\pi}{2}$, it follows that $\sin\dfrac{A}{2} > 0$.

$$2\sin^2\dfrac{A}{2} = 1 - \cos A = 1 - \dfrac{b^2 + c^2 - a^2}{2bc} = \dfrac{a^2 - (b-c)^2}{2bc} \le \dfrac{a^2}{2bc}.$$

VII. 87. Since $r = 4R\sin\dfrac{A}{2}\sin\dfrac{B}{2}\sin\dfrac{C}{2}$ we obtain $\dfrac{r}{R} = 4\sin\dfrac{A}{2}\sin\dfrac{B}{2}\sin\dfrac{C}{2}.$

Then $\dfrac{1}{8} = 2\sin\dfrac{A}{2}\left[\cos\left(\dfrac{B}{2} - \dfrac{C}{2}\right) - \cos\left(\dfrac{B}{2} + \dfrac{C}{2}\right)\right]$ or

$\frac{1}{8} = 2\sin\frac{A}{2}\left[\cos\frac{\pi}{3} - \cos\left(\frac{\pi - A}{2}\right)\right]$ or $\frac{1}{8} = 2\sin\frac{A}{2}\left(\frac{1}{2} - \sin\frac{A}{2}\right)$. Denoting

$\sin\frac{A}{2} = t$, then the quadratic equation $16t^2 - 8t + 1 = 0$ has the root $t = \frac{1}{4}$.

Thus $A = 2\arcsin\frac{1}{4} = 28.955°$.

Orthocenter of a Triangle

VII.

$AA' = h_a$

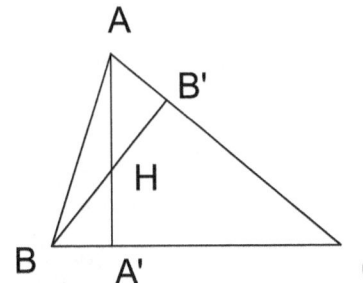

a) In $\triangle ABA'$:

$\sin B = \frac{h_a}{c}$, thus

$h_a = 2R\sin B \sin C$

a) a) In $\triangle ABB'$: $AB' = c \cos A$.

In $\triangle AHB'$: $\cos HAB' = \frac{AB'}{AH}$, thus

$AH = \frac{AB}{\cos\left(90° - C\right)} = \frac{c\cos A}{\sin C} = \frac{2R\sin C \cos A}{\sin C} = 2R\cos A$

c) $aAH + bBH + cCH = 4R^2\left(\sin A \cos A + \sin B \cos B + \sin C \cos C\right) =$

$= 2R^2\left(\sin 2A + \sin 2B + \sin 2C\right) =$.

Using $\sin 2A + \sin 2B + \sin 2C = 4\sin A \sin B \sin C$ (see the problem **VII. 71.**

g)) $= 8R^2\left(\sin A \sin B \sin C\right) = 2\left(2R\sin A\right)\left(2R\sin B \sin C\right) = 2ah_a = 4S$.

VII. 89.

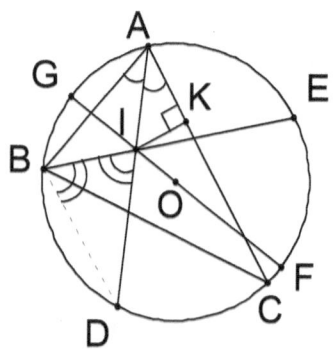

$AO = BO = CO = R$, $IK = r$.

Using the power of a the point I with respect to the circle of radius R and center O, we obtain

$IG \cdot IF = IA \cdot ID$ (1), on the other hand

$IG \cdot IF = (R - IO)(R + IO) = R^2 - IO^2$ (2).

In the right triangle AIK, $IA = \dfrac{r}{\sin\dfrac{A}{2}}$ (3).

$\angle EBD = \dfrac{arc(DF) + arc(FE)}{2} = \dfrac{A+B}{2}$ and in triangle ABI the angle BID

is an exterior angle, thus $\angle BID = \dfrac{A+B}{2}$, therefore the triangle BID is

isosceles, $BD = DI$ (4). Applying Law of Sines in triangle ABD, we

obtain $2R = \dfrac{BD}{\sin\dfrac{A}{2}}$. Using (4) it follows that $ID = 2R\sin\dfrac{A}{2}$ (5). From (1),

(2), (3), and (5) we obtain $R^2 - IO^2 = IA \cdot ID = \dfrac{r}{\sin\dfrac{A}{2}} 2R\sin\dfrac{A}{2} = 2rR$.

VII. 90. $r = 4R\sin\dfrac{A}{2}\sin\dfrac{B}{2}\sin\dfrac{C}{2} = 2R\sin\dfrac{A}{2}\left(\cos\dfrac{B-C}{2} - \cos\dfrac{B+C}{2}\right) =$

$= 2R\sin\dfrac{A}{2}\left(\cos\dfrac{B-C}{2} - \sin\dfrac{A}{2}\right) = -2R\sin^2\dfrac{A}{2} + 2R\sin\dfrac{A}{2}\cos\dfrac{B-C}{2}$, then

the quadratic equation $2R\sin^2\dfrac{A}{2} - 2R\sin\dfrac{A}{2}\cos\dfrac{B-C}{2} + r = 0$, with the

unknown $\sin\dfrac{A}{2}$ has real solutions, then $\Delta \geq 0$ and the result follows.

VII. 91. Prove that, in any triangle ABC, $\dfrac{\sin^n A + \sin^n B + \sin^n C}{3} \geq \left(\dfrac{p}{3r}\right)^n$

where $n \in (-\infty, 0] \cup [1, \infty)$.

Solution

We know that $\dfrac{x_1^n + x_2^n + \ldots + x_m^n}{m} \geq \left(\dfrac{x_1 + x_2 + \ldots + x_m}{m}\right)^n$ where

$n \in (-\infty, 0] \cup [1, \infty)$. Denote $x_1 = \sin A$, $x_2 = \sin B$, and $x_3 = \sin C$.

Using $\dfrac{\sin^n A + \sin^n B + \sin^n C}{3} \geq \left(\dfrac{\sin A + \sin B + \sin C}{3}\right)^n$ and

$\sin A + \sin B + \sin C = \dfrac{p}{R}$, we obtain the result.

VII. 92. Using $\dfrac{x_1^n + x_2^n + \ldots + x_m^n}{m} \geq \left(\dfrac{x_1 + x_2 + \ldots + x_m}{m}\right)^n$ for any

$n \in (-\infty, 0] \cup [1, \infty)$. Denote $x_1 = \sin A$, $x_2 = \sin B$, and $x_3 = \sin C$.

Using $\dfrac{\sin^n A + \sin^n B + \sin^n C}{3} \geq \left(\dfrac{\sin A + \sin B + \sin C}{3}\right)^n$ and

$\sin A + \sin B + \sin C = 4\cos\dfrac{A}{2}\cos\dfrac{B}{2}\cos\dfrac{C}{2}$ the result follow.

UNIT VIII

Complex Numbers

VI. 1. a) $z_1 = 5 + 2i$; **b)** $z_2 = 1 - 3i$; **c)** $z_3 = -5 + 14i$;

d) $z_4 = \dfrac{7}{5} - \dfrac{1}{5}i$; **e)** $z_5 = \dfrac{1-i}{1+2i} = \dfrac{(1-i)(1+2i)}{1^2 - (2i)^2} = \dfrac{3+i}{5}$;

f) $z_6 = \dfrac{3}{5} - \dfrac{1}{5}i$. **VI. 2. a)** $-32i$; **b)** $-1-i$; **c)** 0; **d)** 0; **e)** 0;

f) $\left(i^4\right)^{502} \cdot i^2 + \left(i^4\right)^{502} \cdot i^3 + \left(i^4\right)^{502} \cdot i^4 + \left(i^4\right)^{502} \cdot i^5 = i^2 + i^3 + i^4 + i^5 = 0$;

g) $2^{2014} i$. **VI. 3. a)** $\mathrm{Re}(z_1) = -4$, $\mathrm{Im}(z_1) = 3$, $|z_1| = 5$; **b)** $\mathrm{Re}(z_2) = 3$, $\mathrm{Im}(z_2) = -2$, $|z_2| = \sqrt{13}$;

c) $\mathrm{Re}(z_3) = -2$, $\mathrm{Im}(z_3) = 1$, $|z_3| = \sqrt{5}$;

d) $\mathrm{Re}(z_4) = 1$, $\mathrm{Im}(z_4) = -2$, $|z_4| = \sqrt{5}$.

VI. 5. a) $\dfrac{7}{9} - \dfrac{3}{9}i$, **b)** $-\dfrac{3}{13} + \dfrac{11}{13}i$, **c)** $-\dfrac{4}{5} + \dfrac{7}{5}i$, **d)** $\dfrac{11i}{2}$,

e) $\dfrac{\left(1 + 2i + i^2\right)^2 + \left(1 - 2i + i^2\right)^2}{\left(1 + 4i + 4i^2\right)^2 + \left(1 - 4i + 4i^2\right)^2} = \dfrac{4i^2 + 4i^2}{(-3 + 4i)^2 + (-3 - 4i)^2} = \dfrac{4}{7}$;

f) $z = \dfrac{4}{3}(1+i)$; **g)** $z = -1 + i$; **h)** $z = -2i$; **i)** $\dfrac{3 + 7i}{10}$; **j)** $\dfrac{-15 - 8i}{17}$.

VI. 6. a) $\begin{cases} x + 2y = 3 \\ 3x - y = 2 \end{cases}$, $x = 1$, $y = 1$; **b)** $\begin{cases} x^2 + y = 2 \\ x + 3y^2 = 4 \end{cases}$, $x = 1$, $y = 1$;

c) $\begin{cases} 3\sqrt{x^2-2y}+x^2=2y \\ -x^2=4y-12 \end{cases}$, $x=\pm 2$, $y=2$; d) $x=-6$, $y=7$;

e) $x=0$, $y=3$;

f) $x-3+(y-3)i=(x+2)i-y-4$, $\begin{cases} x-3=-y-4 \\ y-3=x+2 \end{cases} \Rightarrow y=2$, $x=-3$.

g) $x=1$, $y=4$; h) $x=-1$, $y=-2$.

VI. 7. a) $a=0$, $b=1$; **b)** $a=0$, $b=1$; **c)** $a=\dfrac{1}{13}$, $b=-\dfrac{21}{13}$;

d) $a=\dfrac{6}{5}$, $b=0$; **e)** $a=\dfrac{1}{2}$, $b=0$. **VI. 8.** $z=-8$.

VI. 9. a) $x_{1,2}=\dfrac{-1\pm 3i}{3}$; **b)** $x_{1,2}=\dfrac{-5\pm i\sqrt{3}}{2}$;

c) $x_{1,2}=\dfrac{3\pm i\sqrt{7}}{4}$; **d)** $x_{1,2}=\dfrac{3\pm i\sqrt{23}}{2}$.

VI. 10. The complex numbers $-\dfrac{1}{2}+i\dfrac{\sqrt{3}}{2}$ and $1-3i$ are the roots of the

equation $x^2-2x+10=0$, $x^3-4x^2+14x-20=(x^2-2x+10)(x-2)$.

VI. 11. $\left(x+\dfrac{1}{2}-i\dfrac{\sqrt{3}}{2}\right)\left(x+\dfrac{1}{2}+i\dfrac{\sqrt{3}}{2}\right)=x^2+x+1$,

$x^4+x^2+1=(x^2+x+1)(x^2-x+1)$.

VI. 12. We have $x_1^2+x_1+1=0$ and $x_2^2+x_2+1=0$; $x_1+x_2=-1$ and

$x_1 x_2=1$. On the other hand $x_1^3-1=(x_1-1)(x_1^2+x_1+1)=0 \Rightarrow x_1^3=1$,

similarly $x_2^3=1$.

a) $\left(1+x_1^2-1\right)^{2002}+\left(1+x_2^2-1\right)^{2002}=\left(x_1^3\right)^{1334}x_1^2+\left(x_2^3\right)^{1334}x_2^2=$

$=x_1^2+x_2^2=(x_1+x_2)^2-2x_1 x_2=1-2=-1$

b) -1; c) $\left[x_1^2\left(x_1^2+x_1+1\right)+1\right]^{2013}+\left[x_2^2\left(x_2^2+x_2+1\right)+1\right]^{2013}=2$.

VI. 13. $\varepsilon^3=1$, $\varepsilon^3-1=(\varepsilon-1)(\varepsilon^2+\varepsilon+1)=0 \Rightarrow \varepsilon^2+\varepsilon+1=0$

a) $z_1=\varepsilon^4+\varepsilon^2+1=\varepsilon\cdot\varepsilon^3+\varepsilon^2+1=\varepsilon\cdot 1+\varepsilon^2+1=\varepsilon^2+\varepsilon+1=0$

b) $z_2 = \dfrac{1+\varepsilon}{1-2\varepsilon+\varepsilon^2} + \dfrac{1-\varepsilon}{1+2\varepsilon+\varepsilon^2} = \dfrac{1+\varepsilon}{-3\varepsilon} + \dfrac{1-\varepsilon}{\varepsilon} = \dfrac{-2+4\varepsilon}{-3\varepsilon}.$

If $\varepsilon_1 = \dfrac{-1+i\sqrt{3}}{2} \Rightarrow z_2 = \dfrac{-2+4\varepsilon}{-3\varepsilon} = \dfrac{2}{3\varepsilon} - \dfrac{4}{3} = \dfrac{2}{3} \cdot \dfrac{2}{-1+i\sqrt{3}} - \dfrac{4}{3} = \dfrac{-5-i\sqrt{3}}{3}$,

If $\varepsilon_2 = \dfrac{-1-i\sqrt{3}}{2} \Rightarrow z_2 = \dfrac{-2+4\varepsilon}{-3\varepsilon} = \dfrac{2}{3\varepsilon} - \dfrac{4}{3} = \dfrac{2}{3} \cdot \dfrac{2}{-1-i\sqrt{3}} - \dfrac{4}{3} = \dfrac{-5+i\sqrt{3}}{3}$;

c) $z_3 = 0$;   **d)** $z_4 = a^3 + b^3 + c^3 - 3abc$;   **e)** $z_5 = 4$;   **f)** $z_6 = 9$.

VI. 15. $z^2 - z + 1 = 0$ then $z^3 + 1 = (z+1)(z^2 - z + 1) = 0 \Rightarrow z^3 = -1$,

$z^{13} + \dfrac{1}{z^{13}} = (z^3)^4 z + \dfrac{1}{(z^3)^4 z} = z + \dfrac{1}{z} = 1.$

VI. 16. $i^4 = 1$, $\dfrac{1+i}{1-i} = \dfrac{(1+i)(1+i)}{1^2 - i^2} = i$,

$z = (i^4)^{502} \cdot i^2 + (i^4)^{502} \cdot i^3 + (i^4)^{503} + (i^4)^{503} \cdot i = -1 - i + 1 + i = 0.$

VI. 17. $x_1^2 + x_1 + 1 = 0$ and $x_1^3 - 1 = (x_1 - 1)(x_1^2 + x_1 + 1) = 0 \Rightarrow x_1^3 = 1$,

similarly $x_2^3 = 1$, $E = \dfrac{(-x_1^2)^{2014}}{-x_1} + \dfrac{(-x_2^2)^{2014}}{-x_2} =$

$= -(x_1^3)^{1342} x_1 - (x_2^3)^{1342} x_2 = -x_1 - x_2 = -1.$

VI. 18. *Two solutions:* **i)** A complex number is a real number,

if only if $z = \bar{z}$. From $|z_1| = |z_2| = 1$ it follows $z_1 \bar{z}_1 = z_2 \bar{z}_2 = 1$.

$$\overline{\left(\dfrac{z_1 + z_2}{1 + z_1 z_2}\right)} = \dfrac{\bar{z}_1 + \bar{z}_2}{1 + z_1 z_2} = \dfrac{\bar{z}_1 + \bar{z}_2}{1 + \bar{z}_1 \bar{z}_2} = \dfrac{\dfrac{1}{z_1} + \dfrac{1}{z_2}}{1 + \dfrac{1}{z_1 z_2}} = \dfrac{z_1 + z_2}{1 + z_1 z_2}.$$

ii) Consider $z_1 = \cos t_1 + i\sin t_1$, $z_2 = \cos t_2 + i\sin t_2$, $|z_1| = |z_2| = 1$.

$$\dfrac{z_1 + z_2}{1 + z_1 z_2} = \dfrac{\cos t_1 + \cos t_2 + i(\sin t_1 + \sin t_2)}{1 + \cos(t_1 + t_2) + i\sin(t_1 + t_2)} =$$

$$= \frac{2\cos\dfrac{t_1+t_2}{2}\cos\dfrac{t_1-t_2}{2}+2i\sin\dfrac{t_1+t_2}{2}\cos\dfrac{t_1-t_2}{2}}{2\cos^2\dfrac{t_1+t_2}{2}+2i\sin\dfrac{t_1+t_2}{2}\cos\dfrac{t_1+t_2}{2}} = \frac{\cos\dfrac{t_1-t_2}{2}}{\cos\dfrac{t_1+t_2}{2}} \in \mathbf{R}.$$

VI. 19. $|z|=1 \Rightarrow \bar{z}=\dfrac{1}{z}$. Then $\overline{\left(\dfrac{1+z^{2n}}{z^n}\right)} = \dfrac{1+\bar{z}^{2n}}{\bar{z}^n} = \dfrac{1+\left(\dfrac{1}{z}\right)^{2n}}{\left(\dfrac{1}{z}\right)^n} = \dfrac{1+z^{2n}}{z^n}$.

VI. 20. $\overline{\dfrac{1-z+z^2}{1+z+z^2}} = \dfrac{1-z+z^2}{1+z+z^2} \Rightarrow \dfrac{1-\bar{z}+\bar{z}^2}{1+\bar{z}+\bar{z}^2} = \dfrac{1-z+z^2}{1+z+z^2} \Rightarrow$

$(z-\bar{z})(1-z\cdot\bar{z})=0$, $z\neq\bar{z}$, then $\bar{z}\cdot z=1$, which leads to $|z|=1$.

VI. 21. $\bar{z} = \overline{(3+2i)^{4n}+(2+3i)^{4n}} = \overline{(3+2i)^{4n}} + \overline{(2+3i)^{4n}} =$

$= \left(\overline{3+2i}\right)^{4n} + \left(\overline{2+3i}\right)^{4n} = (3-2i)^{4n}+(2-3i)^{4n} =$

$= i^{4n}(2+3i)^{4n} + i^{4n}(3+2i)^{4n} = z$.

VI. 22. a) $\bar{z}_n = z_n$;

b) $2z_{n+1}-2z_n = 2(1+i)^{n+1}+2(1-i)^{n+1}-2(1+i)^n-2(1-i)^n =$

$= 2i(1+i)^n - 2i(1-i)^n = (1+i)^2(1+i)^n + (1-i)^2(1-i)^n = z_{n+2}$.

VI. 23. $OA=|z_1|=\sqrt{10}$, $OB=|z_2|=\sqrt{10}$, $AB=|z_1-z_2|=2\sqrt{5}$,

VI. 24. $|z_1-z_2|=3\sqrt{10}, |z_1-z_3|=2\sqrt{10}, |z_3-z_2|=\sqrt{10}$.

VI. 25. $|z_1-z_2|=3\sqrt{2}$, $|z_1-z_3|=3\sqrt{2}$.   **VI. 26.** $x^2+y^2=5$.

VI. 27. $A(z_A)$, $B(z_B)$, $C(z_C)$, $D(z_D)$

$M\left(\dfrac{z_A+z_B}{2}\right)$, $N\left(\dfrac{z_B+z_C}{2}\right)$,

$P\left(\dfrac{z_C+z_D}{2}\right)$, $Q\left(\dfrac{z_D+z_A}{2}\right)$

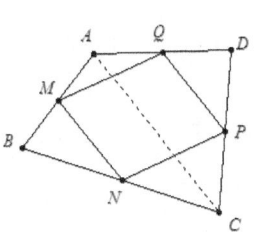

$$m_{MN} = \frac{z_N - z_M}{\overline{z}_N - \overline{z}_M} = \frac{z_C - z_A}{\overline{z}_C - \overline{z}_A},$$

$$m_{PQ} = \frac{z_Q - z_P}{\overline{z}_Q - \overline{z}_P} = \frac{z_C - z_A}{\overline{z}_C - \overline{z}_A} \text{ and so on.}$$

VI. 28. $z = \dfrac{9}{2} + \dfrac{i}{2}$. **VI. 29.** $z_C = -1 - 6i$.

VI. 30. $AB = |z_2 - z_1|$, $MN = |f(z_2) - f(z_1)| = |az_1 + b - az_2 - b| = |a||z_2 - z_1|$.

VI. 31. a) $z = -\dfrac{7}{3} - \dfrac{4}{3}i$; **b)** $z = -\dfrac{7}{6} + \dfrac{5}{6}i$;

c) $z_1 = 2 - 3i$, $z_2 = -4 - 3i$; **d)** $z = a + bi$, $|z| = \sqrt{a^2 + b^2}$,

$$\left|\frac{z+1}{z}\right| = \frac{|z+1|}{|z|} = \frac{\sqrt{(a+1)^2 + b^2}}{\sqrt{a^2 + b^2}}, \quad \left|\frac{z-i}{1-i}\right| = \frac{|z-i|}{|1-i|} = \frac{|z-i|}{\sqrt{2}} = \frac{\sqrt{a^2 + (b-1)^2}}{\sqrt{2}},$$

$$\begin{cases} \dfrac{\sqrt{(a+1)^2 + b^2}}{\sqrt{a^2 + b^2}} = 2 \\ \dfrac{\sqrt{a^2 + (b-1)^2}}{\sqrt{2}} = 1 \end{cases} \Rightarrow \begin{cases} 3(a^2 + b^2) = 2a + 1 \\ a^2 + b^2 = 2b + 1 \end{cases}, \; z_1 = 1,$$

$z_2 = \dfrac{-1 - 2i}{5}$; **e)** $z = 2 - i$.

VI. 32. a) $x^2 + (y-1)^2 = 4$; $C(0,\, 1)$; $r = 2$;

b) $\dfrac{x^2}{4} + \dfrac{y^2}{3} = 1$ (ellipse); **c)** Perpendicular bisector of the segment AO,

where $A(-2,\, 0)$ and $O(0,\, 0)$; **d)** $x^2 + \left(y - \dfrac{1}{2}\right)^2 = \dfrac{1}{4}$; **e)** $x^2 + (y+1)^2 < 4$;

f) $\left(\dfrac{x}{\sqrt{2}}\right)^2 - \left(\dfrac{y}{\sqrt{2}}\right)^2 - 1 < 0$; **g)** $\dfrac{1}{4} < \left(x + \dfrac{3}{2}\right)^2 + (y-1)^2 \le 1$;

h) $x^2 + (y-1)^2 > 16$; **i)** $z = x + iy$,

$z + i - 2 = (x - 2) + i(y + 1) \Rightarrow |z + i - 2| = \sqrt{(x-2)^2 + (y+1)^2}$

$\sqrt{(x-2)^2 + (y+1)^2} \le 2 \Rightarrow (x-2)^2 + (y+1)^2 \le 4$.

VI. 33. a) $x_1 = 1$, $x_{2,3} = \dfrac{-1 \pm i\sqrt{3}}{2}$; **b)** $x_1 = 2$, $x_{2,3} = -1 \pm i\sqrt{3}$;

c) $x_1 = -\dfrac{1}{3}$, $x_{2,3} = \dfrac{1 \pm i\sqrt{3}}{6}$; **d)** $x_1 = \dfrac{3}{4}$, $x_{2,3} = \dfrac{-3 \pm 3i\sqrt{3}}{8}$;

e) ± 2 and $\pm 2i$; **f)** ± 3, $\pm 3i$; **g)** $x_1 = 1$, $x_{2,3} = \dfrac{-1 \pm i\sqrt{3}}{2}$; $x_4 = \sqrt[3]{2}$,

$x_{5,6} = \dfrac{-\sqrt[3]{2} \pm i \cdot \sqrt[6]{108}}{2}$; **h)** $x_1 = -1$, $x_{2,3} = \dfrac{1 \pm i\sqrt{3}}{2}$; $x_4 = -\sqrt[3]{2}$,

$x_{5,6} = \dfrac{\sqrt[3]{2} \pm i \cdot \sqrt[6]{108}}{2}$.

VI. 34. c) $|a+b| \le |a| + |b|$, $\left|(1+2i)z + 2i|z|\right| \le |1+2i||z| + |2i||z| < \sqrt{5} \cdot \dfrac{1}{2} + 2 \cdot \dfrac{1}{2} < 3$;

f) Let $z + \dfrac{1}{z} = u \Rightarrow z^3 + \dfrac{1}{z^3} = u^3 - 3u$. Since $\left|z^3 + \dfrac{1}{z^3}\right| \le 2$, then

$\left|u^3 - 3u\right| \le 2 \Rightarrow |u|\left(|u|^2 - 3\right) \le |u^3 - 3u| \le 2$.

The inequality $|u|\left(|u|^2 - 3\right) \le 2$ is equivalent to $\left(|u| - 1\right)^2 \left(|u| - 2\right) \le 0$.
Therefore $|u| \le 2$.

VI. 35. c) $\varepsilon^2 + \varepsilon + 1 = 0$, $\varepsilon^3 = 1$, $\bar{\varepsilon} = \varepsilon^2$, $\left(\bar{\varepsilon}\right)^2 = \varepsilon$,

(1) $|z - 1|^2 = (z-1)(\bar{z}-1) = |z|^2 - z - \bar{z} + 1$,

(2) $|z - \varepsilon|^2 = |z|^2 - z\varepsilon^2 - \bar{z}\varepsilon + 1$,

(3) $\left|z - \varepsilon^2\right|^2 = |z|^2 - z\varepsilon - \bar{z}\varepsilon^2 + 1$. Add (1), (2) and (3).

VI. 36. $z = \left(\sqrt[4]{a} + i\sqrt[4]{b}\right)\left(\sqrt[4]{a} + i^2\sqrt[4]{b}\right)...\left(\sqrt[4]{a} + i^{4k}\sqrt[4]{b}\right)$.

VI. 37. a) $a = \pm 2$; **b)** $a = \pm\sqrt{6}$.

VI. 38. a) $\dfrac{z-2}{z-6} = \dfrac{x-2+iy}{x-6-iy} = \dfrac{(x-2+iy)(x-6+iy)}{(x-6)^2 - (iy)^2} =$

$= \dfrac{(x-2)(x-6) + y^2}{(x-6)^2 + y^2} + i\dfrac{y(x-2) - y(x-6)}{(x-6)^2 + y^2}$,

$$\frac{(x-2)(x-6)+y^2}{(x-6)^2+y^2} = \frac{y(x-6)-y(x-2)}{(x-6)^2-(iy)^2} = 0 \Rightarrow (x-4)^2+(y+2)^2 = 8.$$

A circle, of center $C(4,-2)$, $r = 2\sqrt{2}$;   **b)** $\left(x-\dfrac{3}{2}\right)^2 + y^2 = \dfrac{1}{4}$, $C\left(\dfrac{3}{2},0\right)$,

$r = \dfrac{1}{2}$;   **c)** The points from the line $x = 0$;

d) From $\left|z^2-i\right| = \left|z^2-2i\right| \Rightarrow \left|x^2-y^2+i(2xy-1)\right| = \left|x^2-y^2+i(2xy-2)\right| \Rightarrow$

$(2xy-1)^2 = (2xy-2)^2$. Finally $y = \dfrac{3}{4x}$, this is a hyperbola.

e) $\dfrac{z+1+i}{2-iz} = \dfrac{x+1+i(y+1)}{2+y-ix} = \dfrac{[x+1+i(y+1)](2+y+ix)}{(2+y)^2+x^2} =$

$= \dfrac{(x+1)(y+2)-x(y+1)+i(x^2+x+y^2+3y+2)}{(2+y)^2+x^2}$, Since $\text{Im}\left(\dfrac{z+1+i}{2-iz}\right) = 0$,

then $x^2+x+y^2+3y+2 = 0 \Rightarrow \left(x+\dfrac{1}{2}\right)^2 + \left(y+\dfrac{3}{2}\right)^2 = \dfrac{5}{2}$, a circle with

center $C\left(-\dfrac{1}{2},-\dfrac{3}{2}\right)$ and radius $r = \sqrt{\dfrac{5}{2}}$.

VI. 39. a) $x_1^2+x_2^2 = m^2-m$, $x_1^3+x_2^3 = m^2(m-1.5)$, $\dfrac{1}{x_1}+\dfrac{1}{x_2} = 2$;

b) $m \in (0,2)$;   **c)** 4 and $1-\sqrt{17}$.

VI. 40. $a = \dfrac{8-3i}{2}$.   **VI. 41.** $a = -3$, $b = 6$.

VI. 42. a) $x(x-1)(x+1)\left(x+\dfrac{1}{2}-i\dfrac{\sqrt{3}}{2}\right)\left(x+\dfrac{1}{2}+i\dfrac{\sqrt{3}}{2}\right)$;

b) $(x+2)(x-2-i\sqrt{3})(x-2+i\sqrt{3})$;   **c)** $x^2(x-i)(x+i)$;

d) $x^2(x+3)\left(x+\dfrac{3}{2}-i\dfrac{3\sqrt{3}}{2}\right)\left(x+\dfrac{3}{2}+i\dfrac{3\sqrt{3}}{2}\right)$.

VI. 43. a) $0+0i$, $1+0i$, $-\dfrac{1}{2}\pm\dfrac{i\sqrt{3}}{2}$;

b) $z = 0$ is a solution. For $z \neq 0$, $r^2 = r \Rightarrow r = 1$. Then $z^3 = \bar{i}zz \Rightarrow z^3 = i$

or $z^3 = -i^3$ or $(z+i)(z^2 - iz - 1) = 0$. The solutions are $z = 0$, $z = -i$,

$z = \dfrac{\pm\sqrt{3}+i}{2}$;   c) $z = 0$, $z = \pm i$;

d) $z_1 = 0$, $z_2 = -3i$, $z_3 = 3i$;   e) If $z = a + bi$, then the solutions are

$z = a(1 \pm i\sqrt{3})$, where $a \le 0$.

VI. 44. a) $z_1 = 1 - 2i$, $z_2 = -1 + 2i$;   b) $z_1 = 2 - \sqrt{3}i$, $z_2 = -2 + \sqrt{3}i$;

c) $z_1 = 2 + i$, $z_2 = -2 - i$;   d) $z_1 = 4 + 3i$, $z_2 = -4 - 3i$.

VI. 45. a) $z_1 = i$, $z_2 = 3i$;

b) $z_{1,2} = \dfrac{1 - i \pm \sqrt{-8-6i}}{2} = \dfrac{1 - i \pm \sqrt{(1-3i)^2}}{2} = \dfrac{1 - i \pm (1 - 3i)}{2}$, $z_1 = i$, $z_2 = 1 - 2i$;

c) $z_{1,2} = \dfrac{2 - 6i \pm \sqrt{-48+14i}}{2-i} = \dfrac{2 - 6i \pm (1 + 7i)}{2-i}$, $z_1 = 1 + i$, $z_2 = \dfrac{-11+27i}{5}$;

d) $z_1 = -1 + 3i$, $z_2 = 1 + 2i$;   e) $z_1 = -1 - i$, $z_2 = i$;

f) $z_1 = 2$, $z_2 = 3 - 2i$;   g) $z_1 = 1 - i$, $z_2 = 1 + 2i$;   h) $z_1 = 2 - i$, $z_2 = 1 + 2i$;

i) $z_1 = 3 - i$, $z_2 = -1 + 2i$;   j) $z_1 = 2 + 3i$, $z_2 = 3 - i$;

k) $z_1 = 5 + i$, $z_2 = 3 + 2i$.

VI. 46. a) $(1+i)(x+iy) + (5-3i)(x-iy) = 20 - 4i \Rightarrow 6x - 4iy - 2ix - 4y = 20 - 4i$.

Then $\begin{cases} 6x - 4y = 20 \\ -4y - 2x = -4 \end{cases}$ whose solution is $\begin{cases} x = 3 \\ y = -\dfrac{1}{2} \end{cases}$. Therefore $z = 3 - \dfrac{1}{2}i$;

b) $z = 2 + 8i$;   c) $z = 1 + i$;   d) $z = -1 - i$;

e) $(1+3i)z - (3-2i)z = 11i + 13 \Rightarrow z(-2+5i) = 11i + 13 \Rightarrow z = 1 - 3i$.

f) $2(a+bi) + i(a-bi) = 9 + 3i$, $\begin{cases} 2a + b = 9 \\ 2b + a = 3 \end{cases}$, $z = 5 - i$.

VI. 47. a) $z_1 = -\sqrt{2} - i\sqrt{2}$, $z_2 = \sqrt{2} + i\sqrt{2}$, $z_3 = -i\sqrt{2}$, $z_4 = i\sqrt{2}$;

b) $z_1 = 1 - i$, $z_2 = -1 + i$, $z_3 = 2 - i$, $z_4 = -2 + i$;

c) $(z^2+1)(z^2 - 2z + 3) = 0$, $z_1 = -i$, $z_2 = i$, $z_3 = 1 - i\sqrt{2}$, $z_4 = 1 + i\sqrt{2}$;

d) $z_1 = -1-i$, $z_2 = 1+i$, $z_3 = -1-2i$, $z_4 = 1+2i$;

e) Denote $\dfrac{5z-2i}{2z-5i} = a$, $a^3 + a^2 + a + 1 = 0$. Thus $a_1 = -1$, $a_2 = -i$, and

$a_3 = i$. Finally $z_1 = i$, $z_2 = \dfrac{-29+20i}{29}$, and $a_3 = \dfrac{29+20i}{29}$.

VI. 48. a) From $(a+bi)^3 = -11 + 2i$ we obtain

$(a^3 - 3ab^2) + (3a^2 b - b^3) = -1 + 2i$ or

$\begin{cases} a^3 - 3ab^2 = -1 \\ 3a^2 b - b^3 = 2 \end{cases}$. In the equation $2(a^3 - 3ab^2) = -1 \,(3a^2 b - b^3)$ setting

$b = t\,a$, yields $2(1 - 3t^2) = -11(3t - t^3)$ or $(t+2)(11\,t^2 - 16t - 1) = 0$.

The only convenient solution of this equation is $t_1 = -2$; hence

$z = 1-2i$; **b)** $z = 3-2i$; **c)** $z_1 = 2i$, $z_2 = \sqrt{3} - i$, $z_3 = -\sqrt{3} - i$;

d) $z_1 = \sqrt{3} + 3i$, $z_2 = -\sqrt{3} - 3i$; **e)** $z_1 = -\sqrt{3} + i$, $z_2 = \sqrt{3} - i$,

$z_3 = 1 + i\sqrt{3}$, $z_4 = -1 - i\sqrt{3}$; **f)** $z_1 = -\sqrt{3} + i$, $z_2 = \sqrt{3} - i$, $z_3 = \sqrt{3} + i$,

$z_4 = \sqrt{3} - i$, $z_5 = 2i$, $z_6 = -2i$.

VI. 49. a) $\dfrac{z-5+12i}{z-3+4i}$; **b)** $\dfrac{z+1}{z-1+i}$; **c)** $\dfrac{z-i}{z+i}$.

VI. 50. a) $x = 1-i$, $y = i$; **b)** $x = i$, $y = 2+i$; **c)** $x = 1+i$, $y = 1-i$;
d) $x = 1+i$, $y = 2+i$.

VI. 51. a) $x_1 = \dfrac{3+i\sqrt{7}}{2}$, $y_1 = \dfrac{3-i\sqrt{7}}{2}$, $x_2 = \dfrac{3-i\sqrt{7}}{2}$, $y_2 = \dfrac{3+i\sqrt{7}}{2}$;

b) $x_1 = \dfrac{\sqrt{2}}{2} + i$, $y_1 = -\dfrac{\sqrt{2}}{2} + i$; $x_2 = -\dfrac{\sqrt{2}}{2} + i$, $y_2 = \dfrac{\sqrt{2}}{2} + i$;

c) $x_1 = i\sqrt{2}$, $y_1 = 1$; $x_2 = i\sqrt{\dfrac{41}{15}}$, $y_2 = \dfrac{4}{15}$;

d) $x_1 = \dfrac{1}{2} + i\dfrac{\sqrt{39}}{6}$, $y_1 = 1 - i\dfrac{\sqrt{39}}{3}$; $x_2 = \dfrac{1}{2} - i\dfrac{\sqrt{39}}{6}$, $y_2 = 1 + i\dfrac{\sqrt{39}}{3}$.

VI. 52. a) $f(-1) = -3$, $f(i) = -1 + 4i$, $f(2-i) = 7 - 6i$, $f(\bar{z}) = \bar{z}^2 + 3\bar{z} - z$,

$f(|z|) = |z|^2 + 2|z|$.

VI. 53. $A = \{(x, 0) \mid x \in \mathbf{R} \setminus \{1\}\}$.

VI. 54. Switching x and $\varepsilon^2 x$ we obtain $f(x) = \varepsilon^2 x$.

VI. 55. $f(x)f(ix) = x^2$ also $f(ix)f(-x) = -x^2$, hence $f(-x) = -f(x)$. We
have $g(x) + g(\varepsilon x) = x$ (1), $g(\varepsilon x) + g(\varepsilon^2 x) = \varepsilon x$ (2), $g(\varepsilon^2 x) + g(x) = \varepsilon^2 x$ (3).
Adding (1), (2), and (3) we obtain
$g(x) = -\varepsilon x$.

VI. 56. Clearly $b = \dfrac{b}{1+\varepsilon} + \varepsilon \cdot \dfrac{b}{1+\varepsilon}$, therefore the relation (1)

becomes (2) $f(x+a) - \dfrac{b}{1+\varepsilon} = -\varepsilon\left(f(x-a) - \dfrac{b}{1+\varepsilon} \right)$. Denoting

$g(x) = f(x-a) - \dfrac{b}{1+\varepsilon}$, (2) becomes $g(x+2a) = -\varepsilon \cdot g(x)$ and in general

(3) $g(x + 4an) = (-\varepsilon)^{2n} \cdot g(x)$. Since $\varepsilon^n = 1$, (3) becomes

(4) $g(x + 4an) = g(x)$. Using (2) and (4)

$f(x + 4an - a) - \dfrac{b}{1+\varepsilon} = f(x-a) - \dfrac{b}{1+\varepsilon} \Leftrightarrow f(x + a(4n-1)) = f(x-a)$.

Taking $x - a = t$, we obtain $f(t + 4an) = f(t)$, $\forall t \in \mathbf{R}$. Therefore the
period is $T = 4an$.

VI. 57. For $\alpha = 1$, $E = 1 + 2 + 3 + \ldots + n = \dfrac{n(n+1)}{2}$. For $\alpha \neq 1$,

$E = (\alpha + \alpha^2 + \alpha^3 + \ldots + \alpha^n) + (\alpha^2 + \alpha^3 + \ldots + \alpha^n) + (\alpha^3 + \alpha^4 + \ldots + \alpha^n) +$

$\ldots + \alpha^n = \alpha\dfrac{1-\alpha^n}{1-\alpha} + \alpha^2\dfrac{1-\alpha^{n-1}}{1-\alpha} + \alpha^3\dfrac{1-\alpha^{n-2}}{1-\alpha} + \ldots + \alpha^n\dfrac{1-\alpha}{1-\alpha} =$

$= \dfrac{(\alpha - \alpha^{n+1}) + (\alpha^2 - \alpha^{n+1}) + (\alpha^3 - \alpha^{n+1}) + \ldots + (\alpha^n - \alpha^{n+1})}{1-\alpha} =$

$$= \frac{\left(\alpha + \alpha^2 + \alpha^3 + \ldots + \alpha^n\right) - \left(n\alpha^{n+1}\right)}{1 - \alpha} = \frac{\left(\alpha + \alpha^2 + \alpha^3 + \ldots + \alpha^n\right) - \left(n\alpha^{n+1}\right)}{1 - \alpha} =$$

$$= \frac{\alpha \dfrac{1 - \alpha^n}{1 - \alpha} - n\alpha^{n+1}}{1 - \alpha} = \frac{-n\alpha}{1 - \alpha}.$$

VI. 58. Let $z = \cos t + i\sin t$, then $z^n = \cos nt + i\sin nt$, $n \in \mathbf{N}$

(De Moivre's theorem) also $z^{-n} = \cos nt - i\sin nt$. Therefore

$$\left(\cos nt + i\sin nt\right) + \left(\cos nt - i\sin nt\right) = 2\cos nt = z^n + z^{-n} = \frac{z^{2n} + 1}{z^n} \text{ and}$$

$$\left(\cos nt + i\sin nt\right) - \left(\cos nt - i\sin nt\right) = 2i\sin nt = z^n - z^{-n} = \frac{z^{2n} - 1}{z^n}.$$

Let $z = \cos 20° + i\sin 20°$, then $z^9 = \cos 180° + i\sin 180° = -1$.

We have $\sin 70° = \cos 20° = \dfrac{z^2 + 1}{2z}$,

$\sin 50° = \cos 40° = \dfrac{z^4 + 1}{2z^2}$ and $\sin 10° = \cos 80° = \dfrac{z^8 + 1}{2z^4}$. So

$$A = \frac{z^2 + 1}{2z} \frac{z^4 + 1}{2z^2} \frac{z^8 + 1}{2z^4} = \frac{1 + z^2 + z^4 + z^6 + z^8 + z^{10} + z^{12} + z^{14}}{8z^7} =$$

$$= \frac{\dfrac{1 - z^{16}}{1 - z^2}}{8z^7} = \frac{1 - z^7 z^9}{8\left(z^7 - z^9\right)} = \frac{1 + z^7}{8\left(z^7 + 1\right)} = \frac{1}{8}.$$

VI. 60. Let $z = \cos t + i\sin t$ then

$$Z_k = \sqrt[n]{z} = \sqrt[n]{\cos t + i\sin t} = \cos \frac{t + 2k\pi}{n} + i\sin \frac{t + 2k\pi}{n}, k = 0, 1, \ldots, n - 1 \text{ . We have}$$

$$z^3 = \frac{3 + i}{2 - i} = 1 + i = \frac{\sqrt{2}}{2}\left(\cos \frac{\pi}{4} + i\sin \frac{\pi}{4}\right), \text{ then } Z_k = \sqrt[3]{\frac{\sqrt{2}}{2}}\left(\cos \frac{\pi}{4} + iso \frac{\pi}{4}\right) \Rightarrow$$

$$Z_k = \frac{1}{\sqrt[6]{2}}\left[\cos \frac{1}{3}\left(\frac{\pi}{4} + 2k\pi\right) + i\sin \frac{1}{3}\left(\frac{\pi}{4} + 2k\pi\right)\right], \ k = 1, 2, 3.$$

VI. 62. Let $M(f(z_1)), N(f(z_2))$, and $P(f(z_3))$ be the geometric images of the numbers $f(z_1)$ $f(z_2)$ and $f(z_3)$, then we have

$$\frac{f(z_3) - f(z_1)}{f(z_2) - f(z_1)} = \frac{az_3 + b - az_1 - b}{az_2 + b - az_1 - b} = \frac{a(z_3 - z_1)}{a(z_2 - z_1)} = \frac{z_3 - z_1}{z_2 - z_1} \in \mathbf{R}^*, \text{ therefore } M,$$

N and P are collinear.

VI. 63. A function is a *bijection*, if and only if it is a injection (one-to one) and a *surjection* (onto).

A function $f : A \to B$ is *one-to-one (injective)*, if and only if $f(z_1) = f(z_2)$ implies that $z_1 = z_2$ for all z_1 and z_2 in the domain of f.

A function $f : A \to B$ is called *onto (surjective)*, if and only if for every element y in B there is an element x in A with $f(x) = y$ ($\forall y \in B \Rightarrow \exists x \in A$ such that $f(x)=y$).

a) For injectivity let $z_1, z_2 \in C, z_1 = x_1 + iy_1$ and

$z_2 = x_2 + iy_2$ if $f(z_1) = f(z_2) \Rightarrow 3z_1 + |z_1| = 3z_2 + |z_2| \Rightarrow$

$3(x_1 + iy_1) + \sqrt{x_2^2 + y_2^2} = 3(x_2 + iy_2) + \sqrt{x_2^2 + y_2^2} \Rightarrow$

$$\begin{cases} 3x_1 + \sqrt{x_2^2 + y_2^2} = 3x_2 + \sqrt{x_2^2 + y_2^2} \\ y_1 = y_2 \end{cases} \Rightarrow \begin{cases} x_1 = x_2 \\ y_1 = y_2 \end{cases} \Rightarrow z_1 = z_2.$$

b) For surjectivity $\forall u \in C, \ u = a + ib \Rightarrow \exists z \in C$ such that $f(z)=u$, z=x+iy. Since $f(z) = 3x + 3iy + \sqrt{x^2 + y^2}$. Therefore the function f is a surjection

(onto) if the system (1) $\begin{cases} 3x + \sqrt{x^2 + y^2} = a \\ 3y = b \end{cases}$ has a unique solution.

Substituting $y = \dfrac{b}{3}$ in the first equation we have

$\sqrt{9x^2 + b^2} = 3a - 9x$. This is an irrational equation, $3a - 9x \geq 0$ and

we get $x \leq \dfrac{a}{3}$. Squaring the both sides of the irrational equation we

get $9x^2 + b^2 = (3a - 9x)^2 \Rightarrow 72x^2 - 54ax + 9a^2 - b^2 = 0$ whose roots are

$$x_1 = \frac{9a + \sqrt{9a^2 + 8b^2}}{24} \text{ and } x_2 = \frac{9a - \sqrt{9a^2 + 8b^2}}{24}.$$

Notice that $x_1 = \dfrac{9a + \sqrt{9a^2 + 8b^2}}{24} > \dfrac{9a + 3|a|}{24} > \dfrac{a}{3}$ and

$x_2 = \dfrac{9a - \sqrt{9a^2 + 8b^2}}{24} < x_1 = \dfrac{9a - 3|a|}{24} < \dfrac{a}{3}.$ Finally $x_2 = \dfrac{9a - \sqrt{9a^2 + 8b^2}}{24}.$

So the solution of the system (1) is $x = \dfrac{9a - \sqrt{9a^2 + 8b^2}}{24}$ and $y = \dfrac{b}{3}.$

Since the function is injective and surjective, it is a bijection. Therefore the function f is invertible and its inverse is $f^{-1}(u) = z$. Finally

$$f^{-1}(u) = \frac{9\,\mathrm{Re}\,u - \sqrt{9(\mathrm{Re}\,u)^2 + 8(\mathrm{Im}\,u)^2}}{24} + i\frac{\mathrm{Im}\,u}{3}.$$

VIII. 64. a) $(x + yi)^2 = a(1 - i) + 1$. Then $x^2 - y^2 = a + 1$ and $xy = -a$. Eliminating a, we obtain $x^2 + xy - y^2 = 1$ (*hyperbola*);
b) The circle $x^2 + y^2 = 1$ and the line $x = 0$.

VIII. 65. a) $x^2 + 1 = 0$, then $x = \pm i$. We have
$$f(i) = (i^2 + i + 1)^{8n+1} - i = i^{8n+1} - i = (i^4)^{2n} \cdot i - i = i - i = 0. \text{ Similarly, } f(-i) = 0.$$

d) *Solution* 1: $x^2 - x + 1 = 0$, then $x = \dfrac{1 \pm \sqrt{3}}{2}$. Let ε be one of the roots, that is, $\varepsilon^2 - \varepsilon + 1 = 0$ or $\varepsilon^3 = 1$ and $\varepsilon \neq 1$. Since $\varepsilon^2 = \varepsilon - 1$, we have
$$f(\varepsilon) = (\varepsilon - 1)^{2n+1} - \varepsilon^{n+2} = (\varepsilon^2)^{2n+1} - \varepsilon^{n+2} = (\varepsilon^3)^n \cdot \varepsilon^n \cdot \varepsilon^2 - \varepsilon^n \cdot \varepsilon^2 =$$
$$= \varepsilon^n(\varepsilon^2 - \varepsilon^2) = 0.$$

Solution 2: $x^2 - x + 1 = 0$, then $x = \cos\dfrac{\pi}{3} \pm i\sin\dfrac{\pi}{3}$.

$$f\left(\cos\frac{\pi}{3} + i\sin\frac{\pi}{3}\right) = \left(\cos\frac{\pi}{3} + i\sin\frac{\pi}{3} - 1\right)^{2n+1} - \left(\cos\frac{\pi}{3} + i\sin\frac{\pi}{3}\right)^{n+2} =$$

$$= \left(-2\sin^2\frac{\pi}{6} + 2i\sin\frac{\pi}{6}\cos\frac{\pi}{6}\right)^{2n+1} - \left(\cos\frac{(n+2)\pi}{3} + i\sin\frac{(n+2)\pi}{3}\right) =$$

$$= \left(2\sin\frac{\pi}{6}\right)^{2n+1}\left(-\sin\frac{\pi}{6}+i\cos\frac{\pi}{6}\right)^{2n+1} - \left(\cos\frac{(n+2)\pi}{3}+i\sin\frac{(n+2)\pi}{3}\right) =$$

$$= -\sin\frac{(2n+1)\pi}{6}+i\cos\frac{(2n+1)\pi}{6} - \left(\cos\frac{(n+2)\pi}{3}+i\sin\frac{(n+2)\pi}{3}\right) =$$

$$= -\cos\frac{(n-1)\pi}{3}-\cos\frac{(n+2)\pi}{3}-i\left(\sin\frac{(n-1)\pi}{3}+\sin\frac{(n+2)\pi}{3}\right) = 0 .$$

VIII. 66. a) Let α be the real root, then

$\alpha^3 - (4+i)\alpha^2 + (7+3i)\alpha - 2i - 6 = 0$ or $\alpha^3 - 4\alpha^2 + 7\alpha - 6 - i\left(\alpha^2 - 3\alpha + 2\right) = 0$.

Therefore $\alpha = 2$ and $z_2 = 1-i$, $z_3 = 1+2i$.

b) Let $x_1 = \alpha$ be a real root, then $\alpha^3 - 4\alpha^2 + (6-i)\alpha - 3 + i = 0$ or

$\alpha^3 - 4\alpha^2 + 6\alpha - 3 + i(-\alpha+1) = 0 \Rightarrow -\alpha+1 = 0$ and $\alpha^3 - 4\alpha^2 + 6\alpha - 3 = 0$

$\Rightarrow \alpha = 1$. Therefore $f(x) = (x-1)(x^2 - 3x + 3 - i)$, $x_2 = 2+i$ and $x_3 = 1-i$.

VIII. 67. *Three solutions*: **i)** $x_2 = \dfrac{1+i\sqrt{3}}{2}$ is also a root, therefore there is p

$\in \mathbf{R}$ such that $f(x) = (x^2 - x + 1)(x + p) = x^3 + (p-1)x^2 + (1-p)x + p$, then

equating the coefficients $p-1 = a$, $1-p = 4$ and $p = b$.

Finally $p = -3$, $a = -4$ and $b = -3$.

ii) Using long division

$x^3 + ax^2 + 4x + b = (x^2 - x + 1)(x + a + 1) + x(a+4) + a + b + 1$. Since the

remainder is zero, $a = -4$ and $b = -3$.

iii) x_1 is a root, then $x_1^2 - x_1 + 1 = 0 \Rightarrow$

$x_1^3 + 1 = (x_1 + 1)(x_1^2 - x_1 + 1) = 0 \Rightarrow x_1^3 = -1$. Therefore

$f(x_1) = x_1^3 + ax_1^2 + 4x_1 + b = -1 + a(x_1 - 1) + 4x_1 + b = 0 \Rightarrow$

$-1 + a\cdot\left(\dfrac{1-i\sqrt{3}}{2} - 1\right) + 4\cdot\dfrac{1-i\sqrt{3}}{2} + b = 0$ or

$b + 1 - \dfrac{a}{2} + i\cdot\left(\dfrac{-\sqrt{3}}{2}a - 2\sqrt{3}\right) = 0 \Rightarrow a = -4$ and $b = -3$.

VIII. 68. $a = -6$ and $b = -20$.

VIII. 69. $x_2 = 1 + i$ is also a root, therefore $f(x) = (x^2 - 2x + 2)(x^2 + ax + b)$.
Equating the coefficients
$a = -4$, $b = 3$, $m = -6$, $n = 6$, $x_3 = 1$, $x_4 = 3$.

VIII. 70. a) $-1 - i\sqrt{3}$, $-1 - i$, $f(x) = x^4 + 4x^3 + 10x^2 + 12x + 8$;
b) $3 - 2i$, $1 + i$, $f(x) = x^4 - 8x^3 + 27x^2 - 38x + 26$.

VIII. 71. If the geometric imagine of z, with $\arg z = t$ is $M(a,b)$,
then the geometric imagine of $-z$ is the point $M(-a,-b)$, therefore
$\arg(-z) = \pi + \arg z$

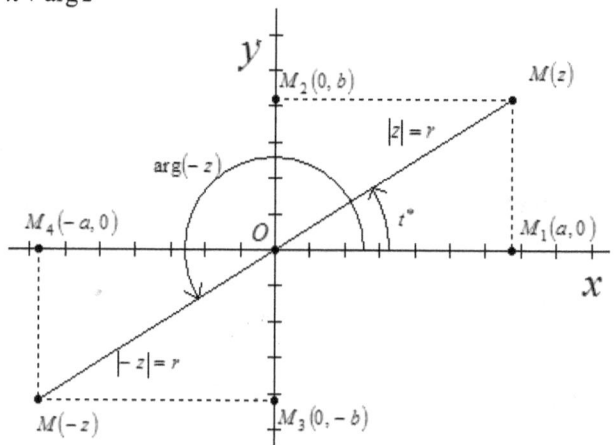

VIII. 72. $M_1\left(2, \dfrac{\pi}{3}\right)$, $M_2\left(\sqrt{2}, \dfrac{7\pi}{4}\right)$, $M_3\left(\sqrt{5}, \pi + \arctan 2\right)$, $M_4\left(1, \dfrac{3\pi}{2}\right)$,
$M_5(1,0)$, $M_6\left(3, \dfrac{\pi}{2}\right)$.

VIII. 73. a) $z_1 = \sqrt{2}\left(\cos\dfrac{5\pi}{4} + i\sin\dfrac{5\pi}{4}\right)$;

b) $z_2 = \sqrt{5}[\cos(\pi - \arctan 2) + i\sin(\pi - \arctan 2)]$; **c)** $z_3 = 2\left(\cos\dfrac{2\pi}{3} + i\sin\dfrac{2\pi}{3}\right)$;

d) $z_4 = 2\left(\cos\dfrac{7\pi}{4} + i\sin\dfrac{7\pi}{4}\right)$; **e)** $z_5 = \sqrt{2}\left(\cos\dfrac{3\pi}{2} + i\sin\dfrac{3\pi}{2}\right)$;

f) If $\theta \in (0, \pi)$, the point $M(1+\cos\theta, \sin\theta)$ lies on the first quadrant and if $\theta \in (\pi, 2\pi)$, the point $M(1+\cos\theta, \sin\theta)$ lies on the fourth quadrant.

$$z_6 = \begin{cases} 2\cos\dfrac{\theta}{2}\left(\cos\dfrac{\theta}{2} + i\sin\dfrac{\theta}{2}\right) & \text{if} \quad \theta \in (0, \pi] \\[3mm] -2\cos\dfrac{\theta}{2}\left(\cos\left(\dfrac{\theta}{2} + \pi\right) + i\sin\left(\dfrac{\theta}{2} + \pi\right)\right) & \text{if} \quad \theta \in (0, \pi) \end{cases}$$

VIII. 74. a) $z = \sqrt{3}$;   **b)** $z = \sqrt{2}$;   **c)** $z = -\sqrt{3}$;   **d)** $z = \dfrac{1}{2} + i\dfrac{\sqrt{3}}{2}$;

e) $z = -\dfrac{\sqrt{2}}{2} + i\dfrac{\sqrt{2}}{2}$;   **f)** $z = -i$;   **g)** $z = -\dfrac{1}{2} - i\dfrac{\sqrt{3}}{2}$;   **h)** $z = \sqrt{3} - i$;

i) $z = \dfrac{\sqrt{6}+\sqrt{2}}{2} + i\dfrac{\sqrt{6}-\sqrt{2}}{2}$;

j) $z = \dfrac{\sqrt{2+\sqrt{2}}}{2} + i\dfrac{\sqrt{2-\sqrt{2}}}{2}$;   **k)** $z = \sqrt{3} + 1 + i\left(\sqrt{3} - 1\right)$.

VIII. 75. a) $(1-i)(\sqrt{3}+i) = \sqrt{2}\left(\cos\dfrac{7\pi}{4} + i\sin\dfrac{7\pi}{4}\right)2\left(\cos\dfrac{\pi}{6} + i\sin\dfrac{\pi}{6}\right) =$

$= 2\sqrt{2}\left(\cos\dfrac{23\pi}{12} + i\sin\dfrac{23\pi}{12}\right)$;   **b)** $8\sqrt{2}\left(\cos\dfrac{5\pi}{4} + i\sin\dfrac{5\pi}{4}\right)$;

c) $\dfrac{(1-i)^8(\sqrt{3}+i)^5}{(-1-i\sqrt{3})^5} = \dfrac{\sqrt{2}^8\left(\cos\dfrac{7\pi}{4} + i\sin\dfrac{7\pi}{4}\right)^8 2^5\left(\cos\dfrac{\pi}{6} + i\sin\dfrac{\pi}{6}\right)^5}{2^5\left(\cos\dfrac{4\pi}{3} + i\sin\dfrac{4\pi}{3}\right)^5} =$

$= \dfrac{2^4\left(\cos\dfrac{56\pi}{4} + i\sin\dfrac{56\pi}{4}\right)2^5\left(\cos\dfrac{5\pi}{6} + i\sin\dfrac{5\pi}{6}\right)}{2^5\left(\cos\dfrac{20\pi}{3} + i\sin\dfrac{20\pi}{3}\right)} = 2^4\left(\cos\dfrac{\pi}{6} + i\sin\dfrac{\pi}{6}\right)$;

d) $\left(\dfrac{\sqrt{3}+i}{1+i}\right)^{10} = (\sqrt{2})^{10}\left[\cos\left(-\dfrac{\pi}{12}\right) + i\sin\left(-\dfrac{\pi}{12}\right)\right]^{10} = 2^5\left(\cos\dfrac{7\pi}{6} + i\sin\dfrac{7\pi}{6}\right)$;

e) $\left(\dfrac{3-i}{1-2i}\right)^8 + \left(\dfrac{3+i}{2-i}\right)^8 = (1+i)^8 + (1-i)^8 =$

$= \sqrt{2}^4\left(\cos\dfrac{8\pi}{4} + i\sin\dfrac{8\pi}{4}\right) + \sqrt{2}^4\left[\cos\left(-\dfrac{8\pi}{4}\right) + i\sin\left(-\dfrac{8\pi}{4}\right)\right] = 2^5$;

f) $2\cos\dfrac{n\pi}{3}$; **g)** $\dfrac{\sqrt{2}}{\cos\theta}\left[\cos\left(\dfrac{\pi}{4}-\theta\right)+i\sin\left(\dfrac{\pi}{4}-\theta\right)\right]$;

h) $-i\tan\dfrac{\theta}{2}$; **i)** $\cos 2\theta+i\sin 2\theta$;

j) $2^n\cdot\sin^n\dfrac{x}{2}\cdot\left[\cos\dfrac{n(\pi-x)}{2}+i\sin\dfrac{n(\pi-x)}{2}\right]$.

VIII. 77. a) $z_k=\sqrt[3]{2(\cos\pi+i\sin\pi)}=\sqrt[3]{2}\left(\cos\dfrac{\pi+2k\pi}{3}+i\sin\dfrac{\pi+2k\pi}{3}\right)$,

$k=0,1,2$; **b)** $z^6=1-i$,

$$z_k=\sqrt[6]{\sqrt{2}\left(\cos\dfrac{7\pi}{4}+i\sin\dfrac{7\pi}{4}\right)}=\sqrt[12]{2}\left(\cos\dfrac{\dfrac{7\pi}{4}+2k\pi}{6}+i\sin\dfrac{\dfrac{7\pi}{4}+2k\pi}{6}\right)$$

$k=0,1,2,3,4,5$; **c)** $z_k=\cos\dfrac{1+4k}{12}\pi+i\sin\dfrac{1+4k}{12}\pi$, $k=0,1,2,3,4,5$;

d) $(z^3-2i)(z^3+i)=0$; **e)** $(z^4-1+i)(z^4+2-i)=0$;

f) $(z^5-1-i)(z^5+1-2i)=0$;

g) $z^5-1=(z-1)(z^4+z^3+z^2+z+1)=0\Rightarrow z^5=1\Rightarrow$

$z^5=\cos 0+i\sin 0$, $z_k=\cos\dfrac{2k\pi}{5}+i\sin\dfrac{2k\pi}{5}$, $k=1,2,3,4,5$;

h) $z^4=\cos\dfrac{\pi}{6}+i\sin\dfrac{\pi}{6}$, $z_k=\cos\dfrac{\dfrac{\pi}{6}+2k\pi}{4}+i\sin\dfrac{\dfrac{\pi}{6}+2k\pi}{4}$, $k=0,1,2,3$;

i) $z^8=\dfrac{\sqrt{2}}{2}\dfrac{\dfrac{1+i}{\sqrt{2}}}{\dfrac{\sqrt{3}-i}{2}}=\dfrac{\sqrt{2}}{2}\dfrac{\cos\dfrac{\pi}{4}+i\sin\dfrac{\pi}{4}}{\cos\dfrac{1}{6}\pi+i\sin\dfrac{1}{6}\pi}=\dfrac{\sqrt{2}}{2}\left(\cos\dfrac{-19\pi}{12}+i\sin\dfrac{-19\pi}{12}\right)$;

j) $(z^4-z^2+1)(z-i)=0$, $\dfrac{z^6+1}{z+i}=0$, $z^6=\cos\pi+i\sin\pi$,

$z_k=\cos\dfrac{\pi+2k\pi}{6}+i\sin\dfrac{\pi+2k\pi}{6}$, where $k=1,2,3,4,5$.

VIII. 78. a) Consider $z_1=\sin x+i\cos x$, $z_2=\sin y+i\cos y$,

$z_3=\sin z+i\cos z$. We have

$$z_1 + z_2 + z_3 = 0 = \overline{z_1 + z_2 + z_3} = \overline{z}_1 + \overline{z}_2 + \overline{z}_3 = \frac{1}{z_1} + \frac{1}{z_2} + \frac{1}{z_3} =$$

$$= \frac{z_1 z_2 + z_2 z_3 + z_3 z_1}{z_1 z_2 z_3}.$$

Then $z_1^2 + z_2^2 + z_3^2 = (z_1 + z_2 + z_3)^2 - 2(z_1 z_2 + z_2 z_3 + z_3 z_1) = 0$;

b) Use mathematical induction;

c) $(z_1 + z_2 + z_3)^3 = z_1^3 + z_2^3 + z_3^3 + 3(z_1 + z_2)(z_2 + z_3)(z_3 + z_1) = 0$. Therefore
$\cos 3x + \cos 3y + \cos 3z + i(\sin 3x + \sin 3y + \sin 3z) = 3z_1 z_2 z_3 =$
$3[\cos(x + y + z) + i \sin(x + y + z)].$

VIII. 79. a) $z^2 - 2z \cos a + 1 = 0$, $z = \cos a \pm i \sin a$. Then
$$z^n + \frac{1}{z^n} = (\cos a \pm i \sin a)^n + (\cos a \pm i \sin a)^{-n} = 2 \cos na.$$

VIII. 80. $\cos 5x + i \sin 5x = (\cos x + i \sin x)^5 =$

$$= \cos^5 x - 10 \cos^3 x \sin^2 x + 5 \cos x \sin^4 x + i(5 \cos^4 x \sin x - 10 \cos^2 x \sin^3 x + \sin^5 x)$$

Therefore $\tan 5x = \dfrac{5 \tan x - 10 \tan^3 x + \tan^5 x}{1 - 10 \tan^2 x + 5 \tan^4 x}.$

VIII. 81. $z_1 z_2 = \cos(\alpha + \beta) + i \sin(\alpha + \beta)$, $\overline{z}_1 \overline{z}_2 = \cos(\alpha + \beta) - i \sin(\alpha + \beta)$.

Therefore $2 \cos(\alpha + \beta) = z_1 z_2 + \overline{z}_1 \overline{z}_2 = z_1 z_2 + \dfrac{1}{z_1 z_2} = \dfrac{z_1^2 z_2^2 + 1}{z_1 z_2}.$

VIII. 82. $z_1 = \cos \alpha + i \sin \alpha$, $z_2 = \cos \beta + i \sin \beta$ be and $z_3 = \cos \theta + i \sin \theta$
and $z_4 = \cos \omega + i \sin \omega$ be four complex numbers.

We have $\dfrac{z_1^2 z_2^2 - 1}{2 z_1 z_2} \cdot \dfrac{2 z_1 z_2}{z_1^2 - z_2^2} = \dfrac{z_3^2 z_4^2 - 1}{2 z_3 z_4} \cdot \dfrac{2 z_3 z_4}{z_3^2 - z_4^2} \Rightarrow$

$z_1^2 z_2^2 z_3^2 - z_1^2 z_2^2 z_4^2 - z_1^2 z_3^2 z_4^2 + z_2^2 z_3^2 z_4^2 + z_1^2 - z_2^2 - z_3^2 + z_4^2 = 0 \Rightarrow$

$\dfrac{z_1^2 z_4^2 + 1}{2 z_1 z_4} \cdot \dfrac{2 z_1 z_4}{z_1^2 + z_4^2} = \dfrac{z_2^2 z_3^2 + 1}{2 z_2 z_3} \cdot \dfrac{2 z_2 z_3}{z_2^2 + z_3^2} \Rightarrow \dfrac{\cos(\alpha + \omega)}{\cos(\alpha - \omega)} = \dfrac{\cos(\beta + \theta)}{\cos(\beta - \theta)}.$

VIII. 83. Let z, $z_A = -3 + 2i$, and $z_B = -1 + i$ be three complex numbers
and $M(z)$, $A(-3 + 2i)$, and $B(-1 + i)$ their geometric image on the

complex plane. The given relation becomes $\left|z - z_A\right| = \left|z - z_B\right|$, which means $MA = MB$. Therefore the geometric imagine of the complex numbers z is the set of points on the perpendicular bisector of the segment AB.

VIII. 84. $IA = \left|z_A - z_I\right| = \left|5 + i - (2 + i)\right| = 3,$

$IB = \left|z_B - z_I\right| = \left|4 + \left(1 - \sqrt{5}\right)i - (2 + i)\right| = \left|2 - i\sqrt{5}\right| = \sqrt{4 + 5} = 3,$

$IC = \left|z_C - z_I\right| = \left|2 - 2\sqrt{2} + 2i - (2 + i)\right| = \left|-2\sqrt{2} + i\right| = \sqrt{8 + 1} = 3.$ Therefore the points A, B, C are the points of the circle of center I and radius 3.

VIII. 85. a) $\left(\dfrac{1 + iz}{1 - iz}\right)^n = -1 \Rightarrow \dfrac{1 + iz}{1 - iz} = \sqrt[n]{\cos\dfrac{3\pi}{2n} + i\sin\dfrac{3\pi}{2n}} \Rightarrow$

$\dfrac{1 + iz}{1 - iz} = \cos\dfrac{3\pi + 4k\pi}{2n} + i\sin\dfrac{3\pi + 4k\pi}{2n}, \ k = 0, 1, 2, \dots, n - 1$

Denoting $\alpha = \dfrac{3\pi + 4k\pi}{4n}$, then $\dfrac{1 + iz}{1 - iz} = \cos 2\alpha + i\sin 2\alpha$. Solving for z we

obtain the solutions $z_k = \tan\dfrac{3\pi + 4k\pi}{4n}, \ k = 0, 1, 2, \dots, n - 1;$

b) Let $\dfrac{1 + ia}{1 - ia} = r(\cos t + i\sin t)$. $\left|\dfrac{1 + ia}{1 - ia}\right| = \dfrac{\left|1 + ia\right|}{\left|1 - ia\right|} = \dfrac{\sqrt{1 + a^2}}{\sqrt{1 + a^2}} = 1 \Rightarrow r = 1$.

So $\dfrac{1 + ia}{1 - ia} = \cos t + i\sin t \Rightarrow a = \tan\dfrac{t}{2}$. From $\left(\dfrac{1 + iz}{1 - iz}\right)^n = \cos t + i\sin t$ we

get $\dfrac{1 + iz}{1 - iz} = \sqrt[n]{\cos t + i\sin t} = \cos\dfrac{t + 2k\pi}{n} + i\sin\dfrac{t + 2k\pi}{n}, k = 0,1,\dots,n - 1$ and

$z = \tan\dfrac{t + 2k\pi}{2n} \ k,= 0,1,\dots,n - 1$ Since $t = 2\arctan a$, therefore $z \in \mathbf{R}$.

c) We have $\left(\dfrac{z + i\sqrt{1 - z^2}}{z - i\sqrt{1 - z^2}}\right)^n = -1 = \cos\pi + i\sin\pi \Rightarrow$

$\dfrac{z + i\sqrt{1 - z^2}}{z - i\sqrt{1 - z^2}} = \cos\dfrac{\pi + 2k\pi}{n} + i\sin\dfrac{\pi + 2k\pi}{n} \Rightarrow$

$\dfrac{z}{\sqrt{1 - z^2}} = \dfrac{i\left[\cos\dfrac{(2k + 1)\pi}{2n} + i\sin\dfrac{(2k + 1)\pi}{2n} + 1\right]}{\cos\dfrac{(2k + 1)\pi}{2n} + i\sin\dfrac{(2k + 1)\pi}{2n} - 1} =$

$$= \frac{2i\cos^2\dfrac{(2k+1)\pi}{2n} - 2\sin\dfrac{(2k+1)\pi}{2n}\cos\dfrac{(2k+1)\pi}{2n}}{-2\sin^2\dfrac{(2k+1)\pi}{2n} + 2i\sin\dfrac{(2k+1)\pi}{2n}\cos\dfrac{(2k+1)\pi}{2n}} = \cot\dfrac{(2k+1)\pi}{2n}\;.$$

Finally $z = \cos\dfrac{(2k+1)\pi}{2n}$ $k = 0,1,...,n-1$.

d) $z_k = n\tan\dfrac{k\pi}{n}$, $k = 0,1,2,...,n-1$;

e) For $a \neq 0$, we have $\dfrac{ax + i\sqrt{1-x^2}}{ax - i\sqrt{1-x^2}} = \cos\dfrac{2k+1}{n}\pi + i\,\sin\dfrac{2k+1}{n}\pi$.

Using the properties of proportions, we obtain

$$\frac{ax}{-i\sqrt{1-x^2}} = \frac{1 + \cos\dfrac{2k+1}{n}\pi + i\,\sin\dfrac{2k+1}{n}\pi}{1 - \cos\dfrac{2k+1}{n}\pi - i\,\sin\dfrac{2k+1}{n}\pi}\ \text{ or }$$

$$\frac{ax}{-i\sqrt{1-x^2}} = \frac{2\cos^2\dfrac{2k+1}{2n}\pi + i\,2\sin\dfrac{2k+1}{2n}\pi\ \cos\dfrac{2k+1}{2n}\pi}{2\sin^2\dfrac{2k+1}{2n}\pi - i\,2\sin\dfrac{2k+1}{2n}\pi\ \cos\dfrac{2k+1}{2n}\pi} \Rightarrow$$

$$\frac{ax}{-i\sqrt{1-x^2}} = ctg\dfrac{2k+1}{2n}\pi \cdot \frac{\cos\dfrac{2k+1}{2n}\pi + i\,\sin\dfrac{2k+1}{2n}\pi}{-i^2\sin\dfrac{2k+1}{2n}\pi - i\,\cos\dfrac{2k+1}{2n}\pi} \Rightarrow$$

$$\frac{ax}{\sqrt{1-x^2}} = \cot\dfrac{2k+1}{2n}\pi \Rightarrow \frac{a^2x^2}{1-x^2} = \cot^2\dfrac{2k+1}{2n}\pi \Rightarrow$$

$$\frac{a^2x^2}{a^2 - a^2x^2} = \frac{\cot^2\dfrac{2k+1}{2n}\pi}{a^2} \Rightarrow x^2 = \frac{\cot^2\dfrac{2k+1}{2n}\pi}{a^2 + \cot^2\dfrac{2k+1}{2n}\pi},\ k = 0,1,2,3,...,n-1.$$

For $a = 0$ n even, $i^n\left(\sqrt{1-x^2}\right)^n + \left(i\sqrt{1-x^2}\right)^n = 0 \Rightarrow$

$\left(\sqrt{1-x^2}\right)^n(i^n + i^n) \neq 0$, $\forall x \in \mathbf{R} \setminus \{-1,1\}$

For $a = 0$, n odd $\left(\sqrt{1-x^2}\right)^{2n+1}(i^{2n+1} - i^{2n+1}) = 0$, $\forall x \in \mathbf{R}$ is a solution.

VIII. 86. $P = \prod_{k=1}^{n}\left(z^k + \overline{z}^k\right) = \prod_{k=1}^{n} 2r^k \cos kt = 2^n (r)^{\frac{n(n+1)}{2}} \cos t \cos 2t \ldots \cos nt$.

VI. 88. Let (1) $\dfrac{x-i}{x+i} = z$, then the initial equation becomes

$z^{n-1} + z^{n-2} + \ldots + z + 1 = 0$.

$z^n - 1 = (z-1)\left(z^{n-1} + z^{n-2} + \ldots + z + 1\right) = 0 \Rightarrow z^n = 1$, with $z \neq 1$. If $z^n = 1$ we

obtain, $z^n = \cos 0 + i \sin 0 \Rightarrow z = \sqrt{\cos 0 + i \sin 0} = \cos\dfrac{0 + 2k\pi}{n} + i \sin\dfrac{0 + 2k\pi}{n}$,

From (1) we obtain $\dfrac{x-i}{x+i} = \cos\dfrac{2k\pi}{n} + i \sin\dfrac{2k\pi}{n} \Rightarrow x = \dfrac{i(1+z)}{1-z}$. So

$$x_k = i \left| \frac{1 + \cos\dfrac{2k\pi}{n} + i\sin\dfrac{2k\pi}{n}}{1 - \cos\dfrac{2k\pi}{n} - i\sin\dfrac{2k\pi}{n}} \right| = \frac{2i\cos^2\dfrac{2k\pi}{2n} - 2i\sin\dfrac{2k\pi}{2n}\cos\dfrac{2k\pi}{2n}}{2\sin^2\dfrac{2k\pi}{2n} - 2i\sin\dfrac{2k\pi}{2n}\cos\dfrac{2k\pi}{2n}} = -\cot\dfrac{k\pi}{n}$$

$k = 1,2,\ldots,n-1$.

VIII. 89.

$A = \left\{ \cos\dfrac{n\pi}{3} + i\sin\dfrac{n\pi}{3} \mid n \in \mathbf{N} \right\} = \left\{ 1, \dfrac{1}{2} + \dfrac{i\sqrt{3}}{2}, -\dfrac{1}{2} + \dfrac{i\sqrt{3}}{2}, -1, -\dfrac{1}{2} - \dfrac{i\sqrt{3}}{2}, \dfrac{1}{2} - \dfrac{i\sqrt{3}}{2} \right\}$,

$B = \left\{ \cos\dfrac{5n\pi}{3} + i\sin\dfrac{5n\pi}{3} \mid n \in \mathbf{N} \right\} = \left\{ 1, \dfrac{1}{2} + \dfrac{i\sqrt{3}}{2}, -\dfrac{1}{2} + \dfrac{i\sqrt{3}}{2}, -1, -\dfrac{1}{2} - \dfrac{i\sqrt{3}}{2}, \dfrac{1}{2} - \dfrac{i\sqrt{3}}{2} \right\}$,

$C = \left\{ \cos\dfrac{2n\pi}{3} + i\sin\dfrac{2n\pi}{3} \mid n \in \mathbf{N} \right\} = \left\{ 1, -\dfrac{1}{2} + \dfrac{i\sqrt{3}}{2}, -\dfrac{1}{2} - \dfrac{i\sqrt{3}}{2} \right\}$,

$D = \left\{ \cos\dfrac{4n\pi}{3} + i\sin\dfrac{4n\pi}{3} \mid n \in \mathbf{N} \right\} = \left\{ 1, -\dfrac{1}{2} + \dfrac{i\sqrt{3}}{2}, -\dfrac{1}{2} - \dfrac{i\sqrt{3}}{2} \right\}$.

Trigonometric Sums

IX. 1. $\sin 91° + \sin 93° + ... + \sin 179° =$

$= \sin(180° - 89°) + \sin(180° - 87°) + ... + \sin(180° - 3°) + \sin(180° - 1°) =$
$= \sin 89° + \sin 87° + ... + \sin 3° + \sin 1° =$

$= \cos(90° - 89°) + \cos(90° - 87°) + ... + \cos(90° - 1°) =$
$= \cos 1° + \cos 2° + ... + \cos 89°.$

IX. 2. The sum has 359 terms.

$\cos 1° + \cos 181° = \cos 2° + \cos 182° = ... = \cos 179° + \cos 359° = 0$. Since

$\cos 180° = -1$, then $S = -1$. **IX. 3.** 0.

IX. 4. Use $\sec k° \sec(k+1)° = \dfrac{1}{\sin 1°}\left(\dfrac{\sin(n+1)°}{\cos(n+1)°} - \dfrac{\sin n°}{\cos n°}\right)$ for

$k = 0, 1, 2, 3, ..., n$. The sum is $\dfrac{\tan(n+1)°}{\sin 1°}$

IX. 6. *Solution 1.* $S_1 + S_2 = n$, $S_1 - S_2 = \cos 2x + \cos 4x + ... + \cos 2nx$,

use **IX. 5.** $S_1 = \dfrac{n}{2} + \dfrac{\cos(n+1)x \sin nx}{2 \sin x}$, $S_2 = \dfrac{n}{2} - \dfrac{\cos(n+1)x \sin nx}{2 \sin x}$.

Solution 2. $S_1 = \cos^2 x + \cos^2 2x + ... + \cos^2 nx =$

$= \dfrac{1 + \cos 2x}{2} + \dfrac{1 + \cos 4x}{2} + ... + \dfrac{1 + \cos 2nx}{2} =$

$$= \underbrace{\frac{1}{2} + \frac{1}{2} + \ldots + \frac{1}{2}}_{n \ times} + \frac{1}{2}(\cos 2x + \cos 4x + \ldots + \cos 2nx) =$$

$$S_1 = \frac{n}{2} + \frac{1}{2}(\cos 2x + \cos 4x + \ldots + \cos 2nx),$$

$$2\left(S_1 - \frac{n}{2}\right) = \cos 2x + \cos 4x + \ldots + \cos 2nx.$$

$$4\sin x \left(S_1 - \frac{n}{2}\right) = 2\sin x \cos 2x + 2\sin x \cos 4x + 2\sin x \cos 6x + \ldots + 2\sin x \cos 2nx =$$

$$= (\sin 3x - \sin x) + (\sin 5x - \sin 3x) + (\sin 7x - \sin 5x) + \ldots + (\sin(2n+1)x - \sin(2n-1)x)$$

Thus $4\sin x \left(S_1 - \frac{n}{2}\right) = \sin(2n+1)x - \sin x = 2\cos(n+1)x \sin nx.$

$$S_1 = \frac{n}{2} + \frac{\cos(n+1)x \sin nx}{2\sin x}.$$

IX. 7. Use $\sin^3 kx = \frac{1}{4}(3\sin kx - \sin 3kx)$, for $k = 1, 2, 3, \ldots, n$.

$$S_1 = \frac{3}{4} \frac{\sin\left(\frac{n+1}{2}\right)x \sin \frac{nx}{2}}{\sin \frac{x}{2}} - \frac{1}{4} \frac{\sin 3\left(\frac{n+1}{2}\right)x \sin \frac{3nx}{2}}{\sin \frac{3x}{2}};$$

Use $\cos^3 kx = \frac{1}{4}(\cos 3kx + 3\cos 3kx)$, for $k = 1, 2, 3, \ldots, n$.

$$S_n = \frac{1}{4} \frac{\sin 3\left(\frac{n+1}{2}\right)x \sin \frac{3nx}{2}}{\sin \frac{3x}{2}} + \frac{3}{4} \frac{\cos\left(\frac{n+1}{2}\right)x \sin \frac{nx}{2}}{\sin \frac{x}{2}}.$$

IX. 8. Use $\sin \frac{x}{3^{k-1}} = 3\sin \frac{x}{3^k} - 4\sin^3 \frac{x}{3^k}$, for $k = 1, 2, 3, \ldots, n$.

$$S_n = \frac{1}{4}\left(3^n \sin \frac{x}{3^n} - \sin x\right).$$

IX. 9. a) Use $\dfrac{1}{\cos x - \cos(2k+1)x} = \dfrac{-1}{2\sin x}[\cot(k+1)x - \cot kx],$

298

$$k = 1, 2, 3, ..., n, \ S_n = \frac{\sin nx}{\sin^2 x \sin(n+1)x};$$

b) Use $\dfrac{1}{\cos x + \cos(2k+1)x} = \dfrac{1}{2\sin x}[\tan(k+1)x - \tan kx], \ k = 1, 2, 3, ..., n,$

$$S_n' = \frac{\sin nx}{\sin 2x \cos(n+1)x};$$

c) Use $\dfrac{1}{\cos kx \cos(k+1)x} = \dfrac{1}{\sin x}[\tan(k+1)x - \tan kx], \ k = 1, 2, 3, ..., n,$

$$S_n'' = \frac{2\sin nx}{\sin 2x \cos(n+1)x}.$$

IX. 10. a) Use $\tan(a+kb) - \tan(a+(k-1)b) = \dfrac{\sin b}{\cos(a+kb)\cos(a+(k-1)b)},$

$$S_n = \frac{\sin nb}{\sin b \cos a \cos(a+nb)}; \ \textbf{b)} \ S_n = \frac{2\sin na}{\sin 2a \cos(n+1)a}.$$

IX. 11. a) Use $\dfrac{1}{2^k}\tan\dfrac{x}{2^k} = \dfrac{1}{2^k}\cot\dfrac{x}{2^k} - \dfrac{1}{2^{k-1}}\cot\dfrac{x}{2^{k-1}}, \ k = 0, 1, 2, 3, ..., n.$

$$S_n = \frac{1}{2^n}\cot\frac{x}{2^n} - 2\cot 2x;$$

b) Use $\tan 2^k x = \cot 2^k x - 2\cot 2^{k+1} x, \ k = 0, 1, 2, 3, ..., n.$

$$S_n' = \cot x - 2^{n+1}\cot 2^{n+1} x.$$

IX. 12. Use $\dfrac{\tan 2^k x}{\cos 2^{k+1} x} = \tan 2^{k+1} x - \tan 2^k x, \ k = 0, 1, 2, 3, ..., n.$

$$S_n = \tan 2^{n+1} x - \tan x.$$

IX. 13. Use $\dfrac{\sin kx}{\cos^k x} = \dfrac{1}{\cos^{k-1} x} \cdot \dfrac{\cos kx}{\sin x} - \dfrac{1}{\cos^k x} \cdot \dfrac{\cos(k+1)x}{\sin x},$

$k = 1, 2, 3, ..., n. \ S_n = \cot x - \dfrac{1}{\cos^n x} \cdot \dfrac{\cos(n+1)x}{\sin x}.$

IX. 14. Use $\dfrac{\cos kx}{\cos^k x} = \cot x\left(\dfrac{\sin(k+1)x}{\cos^{k+1} x} - \dfrac{\sin kx}{\cos^k x}\right), \ k = 1, 2, 3, ..., n+1.$

$$S_n = \frac{\sin(n+1)x}{\sin x \cos^n x}.$$

IX. 15. Use $\dfrac{\sin\dfrac{a}{k(k+1)}}{\cos\dfrac{a}{k}\cos\dfrac{a}{k+1}} = \tan\dfrac{a}{k} - \tan\dfrac{a}{k+1}$, $k = 1,2,3,...,n$.

$$S_n = \tan a - \tan\frac{a}{n+1}.$$

IX. 16. Use $\dfrac{\sin(2k-1)x}{\cos^2 kx \cos^2(k-1)x} = \dfrac{1}{\sin x}\left(\tan^2 kx - \tan^2(k-1)x\right)$, $k = 1,2,3,...,n$.

$$S_n = \frac{\tan^2 nx}{\sin x}.$$

IX. 17. Use $2^{k-1}\tan^2\dfrac{x}{2^k}\tan\dfrac{x}{2^{k-1}} = 2^{k-1}\tan\dfrac{x}{2^{k-1}} - 2^k\tan\dfrac{x}{2^k}$, $k = 1,2,3,...,n$.

$$S_n = \tan x - 2^n\tan\frac{x}{2^n}.$$

IX. 18. Use $\csc 2^k x = \cot 2^{k-1}x - \cot 2^k x$, for $k = 1,2,3,...,n$.

$S_n = \cot x - \cot 2^n x$.

IX. 19. Use $\dfrac{1}{2^{2k}}\sec^2\dfrac{x}{2^k} = \dfrac{1}{2^{2k-2}}\csc^2\dfrac{x}{2^{k-1}} - \dfrac{1}{2^{2k}}\csc^2\dfrac{x}{2^k}$, for

$k = 1,2,3,...,n$. $S_n = \csc^2 x - \dfrac{1}{2^{2n}}\csc^2\dfrac{x}{2^n}$.

IX. 20. $\dfrac{\cos x_i}{\cos x_i + \sin x_i} + \dfrac{\sin x_i}{\sin x_i + \cos x_i} + \dfrac{\cos x_i}{\cos x_i - \sin x_i} + \dfrac{\sin x_i}{\sin x_i - \cos x_i} = 2$,

Then $S_n = 2n$.

VI. 21. Use $\arcsin\dfrac{\sqrt{k+1}-\sqrt{k}}{\sqrt{k+1}\sqrt{k+2}} = \arcsin\dfrac{1}{\sqrt{k+1}} - \arcsin\dfrac{1}{\sqrt{k+2}}$,

$k = 1,2,3,...,n$. $S_n = \dfrac{\pi}{6} - \arcsin\dfrac{1}{\sqrt{n+2}}$.

VI. 22. Denote $a_k = \arctan \dfrac{1}{\sqrt{2k+1}}$ and $b_k = \arcsin \dfrac{\sqrt{2k+1}}{k+1}$.

It is obviously that $b_k \in \left(0, \dfrac{\pi}{2}\right)$ and $a_k \in \left(0, \dfrac{\pi}{4}\right)$. We have

$$\tan b_k = \frac{\sin b_k}{\sqrt{1-\sin^2 b_k}} = \frac{\sqrt{2k+1}}{k} \quad \text{and} \quad \tan 2a_k = \frac{2\tan a_k}{1-\tan^2 a_k} = \frac{\sqrt{2k+1}}{k}.$$

Therefore $\tan 2a_k = \tan b_k$, then $2a_k = b_k$ or $\dfrac{a_k}{b_k} = \dfrac{1}{2}$. Finally

$$S_n = \sum_{k=1}^{n} \frac{a_k}{b_k} = \sum_{k=1}^{n} \frac{1}{2} = \frac{n}{2}.$$

VI. 23. $\displaystyle\sum_{k=1}^{n} \arctan \frac{k^2+k+2}{k^2+k} = \sum_{k=1}^{n}\left[\arctan 1 + \arctan(k+1) - \arctan k\right] =$

$= \arctan(n+1)$. Therefore $S_n = n+1$.

VI. 24. Use $\arctan \dfrac{2k}{k^4+k^2+2} = \arctan\left(k^2+k+1\right) - \arctan\left(k^2-k+1\right)$,

$k = 1, 2, 3, ..., n$. $S_n = \dfrac{n^2+n}{n^2+n+2}$.

IX. 25. Use $\arctan \dfrac{1}{2 \cdot k^2} = \arctan \dfrac{k}{k+1} - \arctan \dfrac{k-1}{k}$,

$k = 1, 2, 3, ..., n$. $S_n = \arctan \dfrac{n}{n+1}$.

IX. 26. Use $\arctan\left(\dfrac{1}{1+k(k+1)x^2}\right) = \arctan(k+1)x - \arctan kx$,

$k = 1, 2, 3, ..., n$. $S_n = \arctan\left(\dfrac{nx}{1+(n+1)x^2}\right)$.

IX. 27. Use $\text{arc}\cot\left(1+k+k^2\right) = \text{arc}\cos(1+k) - \text{arc}\cot k$, $k = 1, 2, 3, ..., n$.

$S_n = \text{arc}\cot 3 - \text{arc}\cot\left(1 + \dfrac{2}{n}\right)$.

IX. 28. Use $\text{arc}\cot\left(2k^2\right) = \text{arc}\cos(2k-1) - \text{arc}\cot(2k+1)$, $k = 1, 2, 3, ..., n$.

$$S_n = \operatorname{arc\,cot} 1 - \operatorname{arc\,cot}(2n+1).$$

IX. 29. Use $\arctan \dfrac{r}{1 + a_{k-1}a_k} = \arctan a_k - \arctan a_{k-1}$. $k = 2,3,...,n$

$$S_n = \arctan a_n - \arctan a_1.$$

IX. 30. Use $\arctan \dfrac{(a_{n+1} - a_k)^2}{a_k^2 - a_k a_{k+1} + a_{k+1}^2} = \arctan \dfrac{r}{a_k} - \arctan \dfrac{r}{a_{k+1}}$,

also $a_k^2 - a_k a_{k+1} + a_{k+1}^2 = (a_k - a_{k+1})^2 + a_k a_{k+1} = r^2 + a_k a_{k+1}$.

$$S_n = \arctan \dfrac{r}{a_1} - \arctan \dfrac{r}{a_{n+1}}.$$

IX. 31. Use $\ln\cos\dfrac{x}{2^k} = \ln\sin\dfrac{x}{2^{k-1}} - \ln\sin\dfrac{x}{2^k} - \ln 2$, for $k = 1,2,3,...,n$.

$$S_n = \ln\sin x - \ln\sin\dfrac{x}{2^n} - n\ln 2.$$

IX. 32. Use $\ln\!\left(1 + 2\cos 3^{k-1}x\right) = \ln\sin\dfrac{3^k x}{2} - \ln\sin\dfrac{3^{k-1}x}{2}$, for $k = 1,2,3,...,n$.

$$S_n = \ln\sin\dfrac{3^n x}{2} - \ln\sin\dfrac{x}{2}.$$

IX. 33. $S_1 + S_2 = \sin(x + 2y) + \sin(2x + 3y) + ... + \sin[(n-1)x + ny]$,

$S_1 + S_2 = \sin(2y - x) + \sin(3y - 2x) + ... + \sin[ny - (n-1)x]$.

Denoting $x + y = h$ and $y - x = a$, we obtain

$$S_1 + S_2 = \sin(y + h) + \sin(y + 2h) + ... + \sin[y + (n-1)h] = \dfrac{\sin\dfrac{n}{2}h \sin\left(y + \dfrac{(n-1)h}{2}\right)}{\sin\dfrac{h}{2}}$$

$$S_1 - S_2 = \sin(y + a) + \sin(y + 2a) + ... + \sin[y + (n-1)a] = \dfrac{\sin\dfrac{n}{2}a \sin\left(y + \dfrac{(n-1)a}{2}\right)}{\sin\dfrac{a}{2}}$$

Finally $S_1 = \dfrac{\sin\dfrac{n}{2}h\sin\left(y+\dfrac{(n-1)h}{2}\right)}{2\sin\dfrac{h}{2}} + \dfrac{\sin\dfrac{n}{2}a\sin\left(y+\dfrac{(n-1)a}{2}\right)}{2\sin\dfrac{a}{2}}$ and

$$S_1 = \dfrac{\sin\dfrac{n}{2}h\sin\left(y+\dfrac{(n-1)h}{2}\right)}{2\sin\dfrac{h}{2}} - \dfrac{\sin\dfrac{n}{2}a\sin\left(y+\dfrac{(n-1)a}{2}\right)}{2\sin\dfrac{a}{2}}.$$

IX. 34. a) Using **IX. 5. a)**, we obtain

$$S_1 = \sum_{k=0}^{n}\cos\frac{(2k+1)\pi}{4n+3} = \sum_{k=0}^{n}\cos\left(\frac{\pi}{4n+3}+k\cdot\frac{2\pi}{4n+3}\right) =$$

$$= \dfrac{\sin\dfrac{n+1}{2}\dfrac{2\pi}{4n+3}\cos\left(\dfrac{\pi}{4n+3}+\dfrac{n}{2}\dfrac{2\pi}{4n+3}\right)}{\sin\dfrac{1}{2}\dfrac{2\pi}{4n+3}} = \dfrac{\sin\dfrac{\pi(n+1)}{4n+3}\cos\dfrac{\pi(n+1)}{4n+3}}{\sin\dfrac{\pi}{4n+3}} =$$

$$= \dfrac{\sin\dfrac{2\pi(n+1)}{4n+3}}{2\sin\dfrac{\pi}{4n+3}} = \dfrac{\cos\left[\dfrac{\pi}{2}-\dfrac{2\pi(n+1)}{4n+3}\right]}{2\sin\dfrac{\pi}{4n+3}} = \dfrac{\cos\dfrac{\pi}{2(4n+3)}}{2\sin\dfrac{\pi}{4n+3}} =$$

$$= \dfrac{\cos\dfrac{\pi}{2(4n+3)}}{4\sin\dfrac{\pi}{2(4n+3)}\cos\dfrac{\pi}{2(4n+3)}} = \dfrac{1}{4\sin\dfrac{\pi}{2(4n+3)}};$$

b) $S_2 = \displaystyle\sum_{k=0}^{n}\cos\frac{2k\pi}{4n+3} = \sum_{k=0}^{n}\cos\left(0+k\cdot\frac{2\pi}{4n+3}\right) =$

$$= \sum_{k=0}^{n}\cos\left(0+k\cdot\frac{2\pi}{4n+3}\right) = \dfrac{\sin\dfrac{n+1}{2}\dfrac{2\pi}{4n+3}\cos\left(0+\dfrac{n}{2}\dfrac{2\pi}{4n+3}\right)}{\sin\dfrac{1}{2}\dfrac{2\pi}{4n+3}} =$$

$$= \dfrac{2\sin\dfrac{(n+1)\pi}{4n+3}\cos\dfrac{n\pi}{4n+3}}{2\sin\dfrac{\pi}{4n+3}} = \dfrac{\sin\dfrac{(2n+1)\pi}{4n+3}+\sin\dfrac{\pi}{4n+3}}{2\sin\dfrac{\pi}{4n+3}} = \dfrac{\cos\left[\dfrac{\pi}{2}-\dfrac{\pi(2n+1)}{4n+3}\right]}{2\sin\dfrac{\pi}{4n+3}}+\dfrac{1}{2} =$$

$$= \frac{\cos\dfrac{\pi}{2(4n+3)}}{4\sin\dfrac{\pi}{2(4n+3)}\cos\dfrac{\pi}{2(4n+3)}} + \frac{1}{2} = \frac{1}{4\sin\dfrac{\pi}{2(4n+3)}} + \frac{1}{2};$$

c) $2\sin\dfrac{\pi}{4n+3}\cdot S_3 = 2\sin\dfrac{\pi}{4n+3}\cos\dfrac{\pi}{4n+3} + 2\sin\dfrac{\pi}{4n+3}\cos\dfrac{3\pi}{4n+3} +$

$+\, 2\sin\dfrac{\pi}{4n+3}\cos\dfrac{5\pi}{4n+3} + ... + 2\sin\dfrac{\pi}{4n+3}\cos\dfrac{(4n+1)\pi}{4n+3},$

$2\sin\dfrac{\pi}{4n+3}\cdot S_3 = \sin\dfrac{2\pi}{4n+3} + \left(\sin\dfrac{4\pi}{4n+3} - \sin\dfrac{2\pi}{4n+3}\right) +$

$+\left(\sin\dfrac{6\pi}{4n+3} - \sin\dfrac{4\pi}{4n+3}\right) + ... + \left(\sin\dfrac{(4n+2)\pi}{4n+3} - \sin\dfrac{4n\pi}{4n+3}\right).$

Therefore $2\sin\dfrac{\pi}{4n+3}\cdot S_3 = \sin\dfrac{(4n+2)\pi}{4n+3} \Rightarrow$

$2\sin\dfrac{\pi}{4n+3}\cdot S_3 = \sin\left[\pi - \dfrac{(4n+2)\pi}{4n+3}\right] \Rightarrow$

$2\sin\dfrac{\pi}{4n+3}\cdot S_3 = \sin\dfrac{\pi}{4n+3} \Rightarrow S_3 = \dfrac{1}{2}.$

d) $S_4 = S_1 - (S_2 - 1) = S_3.$

IX. 35. $\tan(2k+1)x = \dfrac{\tan(2k-1)x + \tan 2x}{1 - \tan(2k-1)x\tan 2x} \Rightarrow$

$\tan(2k-1)x\cdot\tan 2x\cdot\tan(2k+1)x = \tan(2k+1)x - \tan 2x - \tan(2k-1)x.$

Therefore

$$S_n = \sum_{k=1}^{n}\tan(2k+1)x\cdot\tan 2x\cdot\tan(2k-1)x =$$

$$= \sum_{k=1}^{n}\tan(2k+1)x - \sum_{k=1}^{n}\tan(2k-1)x - \sum_{k=1}^{n}\tan 2x =$$

$$= \tan(2n+1)x - n\tan 2x - \tan x.$$

IX. 36. $\tan 1° = \cot(90° - 1°) = \cot 89°,$

$\tan 2° = \cot(90° - 2°) = \cot 8$ °,

....

$\tan 44° = \cot(90° - 44°) = \cot 46°.$ Therefore the product is 1.

IV. 37. For E_1 use $\cos 2^k x = \dfrac{\sin 2^{k+1} x}{2 \sin 2^n x}$, $E_1 = \dfrac{\sin 2^{n+1} x}{2^{n+1} \sin x}$.

$$E_2 = \frac{\sin 2x}{2^{n+1} \sin \dfrac{x}{2^n}}.$$

IX. 38. Use $1 + \sin^{2^k} x = \left(1 - \sin^{2^{k-1}} x\right)\left(1 + \sin^{2^{k-1}} x\right)$ and multiply these relations side by side for $k = 0, 1, 2, 3, ..., n$. $\mathbf{P}_n = \dfrac{1 - \sin^{2^n} x}{1 - \sin x}$.

IX. 39. $\mathbf{P}_n = \dfrac{\cos x \; \cos \dfrac{x}{2} \; \cos \dfrac{x}{2^{n-1}}}{\cos^2 \dfrac{x}{2} \; \cos^2 \dfrac{x}{2^2} \cdots \cos^2 \dfrac{x}{2^n}} = \dfrac{\cos x}{\left(\cos \dfrac{x}{2} \cos \dfrac{x}{2^2} \cdots \cos \dfrac{x}{2^n}\right)\cos \dfrac{x}{2^n}} =$

$$= \frac{2 \sin \dfrac{x}{2^n} \cos x}{\left(\cos \dfrac{x}{2} \cos \dfrac{x}{2^2} \cdots \cos \dfrac{x}{2^n} 2 \sin \dfrac{x}{2^n}\right)\cos \dfrac{x}{2^n}} = \frac{2^n \sin \dfrac{x}{2^n} \cos x}{\sin x \cos \dfrac{x}{2^n}} = 2^n \tan \frac{x}{2^n} \cot x.$$

IX. 40. Use $1 + \sec 2^{k-1} x = \dfrac{\tan 2^{k-1} x}{\tan 2^{k-2} x}$, $k = 1, 2, 3, ..., n$. $\mathbf{P}_n = \dfrac{\tan 2^{n-1} x}{\tan 2^{-1} x}$.

IX. 41. Use $\cos x + \cos y = \dfrac{1}{2} \dfrac{\cos 2x - \cos 2y}{\cos x - \cos y}$. $\mathbf{P}_n = \dfrac{\cos x - \cos y}{2^n \left(\cos \dfrac{x}{2^n} - \cos \dfrac{x}{2^n}\right)}$.

IX. 42. Use $2 \cos 2^{k-1} x - 1 = \dfrac{2 \cos 2^k x + 1}{2 \cos 2^{k-1} x + 1}$, $k = 1, 2, 3, ..., n$.

$$\mathbf{P}_n = \frac{2 \cos 2^n x + 1}{2 \cos x + 1}.$$

IX. 43. Use $2 \cos \dfrac{x}{2^k} - 1 = \dfrac{2 \cos \dfrac{x}{2^{k-1}} + 1}{2 \cos \dfrac{x}{2^k} + 1}$, $k = 1, 2, 3, ..., n$. $\mathbf{P}_n = \dfrac{2 \cos x + 1}{2 \cos \dfrac{x}{2^n} + 1}$.

IX. 44. Use $\dfrac{\cos 2^k a}{1 + \cos 2^k a} = \dfrac{\tan 2^{k-1} a}{\tan 2^k a}$, $k = 1, 2, 3, ..., n$. $\mathbf{P}_n = \dfrac{\tan a}{\tan 2^n a}$.

IX. 45. $P_n = \dfrac{\cos 2^{n+1} a - \cos 2^{n+1} b}{\cos a - \cos b}$.

IX. 46. Use $\dfrac{1 + \tan^2 2^k x}{\left(1 - \tan^2 2^k x\right)^2} = \dfrac{\cos^2 2^k x}{\cos^2 2^{k+1} x}$. Therefore

$$P_n = \prod_{k=1}^{n} \frac{\cos^2 2^k x}{\cos^2 2^{k+1} x} = \frac{\cos^2 2x}{\cos^2 2^{n+1} x} .$$ **IX. 47.** $A_k = \cos k\theta$.

IX. 48. $S_1 + iS_2 = 1 + \cos x + i \sin x + \left(\cos x + i \sin x\right)^2 + \ldots + \left(\cos x + i \sin x\right)^n =$

$$= \frac{1 - \left(\cos x + i \sin x\right)^{n+1}}{1 - \left(\cos x + i \sin x\right)} = \frac{\cos \dfrac{n}{2} x \sin \dfrac{n+1}{2} x}{\sin \dfrac{x}{2}} + i \frac{\sin \dfrac{n}{2} x \sin \dfrac{n+1}{2} x}{\sin \dfrac{x}{2}} .$$

IX. 49. $S_1 + iS_2 = \dfrac{1}{2}\left(\cos x + i \sin x\right) - \dfrac{1}{2^2}\left(\cos x + i \sin x\right)^2 + \ldots - \dfrac{1}{2^n}\left(\cos x + i \sin x\right)^{2n} =$

$$= \frac{1}{2}\left(\cos x + i \sin x\right) \frac{1 - \left[-\dfrac{1}{2}\left(\cos x + i \sin x\right)\right]^{2n}}{1 - \left[-\dfrac{1}{2}\left(\cos x + i \sin x\right)\right]} =$$

$$= \frac{1}{2}\left(\cos x + i \sin x\right) \frac{1 - \dfrac{1}{2^{2n}}\left(\cos 2nx + i \sin 2nx\right)}{1 + \dfrac{1}{2}\left(\cos x + i \sin x\right)} .$$

$$S_1 = \mathrm{Re}\left(S_1 + iS_2\right) = \frac{2 \cos x + 1}{5 + 4 \cos x} - \frac{1}{2^{2n}} \frac{2 \cos(2n+1)x + \cos 2nx}{5 + 4 \cos x} ,$$

$$S_2 = \mathrm{Im}\left(S_1 + iS_2\right) = \frac{2 \sin x}{5 + 4 \cos x} + \frac{1}{2^{2n}} \frac{2 \sin(2n+1)x + \sin 2nx}{5 + 4 \cos x} ,$$

$$S_3 = \frac{\cos nx - 2 \cos(n+1)x - 2^{n+1} \cos x + 2^{n+2}}{2^n \left(5 - 4 \cos x\right)} ,$$

$$S_4 = \frac{\sin nx - 2 \sin(n+1)x + 2^{n+1} \cos x}{2^n \left(5 - 4 \cos x\right)} .$$

IX. 50. Use $z + z^2 + z^3 + \ldots + z^n = z \dfrac{1 - z^n}{1 - z}$, $z \neq 1$, differentiating in both sides

$$1 + 2z + 3z^2 + \ldots + nz^{n-1} = -\frac{nz^n}{1 - z} - \frac{z^n - 1}{\left(1 - z\right)^2} \text{ or}$$

$$z + 2z^2 + 3z^3 + \ldots + nz^n = -\frac{nz^{n+1}}{1-z} - \frac{z^{n+1} - z}{(1-z)^2}. \text{ Taking } z = \cos x + i \sin x, \text{ then}$$

$$S_1 + iS_2 = \cos x + i \sin x + 2(\cos x + i \sin x)^2 + \ldots + n(\cos x + i \sin x)^n =$$

$$= \frac{n \sin \dfrac{2n+1}{2} x}{2 \sin \dfrac{x}{2}} - \frac{\sin^2 \dfrac{n}{2} x}{2 \sin^2 \dfrac{x}{2}} - i \left(\frac{\cos \dfrac{2n+1}{2} x}{2 \sin \dfrac{x}{2}} - \frac{\sin nx}{4 \sin^2 \dfrac{x}{2}} \right).$$

IX. 51. $S_1 + iS_2 = 1 + \dfrac{\cos x + i \sin x}{\cos x} + \dfrac{\cos 2x + i \sin 2x}{\cos^2 x} + \ldots + \dfrac{\cos nx + i \sin nx}{\cos^n x} =$

$$= 1 + \left(\frac{\cos x + i \sin x}{\cos x} \right)^1 + \left(\frac{\cos x + i \sin x}{\cos x} \right)^2 + \ldots + \left(\frac{\cos x + i \sin x}{\cos x} \right)^n =$$

(A geometric progression with $n+1$ terms and ratios $r = \dfrac{\cos x + i \sin x}{\cos x}$)

$$= \frac{1 - \left(\dfrac{\cos x + i \sin x}{\cos x} \right)^{n+1}}{1 - \dfrac{\cos x + i \sin x}{\cos x}} = \frac{1 - \dfrac{\cos(n+1)x + i \sin(n+1)x}{(\cos x)^{n+1}}}{1 - \dfrac{\cos x + i \sin x}{\cos x}} =$$

$$= \frac{\sin(n+1)x}{\sin x \cos^n x} + i \frac{\cos^{n+1} x - \cos(n+1)x}{\sin x \cos^n x}, \quad x \neq \frac{k\pi}{2}, \ k \in \mathbf{Z}.$$

IX. 52. $S_1 = \cos x \cos x + \cos 2x \cos^2 x + \ldots + \cos nx \cos^n x = \displaystyle\sum_{k=1}^{n} \cos kx \cos^k x;$

$$S_2 = \sin x \cos x + \sin 2x \cos^2 x + \ldots + \sin nx \cos^n x = \sum_{k=1}^{n} \sin kx \cos^k x.$$

$$S_1 + i S_2 = \cos x(\cos x + i \sin x) + \cos^2 x(\cos x + i \sin x)^2 + \ldots + \cos^n x(\cos x + i \sin x)^n =$$

$$= \frac{\cos x(\cos x + i \sin x)[1 - \cos^n x(\cos x + i \sin x)^n]}{1 - \cos x(\cos x + i \sin x)} =$$

$$= \frac{\cos x(\cos x + i \sin x)[1 - \cos^n x(\cos nx + i \sin nx)][1 + \cos x(\cos x + i \sin x)]}{\sin^2 x} =$$

$$= \frac{\cos^{n+1} x \sin nx}{\sin x} + i \frac{\cos x(1 - \cos^n x \cos nx)}{\sin x}.$$

Therefore $S_1 = \dfrac{\cos^{n+1} x \sin nx}{\sin x}$ and $S_2 = \dfrac{\cos x\left(1 - \cos^n x \cos nx\right)}{\sin x}$.

IX. 53. Consider $z = \cos\dfrac{\pi}{n} + i\sin\dfrac{\pi}{n}$ and $S = \displaystyle\sum_{k=1}^{n-1}\sin\dfrac{k\pi}{n}$,

$T = \displaystyle\sum_{k=1}^{n-1}\cos\dfrac{k\pi}{n} \Rightarrow T + iS = z + z^2 + \ldots + z^{n-1} = z\cdot\dfrac{z^{n-1} - 1}{z-1}$, $z \neq 1$, hence

$T + iS = i\cot\dfrac{\pi}{2n} \Rightarrow T = 0,\ S = \cot\dfrac{\pi}{2n}$;

IX. 54. Consider the equation $x^{2n} + 1 = 0$. The roots are

$z_k = \cos\dfrac{(2k+1)\pi}{2n} + i\sin\dfrac{(2k+1)\pi}{2n}$, where $k = 0, 1, 2\ldots, 2n-1$.

Since $z_{2n-i} = \overline{z}_{i-1}$, $i = 1, 2, 3\ldots, n$, and

$\left(x - z_p\right)\left(x - \overline{z}_p\right) = x^2 - 2x\cos\dfrac{(2p+1)\pi}{2n} + 1$ we have

$x^{2n} + 1 = \left(x - z_0\right)\left(x - z_1\right)\left(x - z_2\right)\left(x - z_{2n-1}\right) =$

$= \left(x - z_0\right)\left(x - \overline{z}_0\right)\left(x - z_1\right)\left(x - \overline{z}_1\right)\cdot\ldots\cdot\left(x - z_{n-1}\right)\left(x - \overline{z}_{n-1}\right)$.

Therefore $x^{2n} + 1 = \left(x^2 - 2x\cos\dfrac{\pi}{2n} + 1\right)\left(x^2 - 2x\cos\dfrac{3\pi}{2n} + 1\right)\cdot\ldots$

$\cdot\left(x^2 - 2x\cos\dfrac{(2n-1)\pi}{2n} + 1\right)$. For $x = 1$ the last equality becomes

$2 = 4\left(\sin^2\dfrac{\pi}{2n}\right)4\left(\sin^2\dfrac{3\pi}{2n}\right)\cdot\ldots\cdot 4\left(\sin^2\dfrac{(2n-1)\pi}{2n}\right)$.

IX. 55. a) Consider $z_k = \cos\left(x + \dfrac{k\pi}{n}\right) + i\sin\left(x + \dfrac{k\pi}{n}\right)$, $k = \overline{0, n-1}$. We have

$\left(z^2 - z_0^2\right)\cdot\displaystyle\prod_{k=1}^{n-1}\left(z^2 - z_k^2\right) = z^{2n} - z_0^{2n} = \left(z^2 - z_0^2\right)\left(z^{2n-2} + z^{2n-4}z_0^2 + \ldots + z_0^{2n-2}\right)$.

For $z = 1$, we obtain

$\displaystyle\prod_{k=1}^{n-1}\left(1 - z_k\right)^2 = 1 + z_0^2 + z_0^2 + \ldots + z_0^{2n-2}$, thus

$$\frac{z_0^{2n}-1}{z_0^2-1}=\prod_{k=1}^{n-1}\left[1-\cos\left(2x+\frac{2k\pi}{n}\right)-i\sin\left(2x+\frac{2k\pi}{n}\right)\right]\Rightarrow$$

$$\frac{-1+\cos 2nx+i\sin 2nx}{-1+\cos 2x+i\sin 2x}=\prod_{k=1}^{n-1}\left[2\sin^2\left(x+\frac{k\pi}{n}\right)-2i\sin\left(x+\frac{k\pi}{n}\right)\cdot\cos\left(x+\frac{k\pi}{n}\right)\right]\Rightarrow$$

$$\frac{-2\sin^2 nx+2i\sin nx\cdot\cos nx}{\Rightarrow 2\sin^2 x+2i\sin x\cdot\cos x}=\prod_{k=1}^{n-1}\left[2\sin\left(x+\frac{k\pi}{n}\right)\cdot(-i)\cdot\left(\cos\left(x+\frac{k\pi}{n}\right)+i\sin\left(x+\frac{k\pi}{n}\right)\right)\right]$$

$$\frac{\sin nx}{\sin x}\cdot\frac{\cos nx+i\sin nx}{\cos x+i\sin x}=2^{n-1}\cdot(-i)^{n-1}(\cos x+i\sin x)^{n-1}\cdot\left(\cos\frac{\pi}{n}+i\sin\frac{\pi}{n}\right)^{\frac{n\cdot(n-1)}{2}}\cdot P$$

where $P=\displaystyle\prod_{k=1}^{n-1}\sin\left(x+\frac{k\pi}{n}\right)$.

Then $\left(\cos\dfrac{\pi}{n}+i\sin\dfrac{\pi}{n}\right)^{\frac{n\cdot(n-1)}{2}}=i^{n-1}$ and

$(\cos x+i\sin x)^n=\cos nx+i\sin nx\Rightarrow P=\dfrac{\sin nx}{2^{n-1}};$

b) We obtain from **a)** for $x=\dfrac{\pi}{2n}$. **IX. 56 a)** 1; **b)** $(\cosh^2 x-\sinh^2 x)^n=1$.

IX. 57. a) $z=1+i$, $z^n=2^{\frac{n}{2}}\cdot\left(\cos\dfrac{n\pi}{4}+i\sin\dfrac{n\pi}{4}\right)$,

$(1+i)^n=1-\binom{n}{2}+\binom{n}{4}-\binom{n}{6}+...+i\left(\binom{n}{1}-\binom{n}{3}+\binom{n}{5}-\binom{n}{7}+...\right);$

d) $z=1+i\sqrt{3}$, $z^n=2^n\cdot\left(\cos\dfrac{n\pi}{3}+i\sin\dfrac{n\pi}{3}\right)$.

IX. 58. a) $\varepsilon=\dfrac{-1+i\sqrt{3}}{2}$ is the root of the equation $\varepsilon^2+\varepsilon+1=0$, also

$\varepsilon^3=1$. $\varepsilon=\cos\dfrac{2\pi}{3}+i\sin\dfrac{2\pi}{3}$, $1+\varepsilon=\cos\dfrac{\pi}{3}+i\sin\dfrac{\pi}{3}$,

$1+\varepsilon^2=\cos\dfrac{\pi}{3}-i\sin\dfrac{\pi}{3}$. Finally, add the equalities

$2^n=\binom{n}{0}+\binom{n}{1}+\binom{n}{2}+\binom{n}{3}+\binom{n}{4}+...,$

$(1+\varepsilon)^n=\binom{n}{0}+\binom{n}{1}\varepsilon+\binom{n}{2}\varepsilon^2+\binom{n}{3}+\binom{n}{4}\varepsilon+...,$

$$\left(1+\varepsilon^2\right)^n = \binom{n}{0} + \binom{n}{1}\varepsilon^2 + \binom{n}{2}\varepsilon + \binom{n}{3} + \binom{n}{4}\varepsilon^2 + \dots.$$

IX. 59. $z^n = \left(1 + i\dfrac{1}{\sqrt{3}}\right)^n = \left(\dfrac{2}{\sqrt{3}}\right)^n \left(\cos\dfrac{n\pi}{6} + i\sin\dfrac{n\pi}{6}\right).$

IX. 60. $z^{6n} = \left(1 - i\sqrt{3}\right)^{6n} = \left[2\left(\cos\dfrac{5\pi}{3} + i\sin\dfrac{5\pi}{3}\right)\right]^{6n} = 2^{6n}.$

IX. 61. Use $z = (1+i)^n$.

IX. 62. $S_1 + iS_2 = 1 + \binom{n}{1}(\cos x + i\sin x)^1 + \binom{n}{2}(\cos x + i\sin x)^2 + \dots +$

$+ \binom{n}{n}(\cos x + i\sin x)^n = (1 + \cos x + i\sin x)^n = \left(2\cos^2\dfrac{x}{2} + 2i\sin\dfrac{x}{2}\cos\dfrac{x}{2}\right)^n =$

$= 2^n \cos^n\dfrac{x}{2}\left(\cos\dfrac{nx}{2} + i\sin\dfrac{nx}{2}\right).$

IX. 63.

$S_1 + iS_2 = (\cos x + i\sin x) + \binom{n}{1}(\cos x + i\sin x)^2 + \dots + \binom{n}{n}(\cos x + i\sin x)^{n+1} =$

$= (\cos x + i\sin x)(1 + \cos x + i\sin x)^n = (\cos x + i\sin x)\left(2\cos^2\dfrac{x}{2} + 2i\sin\dfrac{x}{2}\cos\dfrac{x}{2}\right)^n =$

$= 2^n \cos^n\dfrac{x}{2}\left(\cos\dfrac{(n+2)x}{2} + i\sin\dfrac{(n+2)x}{2}\right).$

Therefore $S_1 = 2^n \cos^n\dfrac{x}{2}\cos\dfrac{(n+2)x}{2}$ and $S_2 = 2^n \cos^n\dfrac{x}{2}\sin\dfrac{(n+2)x}{2}$.

BIBLIOGRAPHY

*Canadian Mathematics Competition, Problems Problems Problems, Waterloo, On, Canada, Volume 1, 2, 3, 4, 5.

* Functions 11 (MCR3U). Toronto: McGraw-Hill Ryerson, 2008.

* Advanced Functions 12 (MHF4U). Toronto: McGraw-Hill Ryerson, 2008.

* Functions (Grade 11 University) **MCR3U** -Nelson

* R Green, G Nichollas, Harcourt Mathematics 11, Functions/Relations, Harcourt Canada.

*James Stewart, Lothar Redlin, Saleem Watson, *Precalculus Mathematics for calculus 4th*, Brooks/Cole.

*Wesner/Nustad, *Elementary Algebra with applications*, Iowa1983.

*Elliot Mendelson, *Theory and problems of beginning calculus*, 2nd edition, McGra-Hill.

*Titu Andreescu, Dorin Andrica, *Complex Numbers from A to...Z*, Birkhäuser, 2006.

*A. Tsypkin, A. Pinsky, *Methods of solving problems in high school math*, Mir Publishers Moscow, 1983.

*Călugărița Gh., Mangu V., *Probleme de matematică pentru treapta I și a*

II-a de liceu, Ed. Albatros, Bucureşti, 1977.

*Comissaire H., Anzemberger E., *Exercises d'Algebre et de Trigonometrie*, Paris, 1923.

* Iaglom I. M., Iaglom A. M., *Challenging Mathematical Problems with elementary Solutions*, Dover Publications, 1964.

*Kuterov A., Rubanov A., *Zadacnik po algebre i elementarnîm funcţiiam*, Moskva, 1974.

*M. I Skanavi, Sbornik Zadach po matematike, Moskva, 1996.

*Vîşenski V. A., Kartaşov N. V., Mihailovski B. I., Iardenko M. I., *Sbornik zadaci kievskih matematiceskih olimpiad*, Kiev, 1984.

*Nesterenko I. V., Olenik S. N., Potapov M. K., *Zadaci vstupitelnîh eczamenov po matematike*, Moskva, 1986.

*Dorofeev G. V., Potapov M. K., Rozov N. H., *Posobie po matematike dlia postupaiuşcih v vuzî*, Moskva, 1976.

*Sklearski D. O., Cenţov N. N., Iaglom I. M., *Izbrannie zadacii teoremi elementarnoi matematiki (Arifmetica i algebra)*, Moskva, 1965.

*Stamate I., Stoian I., *Culegere de exerciţii şi probleme de algebră*, Ed. Didactică şi Pedagogică, Bucureşti, 1979.

* Nastasescu C., Brandiburu M., C. Nita, D. Joita, *Exercitii si probleme de algebra*, Editura Didactica si pedagogica, Bucuresti.

*Matematică –Algebră (high school textbooks, grades IX - X, 1980 edition), Ed. Didactică şi Pedagogică, Bucureşti, 1980.

*Collection of "Gazeta Matematică - seria B", Bucharest.

* Radu E., Sontea O., Matematica *Exercitii si problemede evaluare pentru clasa a X-a*, Editura All.

*Liliana Nicolescu, *Probleme de matematica pentru liceu*, Editura Cardinal.

*S. Ianus, N. Soare, L. Nicolescu, S. Dragomir, M Tena, *Probleme de* *geometrie si trigonometrie pentru clasele IX-X*, Ed Didactica si Pedagogica, Bucuresti.

*N. Dragomir, C. Dragomir, O. Blag, *Trigonometrie, Exercitii si problemepentru elevii claselor IX-X*, Ed Universal Plan, Bucuresti.

* Oros D., *Trigonometrie Culegere de probleme*, Ed Petrion, Bucuresti.

* Parsan L., *Probleme de algebra si trigonometrie,*Editura Facla, Timisoara.

* Turtoiu F., *Ecuatii si inecuatii trigonometrice*, Ed Tehnica, Bucuresti-1977.

* Turtoiu F., *Probleme de trigonometrie*, Ed Tehnica, Bucuresti-1979.

* Schneider G. A., *Culegere de probleme de trigonometrie*, Ed Hyperion, Craiova, 2012.

* Ganga M., *Ecuatii si inecuatii*, Ed Mathpress, 1998.

*Coța, A., Marta, R., Kurthy, E., Răduțiu, M., Popa, F., E., Vornicescu, F. *Matematică*, manual

pentru clasa a -X-a, *Geometrie* și *trigonometrie*, Editura Didactică și Pedagogică, București, 1987.

*Cuculescu, I. *Culegere de probleme rezolvate pentru admiterea în învățământul superior*

Editura Științifică și Enciclopedică, București, 1984.

*Flondor, D., Donciu, N. *Algebr*ă și *analiz*ă *matematic*ă. *Culegere de probleme*. vol. 2,

Editura Didactică și Pedagogică, București, 1979.

*Barbu, Catalin. Numere Complexe – Aplicatii– Editura Grafit, Bacau, 2005